The Four Cosmic Pillars; The Result Thereof.

In terms of Applying Cosmic Physics

WRITTEN BY P. S. J. Schutte ISBN 978-0-9802725-2-9

All rights are reserved.
No part, parts or the entirety of this book may be reproduced by publishing, electronically copied, duplicated by whatever means that form reproduction or duplication, without the prior written consent of the copy rite owner.

Written by Peet (P.S.J.) Schutte

© KOSMOLOGIESE EN ASTRONOMIESE TEGNIKA

ISBN-13: 978-1539334149
ISBN-10: 1539334147

I do find much pride in my status as being Afrikaner and would like to have my names used by pronouncing it in the manner Afrikaans dictates…therefore I would sincerely appreciate the courtesy when readers will take note that my name and last name are pronounced in Afrikaans, which is originally from Dutch and must be pronounced that way. Peet one would pronounce "here" which is the closest English to the pronouncing of the "ee". The "Sch" in Schutte is pronounced exactly as school is where both actually are pronounced Skutte or "skool". By pronouncing my name in Afrikaans you do me the utmost courtesy any one can.

Being an Afrikaner is what I am most proud of. I submit article to well known physics magazines but my articles are rejected on the most unappeasable grounds and for the most outrageously ridiculous reasons the Newtonians can think of. I explain how gravity forms but I am rejected because they are of the opinion that my work does not meet. One such an article I may use because I said I was going to use the material as an open letter I gladly show.

This book was done with a $25 ^{00} scanner and a $35 ^{00} printer and the reason I explain inside. For the same reason this book was not edited or linguistically checked. I could not because that does not work because I am in the writing business and not the spelling business and while I check spelling the writing gets more and so does the spelling and grammar errors. I had a choice; doing the books with no funding or not doing it at all because while I rubbish Newtonian science and show it is the fake it is, they will never publish my work because I trash Newton. Not having funds and trying to fight science for the truth with the truth was a fight that physically broke my health and still I am not published except in this manner. I apologise for the spelling and language but in poverty that was the best I could do under the prevailing circumstances in which I find myself…. This book is a first in every sense… it unites science and religion because science and religion was separated by human stupidity

Please take note that I sell information and not words or books and therefore the information takes priority and not the spelling or words used to inform the readers. This represents the work of God and not the word of God and so there is no interpretations applying and versus you can learn and sound intellectual but only cold facts you will have to understand.

Please take note that the e-books on offer has not linguistically been edited and with the English language not being my first language, you can bet your pocket money on the fact that there are going to be several language hick-ups that you will encounter as you read along. If you are in search of a language masterpiece without any mistakes then try William Shakespeare and that might satisfy you because I am not William Shakespeare and my language is not there to impress anyone. If you are not prepared to ignore some language and printing mistakes, then please do NOT PURCHASE for you are going to be disappointed.

However if you wish to read about science that no one has ever heard of and you wish to find out why I am prepared to challenge the entire world of physics, as well as the lot of boffins in cosmology and astrophysics then you are holding the correct book. If you purchase this book you will venture into science you have never experienced. You will see a new approach to physics you have never thought were possible. However, I again stress the fact that there are going to be spelling errors and language glitches because I am no language expert, even in Afrikaans, which is my native language and I am too dirt poor to enlist the services of such a person as an language expert because for the past ten years I have been conducting the research you are now about to witness on a full time basis without having an income except that of my wife being a typist.

The books on offer is as follows:

How the Solar System Forms: An Academic Presentation by Peet (P.S.J.) Schutte
ISBN-13: 978-1523217021 (CreateSpace-Assigned)
ISBN-10: 1523217022

A Cosmic Birth as an Academic Presentation Book 1 by Peet (P.S.J.) Schutte
ISBN-13: 978-1517066970 (CreateSpace-Assigned)
ISBN-10: 1517066972

A Cosmic Birth...as a Special Presentation Book 2 by Peet (P.S.J.) Schutte
ISBN-13: 978-1517525460 (CreateSpace-Assigned)
ISBN-10: 1517525462

An Academic Introducing to The Titius Bode Law Book 1 by (P.S.J.) Peet Schutte
ISBN-13: 978-1507845851 (CreateSpace-Assigned)
ISBN-10: 1507845855

An Academic Introducing to The Titius Bode Law Book 2 by Peet (P.S.J.) Schutte
ISBN-13: 978-1507853788 (CreateSpace-Assigned)
ISBN-10: 1507853785

An Academic Introducing to The Titius Bode Law Book 3 by Peet (P.S.J.) Schutte
ISBN-13: 978-1505874884 (CreateSpace-Assigned)
ISBN-10: 1505874882

How the Solar System Forms: a Pre- Script by Peet (P.S.J.) Schutte
ISBN-13: 978-1503023895 (CreateSpace-Assigned)
ISBN-10: 1503023893

Relevant applying literature Go to Google Amazon.com: Peet Schutte: Books
http://www.amazon.com/s?ie=UTF8&page=1&rh=n%3A283155%2Cp_27%3APeet%20Schutte.
Oxford dictionary of Astronomy web site naturescosmicconcept

The Following books are all available from CreateSpace web site.
The Absolute Relevance of Singularity **The Journal**
The Absolute Relevance of Singularity **The Unpublished Article**
The Absolute Relevance of Singularity **The Dissertation**
The Absolute Relevance of Singularity **in terms of** Newton Book 0
The Absolute Relevance of Singularity **in terms of** Cosmic Physics Book 1
The Absolute Relevance of Singularity **in terms of** The Sound Barrier Book 2
The Absolute Relevance of Singularity **in terms of** The Four Cosmic Phenomena Book 3
The Absolute Relevance of Singularity **in terms of** The Cosmic Code Book 4
The Absolute Relevance of Singularity **in terms of** Life Book 5
The Absolute Relevance of Singularity **in terms of** Investigating Kepler Book 6
The Absolute Relevance of Singularity **in terms of** The Thesis Book 7
The Absolute Relevance of Singularity **in terms of** The Cosmic Creation Book 8

peet@naturescosmicconcept.co.za   mail.naturescosmicconcept.co.za

This is not a sales ploy but this warning comes on the ground that I use certain ideas in this book which I explain in the other three in the series and I do not return again to the introducing of the concepts as a whole in *The Cosmic Code as the Absolute Relevancy of Singularity.* Then again...this book shows that cosmology read correctly is so simple that even a person with a simple mind such as I have can understand all information that explains the Universe. I admit that if I could write the work on offer, then any body can read it…but please take note that what you are about to read…you have never read before and in that comes the complexity of understanding. As the books introduce new facts using arguments, which the readers have never come across before, so the understanding about the new concepts will grow in clarity.

Simplifying Complicated Science,

Simplifying Complicated Science Part 1,

Simplifying Complicated Science Part 2,

Articles Revealing Cosmic Unknowns

The Four Cosmic Pillars in mono colour
The Absolute Relevancy of Singularity In Terms of Life In Mono Colour
The Four Cosmic Pillars in colour
Articles Revealing Cosmic Unknowns in colour

Please read on to find more information concerning The Absolute Relevancy of Singularity
My mother tongue is **Afrikaans, which is an African language** and my second language is English, which is the normal British /American/ Canadian / Australian variety used by many if not most. With English, being my second language I am **not boasting** about my verbal skills in English and there is a hidden motive why I am mentioning that at this point, but I shall get to the explaining a little later on.

I have per suiting this theory that I partly present in these books, of which the investigating research was done by me on a part time basis since 1977. Then I decided to formulate my conclusions in a seven part theses I named **Matter's Time In Space: The Theses Vol. 1 to 7** which I then compiled as my presentation of my new cosmic theory and then following that I worked on promoting my theses. This took almost every minute of my life the last past nine years as the promoting required my attention on full time basis whereby I was trying to introduce my findings to many academics without having much joy I should add.

This past almost twenty years saw me go without any income as I tried to get my theorem recognized. Going without a steady income left me almost destitute and in order to find a manner to get my theory across to the attention of influential readers, I decided to publish these books electronically as to try and get around the stranglehold of Newtonian bias controlling science at present worldwide. I decided to publish these articles through LULU.com which I saw as the only manner whereby I could generate funding by which I eventually would be able to have the twenty seven books I already wrote linguistically edited and then to have the funding whereby I could have the books afterwards published on a Print-On-Demand basis and then distributed through the large retail distributors such as Barns and Noble and Amazon.com.

With my first language not being English and the books not linguistically checked by an expert there are bound to be language errors that readers will notice. In the past I tried to check my work myself but after checking say one hundred and fifty pages for language corrections, instead of having corrected work I ended by having four hundred pages of newly written information which is still not language corrected but holds a lot more information. There then are four hundred pages of unchecked work, which exaggerate the first problem. This is because my priorities lie elsewhere. I aim to spend money on correcting the work as far as language goes and then have the books formally printed in ink on paper, as I receive money and in the hope that I will receive money. I will then have all my work including the one you are reading edited professionally and corrected as I find money to do so.

However, the work I present has been introduced for the first time ever via my brain and every concept I offer, as an introduction to the world of science, is entirely a product of my mind. If you are a hardliner Newtonian then brave yourself because you are in for an eye opening experience.

My promise to you is that if ever you are able to prove that the information I present as mine is not completely and altogether new. I shall personally refund your money immediately.

P. S. J. Schutte
P O Box
Town Somewhere in the
Limpopo Province
Rep. South Africa
Tel 27-14-
e-mail gravity@

Dear Reader,
I am Petrus Stephanus Jacobus Schutte going by the name of Peet; the author of the article that I hope would be published. This article is one book of the four (small) books containing the rudimentary basis that forms the pith of my theory I present as **The Absolute Relevancy of Singularity,** which also is part of the title naming each book covering another angle and each having a different I.S.B.N. number. The books are written all explaining my view about **THE ABSOLUTE RELEVANCY OF SINGULARITY** carrying the theme from four different perspectives into one title and the one title I present as four books each titled as **The Absolute Relevancy of Singularity and this one in specific being The Absolute Relevancy of Singularity in terms of Applying Physics** ISBN 978-0-9802725-2-9.

I hope you find your reading of this book presented by this open letter a most fruitful experience. I feel I need to warn you, the person reading this; the work contained herein strays widely from mainstream science and for that there is a very good reason. However, in the least, the content is thought provoking, as it is one hundred percent original in theme and in thought. I researched the work of a man that is most exceptional and therefore should be placed much more prominent in the allocating of the position that his work should have when forming the history of mankind. His contribution in the gathering of information that furthered the entire human species in their accumulation of knowledge as well as the human understanding in cosmic affairs stands second to none in comparison to most others whilst most people are not even aware of the full implication of his work.

Notwithstanding your personal academic qualifications, disregarding your status and achievements and ignoring your abilities however superior they might be, I shall teach you about physics.

Whilst recognising the work of Johannes Kepler, Mainstream science bluntly ignores the impact of his work, and in that they miss the full vastness of the wide influence of his work. Newton shrouded Kepler's work under a blanket of alterations which I show was most unwanted since Kepler's work needs no alterations or corrections and every one since then kept Kepler's work hostage under Newton's changes. It is therefore almost absolutely realistic to say that all information what you are about to read in this letter and article sent to you for your attention was never yet printed in the near or the distant past although Kepler's work has been with us for about four hundred years, during which time it went unnoticed. It seems to me that any research predating Newton never came into use or into practise. My investigation of Kepler's work brought about a conclusion that no one yet arrived at concerning them with the findings of Kepler because no one scrutinised Kepler's formula before. Everyone is satisfied with Newton's version notwithstanding the incorrectness of it. The world seems satisfied with the idea that Kepler found planets rotating around a centre formed by the Sun and because of that Newton saw a circle. Where Newton saw a mathematical circle and was unable to understand $a^3 = T^2 k$, Newton added what he thought is mathematically required to indicate such a circle. Newton added a mathematical $4\Pi^2$ to the formula of Kepler and removed the distance symbolising measure that Kepler introduced using **k**.

On the other side Newton changed the symbol of **k** by using the symbols G ($m + m_p$). This is just a longer and probably a more detailed manner of indicating **k** and better defining of **k** but it symbolises precisely to the point what **k** stands for nonetheless. I wish to draw your attention to the matter of Johannes Kepler's findings that Mainstream science considers as resolved and closed for many a century while it is not. My investigating Kepler helped me too resolve other unresolved matters but it was only possible by using Kepler's work. This brought about the idea that the Universe is in a state of contracting towards a centre of sorts where mass will form this contracting. This was prevailing until a man by the name of E. P. Hubble came to the forefront.

E. P. Hubble (1889-1953) confirmed an expansion through out the Universe, which contradicted all that science thought was known about our Universe. According to the accepted Newtonian cosmology everyone is of thought that the Universe is in a normal state of contraction because that is what

$$F = G \frac{M_1 M_2}{r^2}$$ implies.

The Four Cosmic Pillars; The Result Thereof.

Every person is very aware of the idea that the universal expansion would not last for ever, but has to start with some contracting effort at some point. Then all the heavenly bodies will collide and destruct, without any thought about any wavering on the matter and on the matters reliability there is evidently no doubt.

When $F = G \dfrac{M_1 M_2}{r^2}$ apply, there should not be any force, which is able to keep the mass that is producing all the gravity that contains the Universe apart. Known for almost a century, science has failed to give any explanation about this cosmic phenomenon of a Universal expansion except for some silly notion about dark matter being dormant and not forming gravity, as it should. If the dark matter is present as is claimed, then why doesn't the mass form gravity as it should and contract? What does our ability to see or not to see or the luminosity that the dark matter does not have, got to do with the mass bringing about pulling power, that is if mass brings about any pulling power. If the mass is there, visible or not, then the dark mass has to pull because light has no standing in the forming of gravity and if mass does pull, it has to pull to form gravity. However Hubble's law contradicts this idea of a collective contracting Universe totally. This phenomenon about Hubble's constant finding the cosmos expanding should not occur with Newton's perception about gravity envisaging the contraction that must come by the force created by mass in $F = G \dfrac{M_1 M_2}{r^2}$.

If the Universe is on a contracting as Newton said it has to, we have to first find proof about the location to where such contraction is pointing. In order to locate the contracting we have to locate the centre of the Universe, which means we have to locate singularity. With singularity eternal small, holding the place where the Universe started, we first have to differentiate between singularity and zero, should we wish to find singularity. In modern science the phenomenon we know as the Roche lobe comes more and more to the foreground, indicating an undeniable interaction between orbiting structures sharing a common axis.

That axis science at present does not recognise, notwithstanding the reality and undeniable proof there is behind all evidence. As apparent as it is to me, I went about divorcing $F = G \dfrac{M_1 M_2}{r^2}$ from all ideas forming cosmology and applying the roundness we have in Π to specific positions where one may locate singularity, which we have to locate if we wish to find gravity.

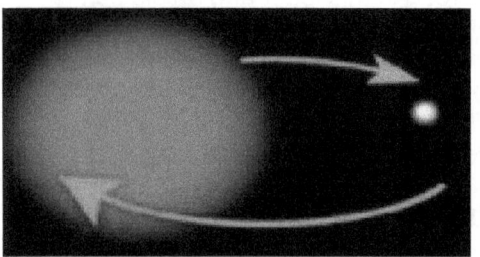 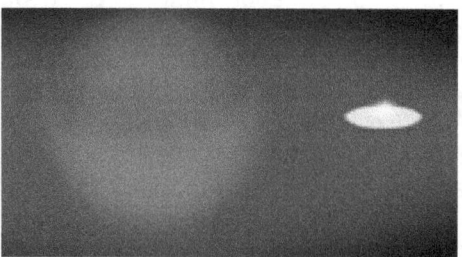

The Roche limit in the practical sense

The formula $F = G \dfrac{M_1 M_2}{r^2}$ cannot explain the comic occurrence shown in the pictures above called the Roche limit, I should find some attention when I say I can explain what is occurring in this instance and this occurrence connects directly to the Roche limit, as explained above. Not only does the Roche limit explain this phenomenon, but also it ties directly to the Titius Bode principle, also being another inexplicable factor in light of the formula $F = G \dfrac{M_1 M_2}{r^2}$.

According to the formula of $F = G \dfrac{M_1 M_2}{r^2}$ all orbiting structures should collide with a bang, but instead they do the tango until one drop, but when dropping it still does not collide with the larger structure, as would the formula $F = G \dfrac{M_1 M_2}{r^2}$ suggest that is used by science. The position where the formula applies is most surprising. Where the formula $F = G \dfrac{M_1 M_2}{r^2}$ applies, one has to find singularity applying because the position of r is pointing to a specific pinpointing of space contracting.

This is not only limited to planets in our solar system. In the Universe, there are giant stars spinning around each other. These stars are binaries, which are also one form of double stars where double stars are another such a form. The difference between the types depends on the distance they remain apart. They keep a certain distance apart and do not collide. In the case of the Sun and its planets, it could be a case that the systems might be to small, or they might be to apart. However, this is not the case with binary stars. They are close, they are big, and they spin around a mean axes called the Roche limit.

The Roche limit is:
The region surrounding each star in a binary system, within which any material is gravitationally bound to that particular star. The boundary of the Roche lobes is an equipotential surface, and the lobes touch at the inner Lagrangian point, L_1, through which mass transfer may occur if one of the components expands to fill its lobe. It names after the French mathematician Edouard Albert Roche (1820-83).

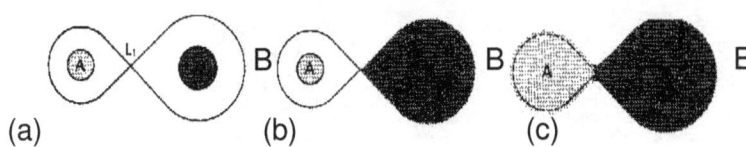

(a) (b) (c)

THE ROCHE LOBE: In a binary system, the Roche lobes of components A and B meet at the L_1 Lagrangian point. (a) In a detached system, neither star fills its Roche lobe. (b) In a semidetached system, one massive component, B, fills its Roche lobe. (c) In a contact binary, both components overfill their Roche lobes and share a common envelope.

LAGRANGIAN POINT:
The Lagrangian points are five equilibrium points in the orbit of one body around another, such as a planet around the Sun

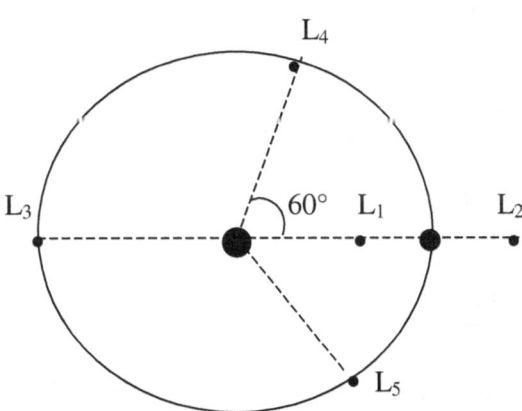

From singularity there comes three values each holding 180^0 and this fact science is familiar with. The straight line is always a potential triangle with on side apparent and the other side in infinity.

Planet	Mercury	Venus	Earth	Mars	Ceres	Jupiter	Saturn	Uranus
Bode's Law distance	4	7	10	16	28	52	100	196
Actual distance	3.9	7.2	10	15.2	28	52	95	192

Bode's Law:
A numerical sequence announced by J.E. Bode in 1772, which matches the distances from the Sun of the six planets then known. It is also known as the Titus-Bode law, as it was first pointed out by the German mathematician Johann Daniel Titius (1729-96) in 1766. It is formed from the sequence 0,3,6,12,24,48,96,

and 192 by adding 4 to each number. The planets were seen to fit this sequence quite well – as did Uranus, discovered in 1781. However, Neptune and Pluto do not conform to the 'law'. Bode's Law stimulated the search for a planet orbiting between Mars and Jupiter that led to the discovery of the first asteroids. It is often said that the law has no theoretical basis, but it does show how orbital resonance can lead to commensurability. The importance that becomes known is the sequence the Ties – Bode law saw in the number arrangement of 3; 6; 12; 24; 48; 96 etc. The incorrect application of the Titus Bode law lies in subtracting the figure of 3 from 10 leaving 7. The other way of reasoning is to add four each time to the firs value of three starting with 3 and so on. The true significance of the Titus-Bode law is that it points directly to a circular growth of 7 stages. The 7 relating to 10 is a precise derogative of the Roche limit or the Roche limit is a precise derogative of the Titus Bode principle because he two systems interlink.

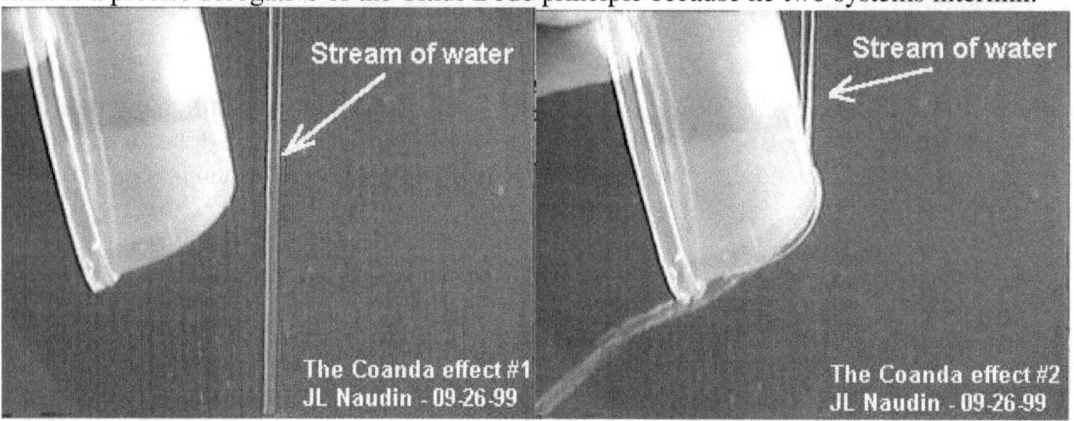

The Coanda effect

The Coanda effect where a liquid concentrates around the surface of a solid and by movement concentrates the density of the liquid to gather and compact while maintaining a relevance to the centre of such a round solid. I discard the idea that mass could be responsible for forming gravity because in almost four hundred years all evidence is indicating that the truth is to the contrary.

In the books I mention (as well as all the others I have written on the matter) I prove that gravity forms in relation to these four cosmic phenomena and mass is a by product resulting from gravity forming and not the factor producing gravity.

Why would a water drop always form a sphere in micro gravity in an outer space environment when free from the earth gravity as part of the natural cosmic free choice?
We all accept that the true cosmic form will be the sphere and most probably is a sphere in the universe as a whole ...but why would the sphere form as the original form when matter is not pre-cast to have any specific form and therefore take on by cosmic pre-cast the sphere as form?

Don't sell the significance of this answer too short, because in the answer hides the fact and the proof why the Universe has gravity.

By merely blaming gravity as BEING a mysterious FORCE that is pulling towards a common point in an allocated general centre is rather avoiding the question with simplicity because the question about how and why remains unanswered. The answer will be unfulfilled because of the following: If gravity pulls towards a centre and gravity holds the Universe attached the question arising from that simplistic answer is then ... where is the centre of the universe?

Where would you, the person that is reading this page, place the centre of the Universe? Whatever your insight into physics might be, it will be unfulfilled because of your inability to accurately place the position of the centre of the Universe. Without such critical knowledge to your disposal, you then have no idea where gravity comes from. Test your thoughts in the following: If gravity pulls towards a centre and gravity holds the Universe attached it has to be pulling to a centre, therefore then the next question arising from that simplistic question must then be... where is the centre of the universe?

I have to resort private with the publishing of my work because from the nature of my work I take Mainstream science head on and am confrontational on most aspects of astronomy. There does not seem to be any publisher that wants to go head bashing with the establishment of science on official science principles, which I have to do to convey my message in a no uncertain language.

ALL QUESTIONS I ASK IN THIS WEB SITE NO PERSON IN SCIENCE PREVIOUSLY WAS EVER ABLE TO ANSWER. WITH ME BEING   THE FIRST ONE IN FOUR HUNDRED YEARS ATTEMPTING TO

STUDY KEPLER I AM THE FIRST ONE SINCE KEPLER PUBLISHED HIS FINDINGS IN THE BEST PART OF SOME FOUR HUNDRED YEARS AGO TO BE ABLE TO ANSWER ALL THE QUESTIONS COMING UP...

If you knew would you tell me what gravity is…yes beside it being a force…well we all know that gravity is a force but other than being a force, what is gravity, which is besides being Newton's force.
If you don't know what gravity is besides knowing that gravity is one of the four forces and it is Newton's original force, then do you know someone that knows what gravity is?

If you still answer in the negative and you still don't know what is the cause of gravity, then have you heard of any one that knows what gravity is?

Maybe there is one Academic professor or a NASA Scientist that knows about some person that knows one who knows what gravity is, other than the fact that gravity is being one of the four forces and being Newton's personal force.

There has to be an Academic professor or a NASA Scientist known to someone somewhere that knows about some person that knows one who knows what gravity is, because there are many Academic professors or even more amongst the many NASA Scientists with super human mathematical abilities in the art of physics.

With their absolute phenomenal calculations as they present physics they are showing such abilities in a class which no one can imitate, which must represent a picture about having super knowledge on gravity in the most precise detail.

When reading what they accomplish it stands to reason that they know gravity to the smallest detail there is to know and know everything anyone ever can know about gravity…after all they can present the cosmos as they are able to explain the cosmos with gravity taking centre stage in the past, present and future of the cosmos.

They can calculate all the matter throughout the entire cosmos, which is responsible for the entirety that is providing all the mass that provides all the gravity. They have gone as far as even calculating the explicit required quantities available, which they then find to be short falling in the mass availability through out the vastness of the Universe and the missing mass is not establishing the matter density throughout the vastness of the Universe required to bring Newton's vision on contracting to reality. The mass they measure that is forming the gravity they require probably falls short of the requirements needed to substantiate the cosmos' effort in rendering a constant supply of gravity that will eventually secure the returning of everything to where the cosmos came from. This eventuality they named The Big Crunch even before locating the Big Crunch. It is like naming a baby even long before knowing how the procreating is taking place that will lead to impregnating of some member of the specie (which member it will be is still unclear) where it later on will lead to conceiving the baby … that is the manner in which science dogma is enunciated but that is how clever those are that knows everything there is to know on gravity.

They know gravity to split detail where the detail goes to such precise extend that they are able to calculate how much gravity the missing dark matter in all the Black stars will provide to allow the cosmos in experiencing the next big implosion that is coming somewhere in the future. That the implosion must come even in the face of insufficient gravity is a certainty otherwise Newtonian physics is completely inadequate in their cosmic vision about the Universal future! With such phenomenal abilities they then would have to know what gravity is!

Well…if you don't know any one that knows anybody that knows someone somewhere that is familiar with the ins and outs of what is causing gravity, I then can ensure you I know about someone that knew all there is to know about gravity. I am the person that knows someone and that someone knew gravity…but he is dead now. He passed away. Still I would like to introduce him to you…and about his work of course, if you would page on.

I present you again with my work and this time I compiled the work into three books. In the three books I introduce for the first time ever into science an explanation for the phenomena not yet explained. They are The Roche limit, 2) The Bode law, 3) The Lagrangian points and The Coanda effect.

These above mentioned phenomena work in harmony to form the principle called gravity. They work as a whole as an assembled unit to compile gravity. Each one has its individual role but unifies their combined principles to form gravity. Each fills a function that finds a position in the unit where it participates in forming that what we regard as keeping the Universe glued. Mass therefore, has nothing whatsoever to do with the generating of gravity. In the books I purposely explain my conclusions in minute detail, which I did not do the previous time when I offered my work for academic investigative evaluation.

This time the information is now so much simplified it is to a level where I believe it will probably be understood by any child still at school on the condition that that scholar has developed an understanding about physics to the level of high school physics standard. In the previous presentations of my work that I submitted my work was rejected at the time by your institution amongst and including other institutions equal in stature to that of yours. At that time I could have been guilty of accepting that there were facts I presented that the reader was unfamiliar with, but I did not realise that because I saw those facts as general knowledge, which at the time I personally would regard as simple everyday information. In the end those facts did prove not to be general knowledge after all and that complicated some of the issues. One such example is where to find singularity. It is so obvious allocated that at the time I could not dream any person did not see it at first glance.

The other possible mistake is not to explain how I conclude how the value of singularity comes about because that too, is so simple to see. One more example I may mention is to explain why singularity has developed space-time by specific measurements according to time and by using time. I might have failed to present the limits of time, which I clearly make an issue of indicating this time round. I took this as general knowledge being located in an obvious position where all can immediately see the presence of eternity and infinity but it turns out not to be that obvious to others. (This is only naming a few.) However, I have corrected such mistakes and this time I have explained every aspect in such great detail that more explaining of any issue with the use of more detail would not help explaining more facts but would become tedious and monotonous. Therefore my explaining at present is so simple a scholar should understand it.

I am under no illusion that you suddenly will find any inspiration to read this letter this time round since you never made any attempt to read any of my correspondence in the past. I am under no illusion that those of you that have incisive influence about recommending what the criteria should be about printing material of published material in major University or stand in as an adviser for any other Publishing Houses which may or may not form a department in your institution, those then will suddenly read my work with inspired interest and ruminate on matters I refer to. I do not envisage that you would this time round find a change of heart where you suddenly would enthusiastically devour every word in this letter, let alone get around to actively read any of my work in any of the books. I am totally aware that when you receive any letter such as this, you would immediately delegated the task of reading the letter to a lesser office since you have no time to deal with such small issues as this letter holds.

I am well aware of all your dismay which I am about to release where your response is dumping this onto a much lesser desk and that is the reaction that this letter charges where it is causing the rejection that stirs in you and I am equally aware that there is a poor chance that any of your members of staff would even care to read this letter to its end. You may well wonder why I then would address this letter to your office. Your reasons for not reading my work in the past was that you were to busy and that did not allow you the time to read it, but it is more to the point of simply being because I reject Newton and you counteract that by rejecting my work from which your reason for doing the rejecting is that in justifying your action as that you simply could not generate any interest to do so.

Sir, Madam, by my sending you this letter is in motivation that I can find solace because you are about to reject what I say again but your doing so notwithstanding my writing you this letter will be my consolation in my conclusion about an academic conspiracy. Reading this statement at this point might seem ridiculous and execrating but if you think in such terms do yourself the favour and keep on reading. If you are surprised in the statement then be even more surprised when I challenge your decency to prove there is no academic conspiracy going on to the effect of deliberately instigated or otherwise being protected because of your feeling vulnerable about academic issues. I see no need to introduce the books in this letter because of the simplicity of the information as it is presented for reading purposes in the books. In the books I am again showing what is incorrect about Newtonian gravity. That is not my personal perception but a fact I established without a doubt, but irrespective of whatever degree of undeniable evidence I bring, notwithstanding the degree of correctness that I deliver, my statements are never good enough, not even to justify an unbiased evaluation thereof.

I know from past experience that this statement where I suggested anything concerning Newton and Newtonian views as being incorrect nailed my coffin shut before you even opened the book. In the past and at present I am aware that such accusations about Newton already bring automatic disqualification to my work. It spurs immediate and total academic rejection of my work by all academic's concerned and I provoke the resentment they experience towards me in person in the most intensity any Academic can experience any resentment. They immediately feel offended by my challenging the system and I am also aware that much of the anger is as a result of their feeling personally offended in their position as guardians of physics. While they are the caretakers and Masters of the physics they guard, I come along and show flaws in what they see as being more perfect than God.

Even those amongst you that declare them to be devoted Christians think more of Newton than of Christ and don't fool yourself on the matter because I have a means to prove this statement. In defence of my condemning Newtonian gravity I ask you which is more important, the ego of the Masters or the truth of the work? In the past it seemed the ego of the Masters about the stature of the absolute Master being Newton carried supremacy above the truth while overshadowing logic and reason. Therefore I say this in total confidence that I am more than aware that I am about to evoke the very same response as I have done in the past in my rejection and disputing of all the Newtonian gravity views. In my saying this which is commanding the anger that I evoke I challenge you when you go onto reject my work again, to have some degree of honesty about why you again reject / ignore or dismiss my work. Be honest to your conscience as to why you are not prepared to analyse and to scrutinize Newton with me and in the manner that I do. However, when you do go on to reject my work once more, then I charge your honesty and your sense of fairness as the bearer of an academic pillar, to go on to prove how mass does bring about gravity. After six years of continues trying and after eight books were presented on the subject and with no clear response yet, what ells must I conclude than that there is some conspiring going on too suppress my findings.

Sir, Madam, I have reached a point where I am beyond diplomacy and I have taken off the gloves of hypocritical politeness. I reached a point where I call a spade by name. I now shout fraud where I detect conspiracy. I will do so even when this is present in the highest circles of the Physics paternity. I feel tested to a point where I am no longer prepared to use cotton wool as a means to avoid confrontation with the highest in all academic intellectuals circles in order to evade embarrassment by hiding honesty, decency and respect behind a cloak of hypocrisy just to save the face of a supposed respected paternity.

In the very same breath I also will admit: Sir, Madam, I feel like Satan incarnated in person while I accuse you because I do not know you in person or even as an individual person. I can only consider you as my superior intellectually and on all other levels and by all other norms and with that then what you represent what I have to regard as being above and beyond reprimand in any way. Yet when no one takes notice of what I say when I show what I say I have to come to a conclusion that there is some conspiracy going on. Why would the members of the academy ignore me rather than act surprised when I show a clear mistake that Newton made. Why trash my work regardless without reading it in serious consideration when I present a new clear definition about views concerning the basics in physics. Your ignoring comes in spite of the fact that no one knows what forms gravity and that includes even Newton in person. I bring a new suggestion to the table that holds all cosmic phenomena dearly and the phenomena are a fact whether academics attribute any reality to their being there or not. From that I define gravity in a new sense but from persons of your status I find rejection because the new definition does not involve mass. If anyone presents an answer to a problem the least one would expect, as an honest reaction is that the party which is informed would be surprised and appreciative of any new suggestion that may solve such a problem and I would think not knowing what mass is would constitute to a problem.

Why would any one dismiss the problem by dismissing the suggestion and ignoring the solution that is presented? The more the lot ignores me the more my suspicions mount because I prove Newton and his mass has no validity. If you are not guilty of what I come to accuse you of when I accuse you of conspiring to suppress evidence about Newtonian incorrectness, then why ignore what I say. Why do you then not just read what I have to say? What I have to say puts the cosmos out there in a completely new perspective and if I may add, a logical and understandable perspective. I bring a perspective where one does not have to go Bohemian to show singularity and to show space-time. It is there for all to see and even a child can understand what there is to see. So far my attempt to use a respectable avenue going through the official channels and following the guidelines on offer got me precisely nowhere and it took me years to get there with all the aid I got from the Academic paternity through the years. I am now at the point where I chuck all the niceties out the window and see what I came to suspect because of the treatment I received. I made excuses on behalf of the Academics and tried to find reasons for their behaviour but the only and last conclusion I could arrive at was that I was very naïve about the honesty and unwavering integrity in which I regarded the Academic world in its complete devoted sincerity to which they strive to achieve about the factual correctness in science.

I feel like the crook in the fairy tale while I accuse you of dirty dealings but what other choice do I have. I prove that Newton does not pan out and you brush me off by telling yourself and me that I do not understand Newton. That is complete and utter rubbish. I understand Newton better than any other person I have come across. Newton does not make sense at all. Mass only has validity when gravity is restrained because gravity is the motion of space that is filled with matter in $a^3 = T^2k$ and when space does not move it cannot act on gravity or motion where it becomes immobile and frustrated. Being immobile and frustrated results in that the space that is filled with matter is forming mass just because it cannot move independently any longer. However gravity is the motion of space in time and mass is the restricting of such independent motion. Read my work and you will see! If you are not prepared to read my work but always delegate the evaluation thereof to a person that is clearly not up to the task then what must I conclude from that? If you do not even try to address the shortfall of Newton then what must I conclude? Then it becomes obvious that those academics being in the status and position that you are either detest what I say or do not believe what I say. If you do not believe what I say, take my challenge serious and prove to your mind why you accept mass has gravity as a result while you scrutinize my ideas. When you detest what I say then find the reason for your detesting me including what I have to say.

Again I have to admit how embarrassed I am in my behaviour by accusing a person in your position but what else is there to conclude. The Critical Density theory makes a mockery of common human intelligence that reminds me of a bunch of drunks going on in an argument about matters they do not understand even when they are sober and in their state of drunkenness they are going mad. I just cannot believe the world's most intellectuals can reason in such an obscene manner that suits the likes of high school boys arguing about their fantasies rather those persons in such position as they have. It fits boys arguing to impress girls much more than it can be associated with the worlds supposed best minds there are. At the end of this letter when and if you have the courage to read it to the end and simply also on the condition that you will read it to the end this time round, then just be honest to your mind and think what I said and how I say what I prove in my books. I honestly do not wish to be insulting by shouting conspiracy but can you find another way that I am supposed to reason when I gauge the way you react?

The following four phenomena are real and are evidence found in the cosmos where they stand undisputed but alas also unaccepted since the phenomena does not match Newton. Newtonians rather accept Newton than would they accept the phenomena since the phenomena do not apply in the manner Newton explained the cosmos works. The Phenomena are as real as outer space is but because it does not match the concept of Newton and his mass by gravity, therefore the Newtonians discard them with many wide ranging excuses. The phenomena are amongst so many other fitting proofs of Newtonian misconceptions and I can explain every one as much as I can fit the lot into the forming of gravity. I can ensure you they have nothing to do with mass and so has mass nothing to do with gravity where gravity has everything to do with mass since mass announces the end of gravity. Mass forms gravity's grave. The one phenomenon is called the Titius Bode law and because it is named a law most astrophysicists scorn at the fact it is referred to as a law. I prove that this is not only a law but this is the Universe. This is gravity compiling a Universe. This is space-time and without this applying stars collapse as we find Black Holes do. To this day I must still find one Newtonian that even had a glance at this proof I bring where they did not tell me to my face the phenomena are coincidental and my proof therefore is also coincidental. Should these apply then Newton cannot apply because all the planets relate to one another as well as the Sun in this precise fashion, which rebuts mass differences. The relation is there and that Newtonians do not dispute but in order to avoid the explaining they rather put it down as a coincidental insignificant phenomenon.

The fact that all the planets show exactly this formation and that all the planets are distributed (all nine of them excluding no one however Neptune does not precisely fit the order but Pluto does and I do explain why Neptune is slightly out of line) is approached in a manner that would rather be degraded as a fanciful thought best left outside the mentioning of good conversation. They ignore a mountainous discovery because of what disgrace Newton's claims would suffer. The true disgrace is the rape Newton committed about the work of Kepler. Newton removed Kepler's formula and suggested that the rotary motion would nullify the radius since a circle does not develop any drive $\frac{dJ}{dt} = 0$. The concept is mathematical fraud as big and as wide as any scam can go. In Kepler's formula the one holding time (T^2) compliment the other factor in time (**k**) to allow space to duplicate as space moves through time. They are not in division as Newton claimed. Secondly there is no mass involved in outer space and in that we find the Titius Bode law that provides us with the evidence of unanimity where all planets are at a specific ratio, which removes any concept of size differentiation from the possibilities present which should be forthcoming where mass is present. One cannot divide the radius to the smallest point and find the radius would diminish to zero as Newton suggested. That is just a mathematical impossibility. The very essence of the Phenomena renders its result onto form. When a line is shortened by dividing the line will keep on reducing until the line becomes infinitely small and the line would have all point that forms the line positioned on one specific spot. That is singularity where 1^0 have no sides. If Newton did suggest that $\frac{dJ}{dt} = 1^0$ then that statement would ring true because I can prove singularity being present at that very location and that is correct!

As I said, I am about to prove that there is four phenomena in the Universe Newtonian science could never explain. They could not explain these four occurrences because they wish to use mass and mass does not exist. Therefore they cannot explain why these for phenomena occur or what happens when they occur and they cannot claim their understanding of how the phenomenon works and why they work. I took the phenomena and trashed mass as the product responsible for gravity. I dissected the inner working of the four phenomena and found that the four forming a unit in combination produces gravity. I prove this mathematically and without leaving scope for doubt. After trying to reveal my academic findings I went all over but found a deafness as good as a stonewall. No one in powerful academic positions wishes to have his work trashed and therefore they will rather put my work in a silent grave. They did it to most others that came up with new ideas in the past and I don't wish to go into detail but it took Kepler eighty years after his death to have his work merited. In the modern world I don't have to take that. I have been to just about every academic publisher and they hardly ever admit to receive my manuscript. Afterwards they never reply and those that do reply, has a lot to say about every aspect except any relative input on why they will not publish my book. Their academic advisers on astrophysics advice them to stay away from my work because I trash Newton and they maintain Newton was never disprove. You yourself can see how easy it is to prove Newton to be incorrect.

Now I decided to go to the press with all the evidence and start to fight them. Up to now I was trying to get their appreciation and accepting but after writing ten books and trying to convince them the past seven years, I came to realize they have not my or the best interest of science at heart. Every one of them are trying to protect their personal research how invalid that might be, and protect their papers and books and do not care for the truth to come out. If I can reach students by the mass media like the Da Vinci files did, and make the students see before the academic power machine can brainwash the students, I can get people to see and to start asking them questions. Those questions is just as simple as the ones I put to you in the start of this letter, but even the most brilliant academic minds are unable to answer anything when those obvious facts are asked. They expect every student studying physics to accept that mass is what produces gravity in spite of knowing well that mass, as a factor has never been proven. Any student insisting on such proof about mass that produces gravity is flunked by test and is shown away from the University.

If I can reach as many students as possible and drum up their anguish in finding proof, I will succeed. If I can reach every student in my reach via any medium I could commandeer and get him or her to insist on proof about how mass produce gravity, then I have those that try to dig my grave in defeat. Then those all so powerful have to come into the open and say why a feather falls the same speed as a rock and why a comet does not collide with the Sun. However, I need to get to students before they become brainwashed and accept deceit.

You and every other person forms the Centre of the Universe...and that is officially recognised by the Universe, no less

Some years ago I was reading of a remark Einstein made about his realisation on the subject of gravity whiles being a patent clerk. Einstein realised that had he, Einstein fell from the window of the patent office Einstein would feel as if he was as weightless as a chair was and a pen was that was also falling in his imagination, alongside Einstein down the building. The mistake Einstein made is to put his perception to his imagination because what he was thinking was as much a reality as physics could be.

With reading that I then realised Einstein would not have to resort to his imagination to feel as weightless as the other items that fell with him because the reality of physics is that if Galileo is correct then he had to feel weightless. It then would be "not as if" but he would feel that way in the true sense. If Einstein was truly falling the part, which was the falling experience, it was he that was experiencing the falling he was going through and with that he was feeling weightless through falling because that is what was happening to him. He was not pretending to fall whereby he then would feel as if...he was really falling and with that there is no as ifs. Being just as weightless as the chair and the pen is the result of physics and not part of his imagination tricking him to feel what he should otherwise not feel. What he then would have experienced came by means of what he was experiencing as a result of physics and not due to some mind game tricking him to imagine... If Einstein was experiencing weightless ness, it would be because he was weightless while falling and that Galileo said in proving his pendulum could hold time. Einstein would not imagine the weightless ness because Einstein was truly falling in a state of being weightless. He was at that moment truly weightless while descending to the earth and that is why all things fall at the same pace. Einstein, the pen, and the chair had the same weight since they were all weighing the same because they were all falling at the same pace. All three items would be equally weightless during the falling...that was what Galileo found because objects of different sizes and different mass travel equal while descending. The bigger objects do not fall quicker than a smaller object and that can only be attributed to one fact; it can only be true if the three weighed the same while falling when the three were falling. While in motion and off of the ground the three weighed the same since they travelled at equal speed downwards. However, when they stopped moving and came to a standstill they then weighed different. Only when being secured to the earth and therefore standing still while the Earth did the moving on their behalf did the objects have mass and by experiencing motion they weighed the same. This was the most eventful day in my life (apart from my wedding and the birth of my children) because from there on I started to conclude my theory on cosmology!

From this one can deduct that gravity is motion or the intent to commit motion and mass is when the motion of gravity is frustrated by some solid structure blocking or preventing the continuing of the motion. Gravity is motion of space and mass is the restricting of the motion of space. Having mass does not bring about gravity but it does restrict gravity's motion. Gravity produces mass but mass does not produce gravity or in fact produce mass. Mass is the restraining of motion and gravity is material moving about by committing gravity. Mass only comes into the application thereof when two objects filled with space moves into a position where both want to claim space the other occupy. It is the motion and the independence they show by having structural integrity and in that to hold on their individuality being true to the form the atoms hold the structures in that prevent the sharing of space, which in turn prevent further motion that causes mass. In essence gravity is still in a tendency but is then in the frustration of motion and is the commitment to move once the blocking of space is relinquished.

I then with this realised that gravity is the motion differentiation between objects. It is the independent motion providing a different speed while sharing a common centre off attracting that allows a discrepancy to establish mass under specific conditions applying between the two in relevancy. While falling the gravity applies as speed that is putting time in relation to the distance travelled. While the object falls the motion confirm gravity. When motion ends mass sets in and becomes the constraining of the object having further motion. The motion is still there but now it is reduced to a tendency while to move thus establishing the object mass comes as a result of the limiting of further motion when the falling body no longer falls but is then resting and is subdued by the motion of the Earth, which the Earth then introduce as mass. The mass is then a substitute of the motion and gravity then would be the motion performing as being the tendency thereof. However mass then restricts motion and becomes motion in a tendency to apply motion. This tendency to which I refer is what Newtonians see as being "pulled down" by the Earth. More accurately stated would be that the Earth captures the body's motion and by the Earth then providing the motion the body will relent independent motion to have mass.

While falling gravity applies and motion neutralizes size, mass or weight. Mass counters motion when the Earth restrains further motion of the falling object and the moving object is stopped from further movement where mass is then preventing or hindering gravity. The falling object remains individual and still tends to move while Earth individuality resists further movement of the falling body's movement. Further movement is disallowed as other material fill space. Remaining independent the falling object will finally come to rest on the ground while the earth takes over the duty of movement and from that motion restriction becomes resistance that becomes mass. While falling the object is experiencing gravity because the object is in gravity but when on the soil the object experience mass which is the restricting of gravity or motion of the space filled with material. It is a fight of objects to secure and retain the position in the **centre of the Universe**.

Moreover, I came to another conclusion of equal importance. When any person is standing on any place anywhere, while viewing the Universe, that person is filling the **centre of the Universe**. Let's get more personal. When you, the person that is reading this, are standing at night and is looking at the Universe you are seeing the Universe from the position that one only can have if that person is filling the specific spot in the **centre of the Universe**. All the light, every single beam that ever left any destiny at any time acknowledges this fact. You are the most important person in the Universe because you are holding the most important position in the Universe. All the light that come across all of the vacant space from any and all possible positions in space runs directly towards your position using a straight line towards you filling the **centre of the Universe**. Not excluding the effort of one photon, all light is heading to meet you where you are in that centre spot and not one photon will pass you by. Not one photon dare miss you because if they do they miss the effort that all light has to accomplish and that is to locate you as the person filling the **centre of the Universe**.

Should you decide to shift your position to any other place in the Universe, you will shift the **centre of the Universe** to that location as well. If you install a camera on Mars, the light is obliged to acknowledge your relocating the **centre of the Universe** at your will to reposition you're choosing that **centre of the Universe**. All the light that ever left its destination crossing the vast spaces of the Universe, excluding no particular light, travelled all the way just to find you filling the **centre of the Universe**, right where you are. By you're standing anywhere, you fill the **centre of the Universe**, and the entire Universe admits to that because all the light comes to meet you there. If you shift from the North Pole to the South Pole you will shift the **centre of the Universe** because all the light travelling throughout the Universe will find you where you then moved the **centre of the Universe**. The light left its destination billion years ago as it travelled through space at the speed of light anxious to acknowledge you're being in the very **centre of the Universe**. No photon will be able to pass you by where you are in the **centre of the Universe** because all light is heading your way from their start. No wonder every person born has the idea they were born to fill the **centre of the Universe**, which we do fill. The Universe is spinning around you or I, which is filling a centre where all motion is connected. That is why most people have this awkward idea that the Universe was created for man and every person sees him or her being the most important person on Earth because that person fills the **centre of the Universe.** That is the Coanda effect on the utter-most grandest scale imaginable; nevertheless it is only a manifestation of the Coanda effect. It implicates gravity as wide as can be… This might be a tongue in the cheek idea but this idea explained to me what the Universe truly is and why life is in the Universe while never being present in the Universe, but merely experiencing a presence in the Universe. Gravity **is to move or apply the intension to move** space a^3 at the distance of **k** while T^2 is the time **it is going to take to apply gravity or move the space filled with material** space a^3 at the distance of **k in** the time period of T^2. That confirms Kepler's attribution to gravity where according to Kepler space a^3 is equal to the movement T^2 (time it takes to move) at the distance k from the centre specific. Every aspect of deliberation about the Universe was never discussed in the manner it is discussed in **An OPEN LETTER TO SELECTED ACADEMICS ISBN 0-9584410-9-X written by Peet Schutte.**

Then I reviewed his vision he received from a vision Einstein received and applied such a vision on the findings that Kepler received from the Cosmos. It puts all aspects of gravity in the Universe in new dimensions. But the visions formed the beginning because the visions unleashed many new questions. If gravity is motion, what causes motion? What stops motion? That answer is in the Black Hole. If a star is about fusing atoms thereby growing, what happen when all the atoms fused into one all collective atom filling the entire star? Then the Black Hole is gravity so strong it fused all the atoms within the star into one unit filling the star. What is the gravity if the star has one all-inclusive atom providing all the gravity that the star had when the star still had massive volumetric space? If all that space that once filled an entire giant star fused into one enormous gravity applying atom and that enormous force has been secures in the space that one atom holds, the atom would then show a force that would pull the surrounding Universe flat. Where does the gravity of the star end when all the atoms in the star became one giant atom? Gravity is smallest where space is least and this observation is the answer to the question: what is a Black Hole? Where space of an entire massive star is left in the size of one atom the gravity coming from that will pull the Universe flat at that point. Coming to the conclusion about gravity being motion and mass being the restriction of motion was the easy part. What produced the motion and what prevented the restriction from overcoming the motion was the tough part. Figuring out why was everything on the move and where did the motion stop that was the part that took some figuring and some explaining. What made gravity move and why does gravity move…the answers are in the four phenomena never yet explained to satisfaction but now turns out to be the cradle of gravity.

"GRAVITY IS DIVIDED IN TWO FACTORS, BEING **LINEAR DISPLACEMENT** (Π) WHICH IS WHAT **NEWTON'S GRAVITY** IS AND,

There is a position that is in motion that is forming the very edge of the outside of whatever spins or moves or mathematically equated as $\Pi^0\Pi\Pi^2=\Pi^3$. This says that to be in motion the position securing the motion must be in relation to a point forming a centre.

CIRCULAR ISPLACEMENT (Π²) WHICH IS THE "GRAVITY" EINSTEIN RECOGNIZED

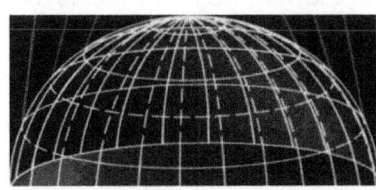

The extension of Π is well received as a dimensional implication to matter holding seven positions from singularity and space having four quarters through out the rotation of singularity forming

Those that are of the opinion that it is friction with the air that heats a spaceship or any object when it enters the atmosphere of the earth, please explain the following: When a spaceship enters the earth's atmosphere by more than 21° it will burn out. If it enters the earth's atmosphere by less than 7°, it would be bounced off into space. The 7° comes about by the concentration of the atmosphere and this bounces the spaceship outwards. However, when a spaceship enters the earth's atmosphere at an angle higher than 21° it will in fact meet less air than in the case of the spacecraft entering at an angle of more than 21°.

the centre to the five dimensions (one side lost to the cube's six sides connecting to the five remaining

sides) making the total sides facing space from the point holding singularity at any given instant at a value of twenty (4 X 5 = 20). **Much of the proof about gravity is part of our perception about gravity because we experience certain conditions being within and being confined by gravity. But are our perceptions about gravity truly correct? We only experience gravity as a factor from the position we have on Earth and that is while we are being forced to be part of the Earth.**
This is what I say what physics is based on...

When one goes about drawing a line the first motion will be to place the pen on the paper. The paper is blank but the blank paper does not represent zero as an option. If it did, there should either be no paper at all or no pen. But that does not bring about the option of zero either because a possibility may arise where one will obtain any instrument to draw the line. It could be a stick or even a finger and the paper may be sand or wood to carve on. The main issue is that there are always many possibilities whereby a line may form and that exclude zero as a possibility. Even by the two stars lining up brings about the forming of a possible line. The shortest line in the realm of possibilities must have a start and finish holding one spot and such a line will also be a dot or a circle. When placing the pen on paper or the chisel on rock or which ever device one may choose to draw a line the line will start with a mark. The mark may be a sizable hole all a chip so small it is barely visible to the naked eye, but it is there, even if it is so small it is only visible by the use of an electronic micro scope, the dot is there.

Humans will tend to favour one specific direction to take the line from the position of the dot, but humans are not creating we are merely duplicating what is already about. **Singularity is a mathematical point at which certain physical quantities reach infinite values for example, according to the general relativity the curvature of space-time becomes infinite in a black hole.** If singularity is a mathematical point we then should discard mans options and look for the option where mathematics will bring about a non-bias flow. Mathematic has no pre conception but man-applying mathematics does and that is where mathematics goes astray. We also have to presume that space-time came from the first dot because where else did it come from? One thing about the Universe is that it holds conformity going out in all directions even-handedly because if it did not there has to be borders, which there are not.

Not favouring one direction puts all directions at equilibrium meaning that any form of what ever might develop from such a spot with the end and the start being in the same position also has to be a sphere because the flow outward will be equal in all directions. This reasoning prompted me to look for singularity in such a spot because if the prime spot from which all came was a spot holding all, then the spot must hold the shortest line but more prominent it will hold the smallest form including the smallest circle or for that matter the smallest sphere. One possibility that the shortest line or smallest spot can never have is having a starting point on the zero mark. If the mark of zero holds the start it must also hold the end because the end and the beginning has the same position. If the position of zero then is the beginning, the end will also be zero leaving the line or spot without an end as well as without a beginning. Such a spot will constitute all of nothing.

The size only depend on the fluctuation of r in the square as a component to the circle or sphere but that does not affect the form by indication of Π in any way there may be. The conclusion from this is that no line can start at zero because that will be a mathematical impossibility. A line or spot starting at zero would therefore be shorter than the shortest line possible. For obvious reasons can no line, or any line grow or extend from zero because such a line must then quit zero and become something, thus abandon its original value. That would mean the start of the line has a different value to the end and a line holds conformity through out. When any line is starting from point zero it can never leave zero because of the influence of being zero disqualifies any possibility of growth. If the line then had to grow in all directions at the same pace the line must therefore be a circle or being three-dimensional, a sphere. Flowing from this fact is that in the Universe there can be no zero point or unfilled space. In the case of the growing sphere the value of the circle is Π, and that is where creation started. That gave me the clue where to start looking for singularity. One would find singularity in the value Π and the value Π will be in all things rotating in a circle. You might wonder how does that apply to the cosmos and moreover to gravity? Let us find the answer from what we know and work our way back to the infinite where we also know we will find singularity as defined by the Brainy Bunch.

By reducing r indefinitely to the tune of half each time, r would become infinitely small, beyond human calculating means, however as mentioned in the case of the smallest dot holding one spot, r would become insignificant beyond human comprehension even, but never reaching zero and still Π would remain intact and dictating form. Reducing r by half would cut the circle by four but that does not mean that Π as such reduces. It is r going in four directions that reduce leaving Π to maintain form. It means the reducing does not affect Π but merely influence Π.

When the circle reduces, the value located to r will become implicated because r determines specific size. Not so in the case of Π, because Π in the true sense only indicate that the circle is a square without corners and therefore Π dictates form and not size. By reducing size only r comes into contest and will point to such reduction. By reducing the circle radius r by half continuously will lead to an infinite small circle but Π will remain because the circle as a form remains even being infinitely small. That then brings about infinity but leaves out zero as an option because by applying infinity, infinity can never be reached because of the qualifications applying to infinity. Being at infinity takes us into the heart of singularity and singularity must therefore be in all round objects since the argument took any or all circles excluding not one.

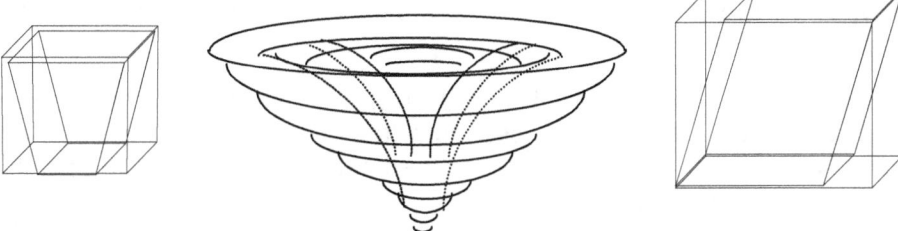

Using: r as reference or using Π as reference only changes form.
An observation coming instinctively to mind one may recognise is that the form reminds rather explicitly of natural phenomenon as hurricanes, water whirls and even the shape most commonly favoured to express the cosmic object referred too as a Black Hole. The definition of singularity specifically places singularity in the centre of the Black Hole and by reducing r we obviously will reach a line of infinity where singularity is by definition. The similarity may be more than coincidental. Let us consider the statement in the reverse. Even by reducing r to the point of hypothetically not existing any longer Π still remains a factor because r becomes the implicated factor and Π merely influenced.

Anything occupying space in the cube will apply r, notwithstanding the name used confirming the shape or r named as length width or height, it is all just a straight line bringing about the cube with all its other names that may find attachment to specific form but nevertheless still remains only a six-sided cube with connecting lines applying different angles changing in some cases.

Since the identification that the star uses hydraulic power and not compressible pneumatic power one must accept that r has little use within the star. In the star the heat surrounding the particles have more density than does the particles themselves. Matter will crack and break under hydraulic pressure, but hydraulics only increase heat and since the liquid is heat, it only brings about that the fluid will become more of what it already is. But when saying pressure it has again more to do with culture in science than with the fact of the matter because heat will flow from hottest to coldest no matter what and therefore the pressure must relate to cold versus hot in relevancy of singularity holding different positions within the stars.

The normal perception is that any circle growing spontaneous would grow by the radius, which is r. That cannot be the case because r is an indication of a straight line. By growing with the aid if a straight line the influence that would have on the circle would result in many circles following one another and not a continuous growth. Gravity is the dimensional changing of space holding r as reference to the sphere holding Π as the reference. In order to generate spin that produces time in matter occupying space, therefore creating dimensional change, Π has to be a factor indicating the possibility of spin. The answer must be in finding Π, and thereby locating singularity.

In considering the spinning motion in the fraction of time in the detailed instant every aspect of rotation will turn in every instant of change in time. Although the points had the same characteristics only seconds before, they oppose the characteristics it had just before and just after the very second in which they are and to which they relate by similar points also in rotation. The fact of the graph proves my point in quarterly opposing dimensions and values,

Due to the spinning nature of such a point with all surrounding the point will be alternating direction favouring change every second and in that the value to such a point can only be Π because of its constant changing. Using r would specifically oppose another r from every angle because the use of r will bring about a static relation to the previous and following instant and therefore it will cancel the constant spin flow.

Space-time is a four dimensional position of the Universe where the position of an object is specified by three coordinates in space and one position in time. According to the theory of special relativity there is no absolute time, which can be measured independently of the observer, so events that are simultaneous as seen from one observer occur at different times when seen from a different place.

From this view and the knowledge that all things being the atom or sub atomic particles of the atom are in motion, therefore all of the Universe are in motion and thus qualifies as space-time.

Locating and finding Singularity in places other than Black Holes must be our next quest because to my view everything connects and that view places a Black Hole in the position of being just another star.

In the beginning there was one spot. It is still there and it is not there. The reason why it is there is because it can never be there and since it can never be there it can never be removed from being there. The spot is always there because only the spot is really they're as an eternal presence but this is only because since the beginning the spot was never there.

The point that I refer too is there by never being present as a part of the Universe. In the centre of all spinning object a line forms that part such a spinning process into four distinct zones in line with three points where one point is in the middle and two more points are to the top and below the centre point. In all that forms seven points and therefore we have a sphere.

However, by the space is not turning the points does not exist because the points form by the spin creating the points. The points can only realise when motion puts the object in relation to the motion and therefore the object has to establish independence by establishing independent time. By not moving the points are not excited and can therefore not have a presence in the Universe. But even when not spinning the points is only one point but is still not present in the Universe. When Kepler's formula on motion only applies in terms of Newton's vision about the top having mass, then the points forms one point that is not serving nor is it representing the top. By coming erect the top dismisses the effect mass should have when pulling the top to the ground.

Locating and finding *Singularity*

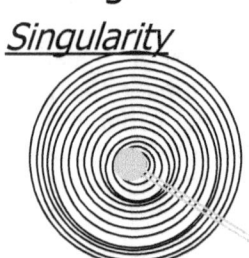

In the **precise middle** of all **objects in rotation** is a precise centre dividing the object in sectors that will **start the spinning initiation** from that centre point. Thus, the spinning object **will have a middle point**, a very specific **centre point that does not spin** and only holds Π as a specific value because no radius can apply. But also the one value such a line **cannot have is zero** because the line **is there and holds contact** to the rest of the material bringing about that **zero does not start any** line and therefore the **value of the line must be infinite**, just as described in **accordance** and by **the definition of singularity**

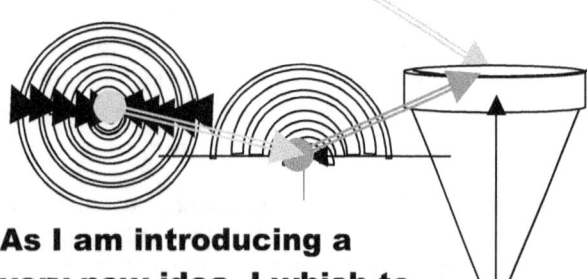

Move the rotating line progressively to the middle by reducing the length the line have from the edge to the middle. At one point all further reducing ends.

As I am introducing a very new idea, I whish to explain in more detail what I try to convey.

As the rotating direction moves inwards, the rings will become smaller and smaller.

That point albeit hypothetical, is also as much a reality none the less and is placed where that point must be standing still because every line running from that point in opposing directions are also in opposing directional spin the other or opposing side.

In considering the spinning motion in the fraction of time in the detailed instant every aspect of rotation will turn in every instant of change in time. Although the points had the same characteristics only one instant before, they oppose the characteristics it had just before and just after the very instant in which they are and to which they relate by similar points also in rotation. The fact of the graph proves my point in quarterly opposing dimensions and values.

The Four Cosmic Pillars; The Result Thereof. Page 19   In Terms Of Applying Cosmic Physics

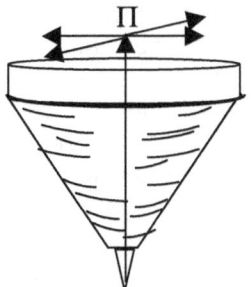

Every quarter provide a distinct value that indicates the progress of the flow of time from the one point Π to the next point Π.

Any changers occurring in Π will lead to a an unequal triangle providing two different values to r and will alternate the link between r and Π² bringing about different form (Π) and time (Π²). When singularity forming the lines of the triangle is not in equilibrium the triangle will destroy the matching of half circle.

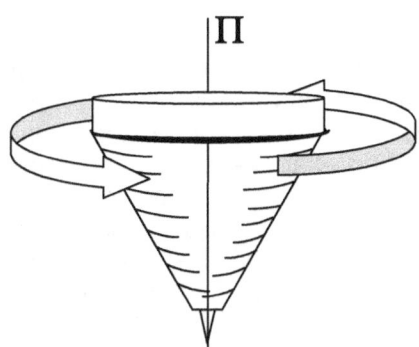

The sectors provide individual singularity as a means in sustaining governing

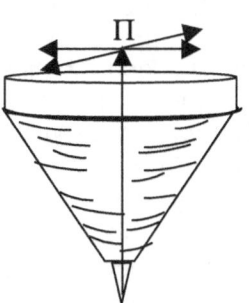

Every quarter provide a distinct value that indicates the progress of the flow of time from the one point Π to the next point Π.

Any changers occurring in Π will lead to a an unequal triangle providing two different values to r and will alternate the link between r and Π² bringing about different form (Π) and time (Π²). When singularity forming the lines of the triangle is not in equilibrium the triangle will destroy the matching of half circle.

In every sector the directional flow will provide a distinct meeting of Π linking r to Π² and this allow the time component in the rotation.

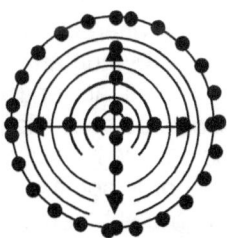

Being motionless on its side portrait the top to the rest of the cosmos as being a smaller part of the larger Earth. In cosmic reality the top not spinning on own incentive is just more Earth as the Earth is just more top. The top does not move and therefore the top is unable to activate singularity through which it can achieve independence. The top forms part of the spin the Earth produces and therefore it forms part of the Earth by having mass in relation to the Earth's structure. The top is merely one point filled with many one points that is in motion around the centre of the Earth forming a line in relation to the Earth roundness that forms singularity. The top is on the one point in singularity of the Earth which, forms part of the four points that rotate around a line in the centre of the Earth which is not there. The top is part of the roundness that the Earth provides in relation to the Earth centre. The top is not spinning but it is the Earth that is spinning…the top as well.

By life giving the top the ability to spin independent it is then Kepler's defining formula that gives the top any notoriety above and beyond being part of the Earth. In order to establish singularity being activated there has to be motion applied to the inertia the top sustains. At this point the overwhelming concept is motion but the most demanding question about motion is what establishes such motion. What will create the motion required to excel motion? To humans having life motion is the most natural and least pondered thought of all. Motion is what we have and motion is what we do. Motion is life and therefore individual motion goes past our thoughts because we cannot think of anything not being in motion. In that we differentiate in terms of our perception and not the relevancy that nature applies. When the top is on the Earth and life never again would intervene to move the top, the top will become as much a stone as the dinosaurs became rocks. There will be no difference between the rock below the top and the soil that will through time cover the top. Eventually the top will be rock. That is if and when no life form ever comes to change the position of the top or to relocate the top as a whole or as parts of the top. If the top remains where it is and never moves the top becomes 1^1 and the Earth in the centre forms 1^0. The top becomes $4\Pi^2$ by motion of Π being relevant to the Earth's centre forming Π^0. The top is the Earth by relevance in the fact it has the motion the top accepts from the Earth. By having mass the top is cosmically absorbed by the motion of the structure of the Earth.

The dot in the centre of all objects is always there all the while the dot cannot leave because it is missing from the Universe and can therefore never be on this side of Creation. From that spot every dot forming the Universe came about and yet not one dot forming the Universe is present on our side of the Universe. There was Π^0, which was α^0 or if you would rather have it be Ω^0 or it maybe was 1^0, but more correctly it was all the above and the beyond because multiplying what ever constitute the mentioned will bring about what is mentioned to a precise equality. It was a spot that was not a spot inside our side of the Universe. It was a line that runs eternally since it not only represents time but it is time but because it runs eternally and keeps repeating exactly what was before to the precise what comes afterwards except for replacing the dot that is not present in the first place the line was there and was eternally running, while never changing in the least or growing by any measure because since time began it was never there. It was not one because before it was one, what was repeated and the process cycled back to before one and before one could be reached. This was when eternity was eternally repeating what were on the spot infinity claims. It was such a continuing of the monotony, no change occurred and therefore never did the running produce progress because the progress was in the perfect repeat of what was before. But then heat changed it all…

When not 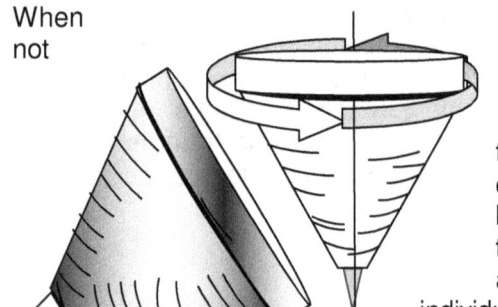 the top is motionless and therefore Kepler's formula $a^3 = T^2k$ is active the space the top holds is invalid although it controls all that is valid. The top moves with and according to the gravity it receives from the Earth, but by moving the top then extend on the movement it receives from the gravity of the Earth. When lying on the ground and not individually spinning the top is not the top but a tiny part of the larger Earth. The motion in space a^3 that the top serves are part of the time the space fills as the Earth moves around its axis and around the axis the Sun provides T^2k. By not individually spinning the top has no defining extension giving the top validity and that is notwithstanding the Human interpretation we give the top whereby such a realisation gained from life then applies to the top being formed by human hand. Only when the top spins, does the top get Universal recognition as an entity within the Universe. We as human minds think of the top as something but the Universe sees the top as part of the Earth. Only when human hand spins the top the top becomes an extension of life because without life being present the top will not be moving. To the Universe the fact that the top is or is not is not because the top has mass but because the top is moving. Even the body the top holds represent life as wood and in that context the top is very connected with the reality of life. There is not and can't be a spinning top on the Moon or on Mars because life is absent. Life as a cosmic presence is not a reality being recognised in the Universe notwithstanding Newtonian madness that wants to put life at a measured preemie of a penny a gross spread wherever the mind may wonder. Until Newtonians find life they have no right to put life in any other place than on Earth where that is the only place life is a fact. Putting life any place other than the Earth is promoting a fairy tail and is not factual science…but then Newtonian science is one big fairy tale. All other places life is not a proven fact or forms a valid factor we can't assume life could be a presence. While it remains unproven life cannot ensure a permanent presence in any other place in the cosmos but here on Earth. Without motion there is no difference putting distinction between the top and the rock it is on. It cannot move as much as the rock it is on cannot move except moving as a joint part of the Earth.

By spinning the top comes erect with an identity the top tries to protect as if the top is fighting for its "life". The top is swaying when going to fast or too slow and the top sings by vibrating when it is in motion. To give the top independence the top has to spin and by spinning the top comes erect. When the top start moving the line of singularity activate within the centre where we think of in terms of an axis forming because the circle parts from the centre. The four position holding time becomes apart from the line that forms and the line that forms never forms. Time with no start parts by motion from time with no end and time that has no space comes loose from time that has eternal space. I can show time with no start as much as I can show time with no end and that, which stands between time with no start and time with no end is a lot of atoms moving. A lot of atoms moving are heat concentrated in a defined space. That means what parts time is heat and heat establishes motion.

This is the most critical realisation about the cosmos man can make. Everyone is all about energy but the term energy is as misleading as the rest of Newtonian science fitted into cosmology. The rock is energy and the Sun is energy and I am energy because my emotions are energy and the chocolate bar is packed with energy while the ship releases energy into the sea, which teams with energy. The term is so widely used it loses all meaning. As soon as the Newtonian wishes to explain what he or she truly has no idea about he or she starts using the term energy. Because no one has the foggiest notion what energy is and every one except that every one knows exactly what energy is, no one asks questions about his true understanding about the concept he has no understanding about. Einstein said that no material could go past the speed of light because when material passes it will become pure energy. How vividly expressed can that statement be. Material is pure energy because unleashing the energy within the atom created energy no man can come to understand. It is called a nuclear bomb and has so much energy it destroys everything in its wake! That makes the atom pure energy in any case so what was Einstein's case about the atom being pure energy when it moves faster than the speed of light? Everything is pure energy because pure energy is unleashed heat. There are heat that is enabling the parting of time in infinity from time in eternity and that parting by heat is the motion that sustains the parting we see as space filled or space formed.

Between the top that is not moving and the top that is moving is one Universe in singularity. A Universe started when the top started moving. Time in eternity separated from time in infinity and in between came heat that is forming motion. When not being active and being on its side the top are not a factor in the Universe but a factor on Earth forming a part as only some small particle of the Earth. In order to receive distinction movement to the space $a^3 = T^2k$ it claims that forms time is required. To all Newtonians that lean so heavy on the energy expression, well, this is the pure energy Einstein referred to because it is the heat or light that parts time. The parting of time by heat creates the motion there are in the Universe.

The spot forming the very centre of what does not spin is not a part of the Universe. It cannot move and in all principles applying it does not exist because it holds no space. As the spot moves to become the dot the motion is not a motion but is merely a reference to a position changing with time. By referring such movement coming about is a reference of motion indicating a change in position that is not but is extending what is not to where what is finds a beginning. In motion by line the motion also moves in a circle since the line not forming a line becomes the line that forms everything and is at the same time also a half circle as well as a triangle. It is within singularity where mathematics just starts! By extending as it expands in overheating the motion places the spot towards the dot but the spot cannot move except to move outwards to form the dot. In that a relevance come into place where the interaction extends to a point beyond the spot held in place by another spot in relation to six points forming the edges of a sphere. When the sphere forms the smallest space in the Universe coming just out of singularity the points are also not in space but merely indicating a start of space but since they hold relevance to the spot they become the dot by six spots held in focus by one centre spot that is not.

Since it is not the spot that moves to the dot but the relevance of associating with six other spots that come together to form a dot and the relevant association is in terms of three point confirming one and forming two. The motion that translates from such expanding is a half circle that is equal to the straight line since going around the spot is equal to going through the spot but at the same time it is bridging the divide and the divide is a circle. The places three point in relation to two where the two in concern bring a change in direction more than merely forming motion.

The motion is not actually taking place because the motion is in the extending of singularity (1 + 1 = 2) and it is in the extending that the circle is equal to the triangle that is equal to the straight line all being 180°. But since the extending takes the relevance from what is not to what should then be the dot where space will begin it forms by seven spots and by extending from the centre, which, is a spot to a point that can only be confirmed by what is not the extending. This action puts the relevance of the motion onto the six sphere – forming points that should form. But such extending is a straight line and therefore the motion forming the circle goes from the dot that was formed by the seven points that are not and the dot extends motion by expanding to another dot formed in the same way. That puts the circle in touch with the next dot by relevance of space formed as a result of motion that never took place in the first instance. Moving from what is not to what should be and having this relate to something formed by the same principle puts the Universe holding 2 forms which is putting the relevance of two in place and going into the fourth dimension $2^4 = 16$ that is associating with the seven of the sphere 7 and that confirms our Universe being at the dimensional standing of $2^4 \times 7 = 112$. For that reason the Universe can retain all that is possible by forming the forth dimension of space from the lagging of time.

All the while it was just a spotted and dotted line running along time as space duplicates with heat surging and cooling as movement brings cold and cold contracts much similar to the actions of stars in the process of pulsating or otherwise known by what ever name one wish to use when referring to a growing and declining star. Time is the moving of space to a new position honouring a direction pointing to a new location in relation to the previous point that will oppose the previous point it had in relation to direction considering the centre point. Looking at the affect of gravity it shows the precise quality of having no distinctive point, as gravity never seems to end at a point but flows all over affecting all that holds a position in its sphere of influence. The gravity coming from China meets the gravity coming from America at no particular spot but intermingles without distinction.

Using the concept that gravity applies Π as the circle factor Π as well as Π^2 replacing r^2 the replacing of Π brings two values as Π and Π^2 this method defines space Π^3 or $\Pi \, \Pi^2 = \Pi^3$. That I found is the case with gravity and will be apparent when explaining the sound barrier as well as the Roche lobe. In order to create a distinction I remained using r as the indicator of the cube or circle that has vacant space and by vacant space I refer to non-solid structure. In the solid structure I use Π as a value for reasons that will become apparent in due time.

To any and to all of my critics of which there will be more than grains of sand on the beach, please take note that it is not humanly possible to supply information contained in these four small books that is representative of the information in the (more or less) twenty five full books I have written on my theory in which I explain my theory. When finding grounds whereby you whish to condemn me then do so after you have investigated most of my work.

Do it on the grounds of the information gathered from most of the books put together as a unit, in which I explain my entire view. Then when ostracizing me, be honest of the motive by which you condemn me, but also please do that when you have read at least the first six books offered as printed material in paper format and then state by what facts you condemn my work.

Moreover, please **PROVE Newton's correctness** and therefore then my incorrectness and do not merely declare it on the grounds of practising physics as a culture that became a religiosity when accepting Newton's correctness.

EXPLAINING THE ABSOLUTE RELEVANCY IN TERMS OF SINGULARITY BY PUTTING SINGULARITY IN PHYSICS

ISBN 978-0-9802725-2-9

All rights are reserved.
No part, parts or the entirety of this book may be reproduced by publishing, electronically copied, duplicated by whatever means that form reproduction or duplication of any description, without the prior written consent of the copy rite owner.

WRITTEN BY PEET SCHUTTE
© KOSMOLOGIESE EN ASTRONOMIESE TEGNIKA

WHOM IT MAY CONCERN,

I do find much pride in my status as being Afrikaner and would like to have my names used by pronouncing it in the manner Afrikaans dictates...therefore I would sincerely appreciate the courtesy when readers will take note that my name and last name are pronounced in Afrikaans, which is originally from Dutch and must be pronounced that way. Peet one would pronounce "here" which is the closest English to the pronouncing of the "ee". The "Sch" in Schutte is pronounced exactly as school is where both actually are pronounced Skutte or "skool". By pronouncing my name in Afrikaans you do me the utmost courtesy any one can. Being an Afrikaner is what I am most proud of. Another point I wish to highlight is that I feel compiled to produce this work in a comic-like format. I have found that the more intellectual and the more educated Academics are, the less they understand the most primitive or classical mistakes in science as well as physics.

As I said my mother tongue is Afrikaans and my second language is English. I have per suiting this theory that I partly present in this book, of which the investigating research was done the past thirty years. Then I compiled my presentation thereof for the past nine years on full time basis whereby I was tying to introduce my findings to many academics without much joy. This past nine years saw me go without any income as I tried to get my theorem recognised. Going without a steady income left me almost destitute and in order to find a manner to get my theory across to the attention of influential readers, I decided to publish these books electronically as to try and get around the stranglehold of Newtonian bias controlling science at present worldwide. I decided to publish these articles through LULU.com which I saw as way the only manner whereby I could generate funding by which I would be able to have the twenty seven books I already wrote linguistically edited and then to have the books published on a Print-On-Demand basis. With my first language not being English and the books not linguistically checked by an expert there are bound to be language errors that readers will notice. In the past I tried to check my work myself but after checking say one hundred and fifty pages for language corrections, instead of having corrected work I ended instead having four hundred pages of new written information which is still not language corrected but holds a lot more information. This is because my priorities lie elsewhere. I aim to spend money on correcting the work as far as language goes, as I receive money and in the hope that I will receive money. I will have all my work including the one you are reading edited professionally and corrected as I find money to do so . . .

I have discovered that the Universe is not employing a Special Relevance of singularity, but there is a state of **_The Absolute Relevancy of Singularity_** that is not only controlling the Universe but is what the Universe constitutes of...it forms the Universe ...it is the Universe.

However, notwithstanding the magnitude in significance **_The Absolute Relevancy of Singularity_** poses to science forming a breakthrough, yet past experience taught me I have no chance that my theory on **_The Absolute Relevancy of Singularity_** will be noticed.

I came to the conclusion that members forming the body of Mainstream science in physics will not care to take any notice of **_The Absolute Relevancy of Singularity_** and I don't believe that it will be read, will be seriously considered and much less be accepted by those with the authority to change physics principles. I hold the opinion that the theory I introduce here and now would never be accepted in my lifetime because science in the Newtonian way is bent on believing in the marvellous, the outrageous and the magic of what can never be explained, although they claim to use facts as a basis. Science has no idea of what a Black Hole is and I can prove what a Black Hole is. Science has no idea what "the sound barrier" is and I can prove what it is. The explaining of science coming from this that I prove is almost endless.

Yet, I feel I need to warn you whom are reading this letter that this work contained in this letter strays widely from mainstream science and for that there is a very good reason, but I should add that in the least it is thought provoking. I researched the work of a man that is most exceptional and even more prominent in the history of mankind and yet the meaning of his work went unnoticed all this time. His role in the gathering of information furthering knowledge accumulating of the human species' efforts stands second to none while most of everyone is not even aware of the full implication of his work.

While recognising his work Mainstream science bluntly ignores his work and in that they miss the full vastness of the wide influencing of his work. It seems to me that any research predating Newton never came into use or in practise. My investigation of Kepler's work brought about a conclusion that no one yet arrived at concerning the findings of Kepler because no one scrutinised Kepler's formula. Kepler found planets rotating around a centre but Newton saw a circle and added what is mathematically required to indicate such a circle. Newton added a mathematical $4\Pi^2$ to the formula of Kepler and removed the distance symbolising measure that Kepler introduced using **k**. On the other side Newton changed the symbol of **k** by using the symbols G $(m + m_p)$. All of this I change and show why it has to change back to Kepler's vision in order to better man's insight into physics, but in that I change the grain and foundation of mainstream physics, I change the total understanding of what forms the basis of cosmology and that part is what mainstream science avoids.

In the book that deals with gravity there are just too many and numerously wide ranging facts that form the complete picture as a whole, which leaves me unable to include a full introduction in a space as small as that which page will allow. The explaining include for instance those phenomena, which I call the four cosmic pillars, but wise as you are, you would not believe me at this point that I have cracked the coconut because I guess in your vast experience you have seen too many idle explanations in the past proving to be senseless and little impressive, therefore my mentioning my success would not matter much either way.

The proof I bring is true about gravity being formed as a result of these phenomena,
1) The Lagrangian system
2) The Roche limit
3) The Titius Bode law
4) The Coanda affect, which I explain by delivering mathematical proof as to how they fit into the overall picture of gravity and which I mention just below. I prove the fact that every individual one of those phenomena is forming a unit that is in total being what we think of as gravity. The phenomena altogether constitutes a unit that forms the process working as gravity. Nevertheless my mentioning these facts will be just completely unbelievable to you without you reading the book, because I guess you have heard some attempt to explain the phenomena before but when I say you have not heard it in the context I put it, you might still be most sceptical because you have never heard it in the correct manner that I explain it and that poses the difference. Still you may not be convinced about my claims and although my explaining the phenomena is correct, does not change the fact that you don't believe me.

The phenomena form an intergraded unit that results in gravity forming where each forms a part of gravity. You may still be you would be sceptical ...but convince yourself that I did manage to:
 1) Find the location, position of singularity as a factor forming space-time
 2) Finding space-time by dissecting Kepler's formula in relation to valuing singularity
 3) Finding and proving space-time and aligning space-time with gravity
 4) Find the working principals behind gravity as a cosmic occurrence.
 5) Find the reason for the Roche limit and explaining the resulting of gravity from that.
 6) Find out why the Lagrangian system, becomes the building form of the Universe.
 7) Find why the Titius Bode law mathematically provides the foundation of gravity
 By proving that the Coanda affect is gravity through activating space-time

By using the above the four cosmic pillars, it enable me to present the proof where I now can explain what conditions bring on the sound barrier. By proving it is gravity that the individual structure generates motion above and beyond the gravity the Earth provide is what is producing individual motion that the independent object earned within the sphere of motion that the Earth's gravity provides where the independent and individual motion put the relevance that gravity has beyond the conserving means gravity has where the space that is serving the independent object is independently in motion.

The adding to the independence on top of the normal structural independence is creating more individualism by the independent motion of the individual structure being apart from the motion that the gravity of the Earth provides. The fact every one misses is that any structure that is not part of the Earth's crust has an independent gravity and the form this gravity applies is stronger than the Earth's gravity which is why the structure maintains its form and this provides the independent individuality the structure has giving the unique structural space. The gravity of the Earth strives to incorporate everything into the Earth's sphere and into the Earth's structure and therefore the fact that the object is not incorporated into the Earth shows defiance and individuality, which gives it, mass.

By applying individual motion on top of the structural individuality that increases by the motion that the Earth provides, the independence of the individual object is becoming further exaggerated by having independent motion, which is further defying the incorporation the Earth strives to achieve. As the motion of the independent object grows more independent by applying more excessive motion to such an extent **where motion creates almost the ultimate independence that may free the individual object with independence from the motion the Earth creates** is what is breaking the restraint gravity has on all objects with independence formed by their structure. The structure show independence at all times by not forming part of the structure of the Earth within the sphere of the Earth's gravity. Moving about shows even more reluctance on the part of the top when spinning allows the top to eventually become part of the Earth. **Breaking the sound barrier is the motion** in space duplicating space by crossing over gravity borders, which is the limit to what constraint the Earth may produce in accordance with what full independence would allow.

These are the definitions underwriting cosmology and while my work is that much ignored; let's see how far I stray from these definitions in comparison of how much Mainstream science underwrites these definitions by bringing indisputable proof in presenting unwavering hardcore facts.

Quoted directly from the Oxford dictionary of Astronomy the following:

The definition of space-time is as follows:

Space-time is a four dimensional position of the Universe where the position of an object is specified by three coordinates in space and one position in time. According to the theory of special relativity there is no absolute time, which can be measured independently of the observer, so events that are simultaneous as seen from one observer occur at different times when seen from a different place. Time must therefore be measured in a relative manner as are positions in three-dimensional Euclidean space, and this is achieved through the concept of space-time. The trajectory of an object in space-time is called world line. General relativity relates to curvature of space-time to the positions and motions of particles of matter.

The definition of singularity is as follows:

Singularity: a mathematical point at which certain physical quantities reach infinite values for example, according to the general relativity the curvature of space-time becomes infinite in a black hole. In the big bang theory the Universe was born from singularity in which the density and temperature of matter were infinite.

The Oxford dictionary of Astronomy defines gravitation as follows

Gravitation is the force of attraction that operates between all bodies. The size of the attraction depends on the masses of the bodies and the distance between them; gravitational force diminishes by the square of the distance apart according to the inverse square law. Gravitation is the weakest of the four fundamental forces in nature. I. Newton formulated the laws of gravitational attraction and showed that a body behaves as though all its mass were concentrated at its centre of gravity. Hence the gravitational force acts along a joining of the centres of gravity of the two masses. In the general theory of relativity gravitation is interpreted as the distortion of space. Gravitational forces are significant between large masses such as stars planets and satellites, and it is this force, which is responsible for holding together the major components of the Universe. However on the atomic scale the gravitational force is about 10^{40} times weaker than the force of electromagnetic attraction.

Singularity: a mathematical point at which certain physical quantities reach infinite values for example, according to the general relativity the curvature of space-time becomes infinite in a black hole.

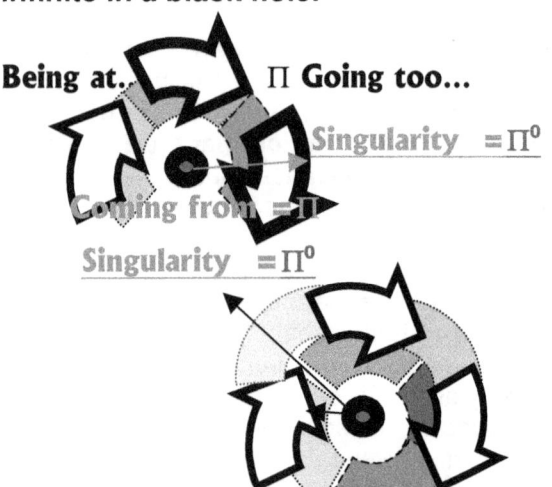

With no line starting from zero because there is no zero as a mathematical fact, then all particles hold the point of infinity and not merely the Black Hole.

here singularity holds position in the centre of any and all rotating objects as a value of Π merely applying movement (in the form of atoms) qualifies all matter to be space-time. It does not only fit the description of space within Black Holes but it fits all stars where singularity becomes part of all the stars from the minute to the largest cluster of matter.

Through rotation encircling the point of singularity and matter is (1) coming from, (2) being at, (3) as it is going too in one movement in relation to the specifics of the centre point being singularity, all matter then qualifies to form space-time.

From that argument one may conclude that all stars will become Black holes depending on the gravity increase they may generate.

I have to give potential readers this fair warning that *The Cosmic Code as the Absolute Relevancy of Singularity* requires a somewhat higher level of understanding and needs a greater degree of insight that the other books in this series does namely

The Absolute Relevancy of Singularity in terms of Applying Physics ISBN 978-0-9802725-2-9
The Absolute Relevancy of Singularity in terms of The Four Cosmic Phenomena ISBN 978-0-9802725-0-5
The Absolute Relevancy of Singularity in terms of The Sound Barrier ISBN 978-0-9802725-3-6
The Absolute Relevancy of Singularity in terms of The Cosmic Code. ISBN 978-0-9802725-5-0
Which all are also available from Lulu.com.

I do realise that have no chance ever that my theory would be well received by the paternity of science just because of the level to which I have been ignored in the past. All my arguments and every concept that I state as my theory on **_The Absolute Relevancy of Singularity_** will not be widely read, or much less that it will be seriously considered and I have not a snowballs hope in hell that it will be accepted by those with the authority to change physics principles...and all because I do not hail Newton for the wonder he was in spite off all the mountains of evidence that I bring with which I show he is totally off the ball with his perception about cosmology. The theory I introduce here and now would never be accepted in my lifetime because science in the Newtonian way is bent on believing in the marvellously fictitious, using all the facts bordering the supernatural, the outrageously inconceivable and the magic of what can never be explained, although they claim to use facts. It is **the marvellous** to think that mass can create gravity. How it is done that mass creates gravity was never explained and the facts put forward works purely on the say so of master Newton and with that no other proof is required. It is **bordering the supernatural** to think that with nothing between stars, yet by the magic of mass, mass has an unexplainable ability to attract another star many astronomical units away. It is the **outrageously inconceivable** to argue that life started on Mars, then overcame the Quite impossible to escape the gravity that Mars holds on all things, and after overcoming the unthinkable, then made a dive for the Earth just to come and evolve over here. Science think they my have the ability to create a Black Hole in a Manmade atom-accelerator because science thinks of the Black Hole as **the magic of what can never be explained** and therefore that proves that science has no idea of what a Black Hole is and I can prove what a Black Hole is.

That fact that I can explain what a Black hole is, that the Wizards of Oz will never allow the explaining I present to be done in as simple manner as I am about to explain the Cosmic Code. However, when I prove what a Black Hole is I am going to destroy the fantasy world everyone makes believe as physics. To science a Black Hole is a world of magic where gravity has the ability to go mad and a Black Hole is something that man could manufacture by creating an atomic accelerator tunnel, or so science thinks. In other words the best science at present can do to explain the gravity in a Black hole is to give gravity a level of superior intellect and then take it away (by allowing gravity to go mad as it seemingly does in Super Novas and in Black Holes). Why can I prove what a Black Hole is...it is because I can prove what gravity is and believe me that is one thing science this far could never get around in proving. The facts they use is as much fiction as Little Red Riding Hood's talking wolf...when it comes to explaining the integrating details of how gravity comes about. In science, when following my theory, everything can be explained by using physics, but using my explanation will make all present science become fiction, make all present science look like a fairy tale and make all present science seem to be good bedtime stories deprived of truth...and the money spent on Newtonian fiction-science will never allow me to have success because that would be too costly for the industry money-wise

Why would I call science a fairy tale...well this is just one of many, many reasons. Science wishes to promote something as impossible as time travel, which I show, is impossible. Science believes in travelling at speeds unlimited that could exceed the speed of light. I prove all such thoughts are impossible because I show that gravity and time is the very same thing. No one can beat gravity because gravity as time maintains the structural integrity of the Universe. In beating gravity one wishes to beat the cosmos that hold us secured. That is why time can manifest as what is known as the Hubble constant. Time is the redeploying of space by extending the absolute relevancy of singularity and that is only one of several factors that serve as time. Every time I declare Newton was mistaken and therefore science is wrong in presenting the most basics of physics, the workings of gravity, I am barraged by rejection and silent ridicule. Every time I challenge the Members of science to either prove Newton correct or to prove me wrong, I am ignored...my challenge goes unmet, so please forgive me for showing much antagonism...it is a result of Mainstream Science rejecting my efforts unfairly for many years. What I write is undeniably and undisputedly correct, but the instant science admits to my work being correct, that admission demotes most of the work science has accepted in the past as correct to the level of science fiction. It will destroy the groundwork of mainstream science and demote what is accepted to become fairy tales, which is what most Newtonian based theories are. Let Newtonian science explain what the cosmic purpose or the function is of a star...of a galactica...of an atom...of gravity...they have no idea. By the time you have finished this book you would have found answers to all the above questions in detail.

Mainstream science has so little idea of what a Black Hole is or what could cause a Black Hole that they devised a "Mini Black Hole" to suit there marvellous misinterpretations of gravity. That is a form of fantasy that fairy tale writers can't compete with. Science is so misguided in understanding life that they put life in all places throughout The Universe without ever finding one shred of evidence of the presence of life. Yet they say they work only with proven facts alone. They hold the opinion that life could have come from Mars but fail miserably in explaining how it will be possible for life to escape the gravity of Mars and then fly all the way, ever so precisely guided; directly to the Earth. How would it be possible for life to escape the gravity of Mars without them when explaining such a possibility by employing realistic physics, going into so much fantasy it leaves the story of the three pigs and the blowing wolf seem real. Science has the explaining of the exploding Super Nova down to the last detail where they explain that a Super Nova is gravity that has gone mad without ever proving how gravity can go mad because the truth of the matter is that gravity has no intellect to "go mad" in any way. Mainstream science always places new object found where their findings prove that the newly found object is on "the edge of the Universe", meaning where the Universe ends by forming an edge. This fantasy they dish up to anyone willing to believe him or her without ever telling what is beyond that edge. All they can see is an end of the Universe but in reality where there is an end there has to be a beginning of something else…this is physics. The Universe I show can't have an edge because I show where the point is that could never start and I show where the point is that could never end. I show that which can go no smaller and I show that which can go no bigger. I am about to introduce a Universe that mathematically can never start and the same Universe can mathematically never end.

I have been on a self-teaching mission that lasted thirty years and now that I have the answers and from which I have drawn the conclusions, I now find so much resistance from mainstream science in getting the findings my research uncovers out in the open. I offer tot academics many books in which I use diagrams, sketches, mathematical explanations and cosmic photos including other tools I employ to promote the required understanding needed to bring the ideas across that I wish to promote. However, publishing in this manner is very costly and money is one thing I do not have and therefore sending it to academics with no reply is an expense I cannot endure. Any academic feeling confronted by my accusation, please show how you prove $F = G \dfrac{M_1 M_2}{r^2}$ is applicable and is true. Show how the use of the formula could be applied meaningfully to present an answer worth of anything. Use the Newton's formula to show when the Moon is going to hit the Earth as the mass of the Earth pulls on the mass of the Moon. Better still, prove that mass does contract to create gravity and then explain how this is done…and please leave out the graviton because that is a joke! The idea that mass draws mass closer $F = G \dfrac{M_1 M_2}{r^2}$ is mathematically proven as an untruth, which means it is not true. What is the truth? …When you have completed this introduction you will have had a peeping view, a tiny glimpse of the truth…but as little as you would gain from reading this introduction alone, when put in comparison to what any person can gain from reading all of my work in total, you will gain endlessly more than what science is to explain about the truth, because what you then have gained by reading this document is much more than what science know about the truth. What I try to convey is that there is a good reason why academics block any and all publishing of my work, and when finishing this book, in comparison to what I offer, you have not even opened a first page of what I offer as new information when judging what my other work uncovers. Still, your effort in reading this document allows you to discover so much more of true science than what previously was known If you think I am boasting I challenge you to show where any of my explaining gravity requires superior intellect to understand... however in my simplistic approach to gravity I prove everything I say by applying the simplest mathematics there is.

The effort that this book represents the informing about an entire new way of cosmic appreciation meant to show that there are grounds for concern in the way science thinks and this book does not even bring all such arguments indicating concern in full. That one can only find when reading the first ten letters forming books named as with a title beginning with **Open Letters…**and those titles are included as books which I mention on my website, having the same name as this book namely www.gravitysveracity.com.

I am about to prove that gravity is **the Coanda effect** and gravity comes about from four cosmic phenomena never yet understood since it was never yet explained. Science doesn't believe there is something such as **the Titius Bode law** but science does believe that mass would generate gravity. Science has no clue about **the Roche limit** but science believes in spite of the Roche limit that big craters on Earth are reminders of massive asteroids that hit the Earth in giant collisions. With the Roche limit in place these crates are the result of something else because it can't be from asteroids colliding with the Earth. We all know how the bicycle rides and we all think we understand how the bicycle rides but having the bicycle ride on two wheels have little to do with balance and everything to do with the Coanda effect.

The bicycle rides forward when peddled but also the bicycle rides downwards when peddled and the two are both linked to gravity. I am going to prove that the Coanda effect forms gravity. I am going to prove that the **Coanda effect** comes as a result of the **Titius Bode law**, **the Roche limit** and the **Lagrangian positioning system** but most of all how these are related to singularity. That means I am going to prove that mass has no effect on gravity but mass comes as a result of gravity. I am going to prove what singularity is and that there are two types of singularity that in the end is only one type of singularity.

Teaching ever since time began forms a pillar on which memory and remembering what you are taught is the most prevalent part of tutoring. One is expected to remember what those coming before and which are tutoring you, wish you to remember. The Tutor lays a foundation by ensuring that everything known and accepted coming from the past are well and truly founded in the mind of the student. In that there is no problem. The problem arises where the information studied is flawed and no one ever realised that. Fortunately this does not occur regularly, but if and when it does, notwithstanding the exceptional part it forms, it then becomes a major problem to deal with. Therefore what comes form the past are carried on into the future as unblemished truth and no person meddles with the thoughts called information given as study material. However, as unlikely as it could be, this did happen and it is part of the basis of physics. When the student is taught, the student is expected to accept without argument. What comes from the past are considered as tested beyond suspicion of inaccuracy and proven to what is absolute unwavering accuracy! It is way beyond doubt. There is this motto that students are mindless and students can only start to think after receiving information that came from the past.

Students are incapable of arguing by reason to introduce new thoughts. This ability to reason only comes after the learning process secured knowledge through the memory process and only when testing shows facts learned by memory is well established and it then forms a solid base for everything the student knows, then the students may form an ability to reason and to argue. This mostly takes about all the time that living one lifetime presents. Well, what happens when that everything that everybody believed in the present, inherited by all from the past, was totally flawed? This has happened to physics and no one in physics so far yet realised it. Not one in physics shows the ability to realise the flaw coming from the past as part of the legacy. Then the mistakes will carry on from forming facts the past, carried over as flaws into the future for as many generations as it takes to realise the mistake that is dragged along and this carrying on of a flaw could continue indefinitely, if there is no clear minds working to recognise the mistake and correct what needs to be corrected. I ask of you not to judge me according to what you have already achieved for in that sense I fall short of receiving your recognition in status. Judge what I present to you, for then you will realise with all my shortcomings, I present you with a truth that exposes short fallings in the basics of physics.

Here and now and before the beginning of what this document may be to any potential reader, all parties reading take note that I state it emphatically that all members forming the community of science in physics judges me being not sufficiently educated and certainly not to the level where I am able to form any opinion on matters concerning **Sir Isaac Newton** or his physics. Any and all of my self-tutoring goes begging in their eyes notwithstanding and regardless of the fact that I did my private and individual studies by which I furthered my insight. That allowed me to show with clarity what destructive force **Sir Isaac Newton** released in order to corrupt the laws of mathematics, contaminating science along the way and mostly raping the work of a great man, Johannes Kepler and what **Sir Isaac Newton** did to derail the truth and disguise scientific correctness where such violation can only be expressed as being blatant incompetence. What his deeds amount to, is to corrupt the laws of mathematics, to render the laws of cosmology useless and to rubbish all of science as far as understanding the truth goes about cosmology. When **Sir Isaac Newton** tried to enter the world of Johannes Kepler he embarked on a journey he had no understanding of. By your reading, you will learn what it is that those academics that are guarding science never wanted published and read by the public at large. What I say is don't run and hide from my attack and coward away from my confrontation as so many of the most intellectuals amongst the Physics Paternity did when I confronted their thinking. On every occasion where I confronted members of the Academic Paternity in the past, those I confronted acted in precisely such a manner, such as cowardly ending all reading by throwing the book down, and then pretending to show the utmost disgust in what I say.

E. P. Hubble (1889-1953) confirmed an expansion through out the Universe, which contradicted all that science thought was known about our Universe. According to the accepted Newtonian cosmology everyone is of thought that the Universe is in a normal state of contraction because that is what $F = G \frac{M_1 M_2}{r^2}$ implies. Every person is very aware of the idea that the universal expansion would not last for ever, but has to start with some contracting effort at some point. Then all the heavenly bodies will collide and destruct, without any thought about any wavering on the matter and on the matters reliability there is evidently no doubt. When $F = G \frac{M_1 M_2}{r^2}$ apply, there should not be any force, which is able to keep the mass that is producing all the gravity that contains the Universe apart. Known for almost a century, science has failed to give any explanation about this cosmic phenomenon of a Universal expansion except for some silly notion about dark matter being dormant and not forming gravity, as it should. If the dark matter is present as is claimed, then why doesn't the mass form gravity as it should and contract? What does our ability to see or not to see or the luminosity that the dark matter does not have, got to do with the mass bringing about pulling power, that is if mass brings about any pulling power. If the mass is there, visible or not, then the dark mass has to pull because light has no standing in the forming of gravity and if mass does pull, it has to pull to form gravity. However Hubble's law contradicts this idea of a collective contracting Universe totally. This phenomenon about Hubble's constant finding the cosmos expanding should not occur with Newton's perception about gravity envisaging the contraction that must come by the force created by mass in $F = G \frac{M_1 M_2}{r^2}$. If the Universe is on a contracting as Newton said it has to, we have to first find proof about the location to where such contraction is pointing. In order to locate the contracting we have to locate the centre of the Universe, which means we have to locate singularity. With singularity eternal small, holding the place where the Universe started, we first have to differentiate between singularity and zero, should we wish to find singularity. In modern science the phenomenon we know as the Roche lobe comes more and more to the foreground, indicating an undeniable interaction between orbiting structures sharing a common axis.

That axis science at present does not recognise, notwithstanding the reality and undeniable proof there is behind all evidence. As apparent as it is to me, I went about divorcing $F = G \frac{M_1 M_2}{r^2}$ from all ideas forming cosmology and applying the roundness we have in Π to specific positions where one may locate singularity, which we have to locate if we wish to find gravity.

$F = G(M \times m)r^2$ may not be able to explain the exploding of double stars, however I can..... by applying $a^3 = T^2 k$...

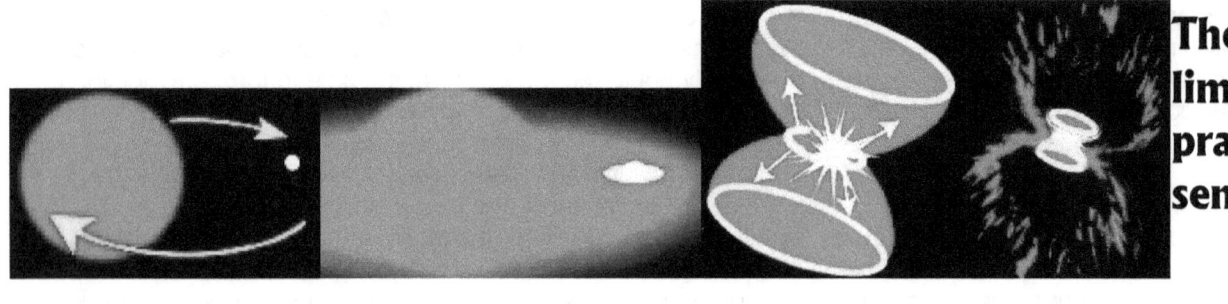

The Roche limit in the practical sense

The formula $F = G \frac{M_1 M_2}{r^2}$ cannot explain the comic occurrence shown in the pictures above called the Roche limit, I should find some attention when I say I can explain what is occurring in this instance and this occurrence connects directly to the Roche limit, as explained above. Not only does the Roche limit explain this phenomenon, but also it ties directly to the Titius Bode principle, also being another inexplicable factor in light of the formula $F = G \frac{M_1 M_2}{r^2}$.

According to the formula of $F = G \frac{M_1 M_2}{r^2}$ all orbiting structures should collide with a bang, but instead they do the tango until one drop, but when dropping it still does not collide with the larger structure, as would the formula $F = G \frac{M_1 M_2}{r^2}$ suggest that is used by science. The position where the formula applies is most surprising. Where the formula $F = G \frac{M_1 M_2}{r^2}$ applies, one has to find singularity applying because the position of r is pointing to a specific pinpointing of space contracting.

The Coanda effect

The Coanda effect where a liquid concentrates around the surface of a solid and by movement concentrates the density of the liquid to gather and compact while maintaining a relevance to the centre of such a round solid. I discard the idea that mass could be responsible for forming gravity because in almost four hundred years all evidence is indicating that the truth is to the contrary.

This is not only limited to planets in our solar system. In the Universe, there are giant stars spinning around each other. These stars are binaries, which are also one form of double stars where double stars are another such a form. The difference between the types depends on the distance they remain apart. They keep a certain distance apart and do not collide. In the case of the Sun and its planets, it could be a case that the systems might be to small, or they might be to apart. However, this is not the case with binary stars. They are close, they are big, and they spin around a mean axes called the Roche limit.

The Roche limit is:
The region surrounding each star in a binary system, within which any material is gravitationally bound to that particular star. The boundary of the Roche lobes is an equipotential surface, and the lobes touch at the inner Lagrangian point, L_1, through which mass transfer may occur if one of the components expands to fill its lobe. It names after the French mathematician Edouard Albert Roche (1820-83).

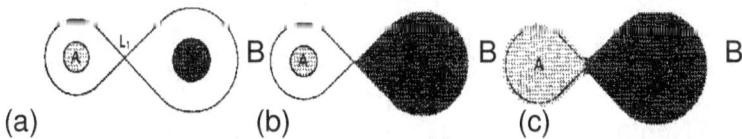

(a)   (b)   (c)

THE ROCHE LOBE: In a binary system, the Roche lobes of components A and B meet at the L_1 Lagrangian point. (a) In a detached system, neither star fills its Roche lobe. (b) In a semidetached system, one massive component, B, fills its Roche lobe. (c) In a contact binary, both components overfill their Roche lobes and share a common envelope.

Bode's Law:

Planet	Mercury	Venus	Earth	Mars	Ceres	Jupiter	Saturn	Uranus
Bode's Law distance	4	7	10	16	28	52	100	196
Actual distance	3.9	7.2	10	15.2	28	52	95	192

Bode's Law:
A numerical sequence announced by J.E. Bode in 1772, which matches the distances from the Sun of the six planets then known. It is also known as the Titus-Bode law, as it was first pointed out by the German mathematician Johann Daniel Titius (1729-96) in 1766. It is formed from the sequence 0,3,6,12,24,48,96, and 192 by adding 4 to each number. The planets were seen to fit this sequence quite well – as did Uranus, discovered in 1781. However, Neptune and Pluto do not conform to the 'law'. Bode's Law stimulated the search for a planet orbiting between Mars and Jupiter that led to the discovery of the first asteroids. It is often said that the law has no theoretical basis, but it does show how orbital resonance can lead to

commensurability. The importance that becomes known is the sequence the Titius – Bode law saw in the number arrangement of 3; 6; 12; 24; 48; 96 etc. The incorrect application of the Titus Bode law lies in subtracting the figure of 3 from 10 leaving 7. The other way of reasoning is to add four each time to the first value of three starting with 3 and so on. The true significance of the Titus-Bode law is that it points directly to a circular growth of 7 stages. The 7 relating to 10 is a precise derogative of the Roche limit or the Roche limit is a precise derogative of the Titius Bode principle because he two systems interlink.

LAGRANGIAN POINT:
The Lagrangian points are five equilibrium points in the orbit of one body around another, such as a planet around the Sun

From singularity there comes three values each holding 180^0 and this fact science is familiar with. The straight line is always a potential triangle with on side apparent and the other side in infinity.

I say this phenomenon called the Coanda effect is gravity. I say mass is a product of gravity whereas Mainstream Science has been saying for centuries that gravity is a product of mass. Science says that gravity is due to mass establishing gravity while not one person could ever explain the least detail as to how it is done. I went on to research Kepler and I discovered gravity through discovering Kepler. I concluded that gravity is the movement of material through space. By following Kepler's guide as Kepler formulated the process in introducing the equation four centuries ago being $a^3 = T^2k$ he gave us an explanation to what gravity is...if only Newton took notice of this important document. This says material holding space moves through space and proves that gravity has nothing to do with mass while mass is the product of space moving.

What is it about gravity that I say which no one wants to know? No one wants to listen to my point because I call Newton a cheat. He defrauded science and took all the other suckers running after him like sheep that are / were unable to think by there own ability. Now no one wants to find out how stupid the entire lot was that came after Newton and followed in his misguided footsteps. Saying this much in the past had every academic rejecting my work at that point. No academic found my work worthwhile to read after reading this much about Newton. No one wants to know that Newton went on lying for almost four hundred years. If you feel annoyed with my remarks concerning Newton, then explain how mass brings about gravity! No one understands the issues of mass and gravity. No one in science clearly distinguishes between gravity and mass and everyone in science tries to confuse the two issues by making them one and the same. They are two distinct different issues never to be confused.

I am Petrus Stephanus Jacobus Schutte going by the name of Peet and who is the author of the above-mentioned book(s). I know very well that you have never heard of me before because I am another nobody living in a backwards, third world African country totally segregated from civilization geographically and culturally and I still have the audacity to criticize everybody living in the sublime world of the mentally superior academic developed Universe of physics. Yet in spite of all the social backwardness and the fact that I am deprived of everything culturally viable, the physics you use is senseless and based on magic. Wish to draw your attention to the fact that science is completely wrong in their approach to cosmology and physics in general.

Whatever gravity is, gravity has to be Π. If gravity is linked to mass as Newton stated, then mass has to be very closely connected to Π. Looking at every aspect that forms gravity, it is formed by a circle. The Earth as much as the Sun as much as all stars and galactica holding gravity is round and the roundness are Π. The curvature of space-time, the fact that gravity bends light into a curve, this bending comes in the form of a circle that is formed by Π. The Sun for instance spins around and that is formed by Π. The Earth holds the Moon captured while the Moon circles around the Earth and the circle is a result of Π. If it is with gravity that the Moon circles around the Earth, then in all of this we must locate gravity holding Π as a value.

Lets find what is in the Universe and what was never part of the Universe. The Universe we see as the Universe is an illusion and has no permanency and therefore it will end as it began; in singularity, but I am not going down that road at this point. If the Universe started in singularity and as it will end in singularity we better look at singularity. Singularity forming one point is never part of the Universe but only holds a relevancy to all other point not being part of the Universe and only this relevancy grants the point by movement to become part of the Universe. Everything we think of in terms of being the Universe is 1^0 or 4^0 or 1^{1000} or whatever could form a singular value. Singularity is the point where the Universe starts and is the point that forms the centre of the Universe.

The first time any person gave a step in the correct direction was when Albert Einstein formulated a concept in 1905 he called The Special Theory of Relativity and in 1915 he introduced his assessment on the principle of The General Theory on Relativity. Although it must be the most accurate document ever printed by man up to this point in time and yet I do not quite agree with his findings. What I discovered goes far beyond the discovery that Albert Einstein formulated because Einstein asserted singularity while I go beyond asserting; I prove and show singularity. I have discovered that the Universe is not employing a general relevance of singularity, but throughout the Universe there is a fixed overall state of **The Absolute Relevancy of Singularity** that is not only **controlling the Universe**, but is what the Universe **constitutes of**...**it forms the Universe**...**it is the Universe**.

Every point in the Universe that is no point in the Universe but only represents that which confirms by forming the Universe is representing as well as represented by singularity. Since I go beyond any idea what Newton could understand when Newton formed his ideas back almost when science began, notwithstanding the magnitude in significance ***The Absolute Relevancy of Singularity*** presents as a breakthrough in science, the influential members of the scientific establishment will not recognise my theory on **The Absolute Relevancy of Singularity** just because it does not glorify Newton.

Science has a mouth full of criticism about the Roman Catholic Church protesting the manner the Roman Catholic Church went on in their conduct with Galileo Galilee, but I am the walking proof that Mainstream Science makes the conduct of the Church look like kinder garden teachers reprimanding toddlers when compared to the way the science world treated me. The Church back then saw their position as the custodians protecting the interest of what they saw was the interest of God, but Mainstream Science see their position as performing in what would be the function of God and therefore they presume the role of God. They see them being untouchable and beyond any approach or reproach on matters conducted by science for they only confirm what their god Newton formulated...and everyone knows, Newton can do no wrong.

I found any body that does not absolutely sacrilegiously confirm the greatness of Sir Isaac Newton; not bowing to his mighty splendour has condemned me into the hell pit of obscurity by those utmost important Academics being filled with contemptuousness had the likes of my views totally ignored in condemnation. Such past encounters where I did not show absolute fascination with the greatness of Sir Isaac Newton brought me the condemnation of all thought to be important in science and their resentment and that brought me their absolute fustigation which then taught me that mainstream science in physics will again ignore my ideas that I formulated as ***The Absolute Relevancy of Singularity***. I don't believe it would even be read, will be seriously considered and much less be accepted by those with the authority to change physics principles. At least the Pope allowed Galileo Galilee to print a book in which he manifested his ideas while Mainstream science in the present has so much control they prevent me from even getting some ideas printed in ink and on paper. I think the theory I introduce would never be accepted during my lifetime because science is fixated on Newtonian ideas, which makes them bent on believing in the outrageously marvellous, and the unexplainable magical powers with gravity working by mass supplying a pulling power, which is a fact never proven and accepted only on Newton's word and Newtonian cultural bias, although they claim to only use proven facts. The way science considers gravity has not grown one inch since Newton...and Newton was still part of the Middle Ages in thinking! Newton actually considered and approved the ideas brought forward by alchemists where amongst many other obscurities it was thought that a machine could run on 100 % efficiency. Newton thought that the idea of introducing nothing, as a mathematical substance that at the time blew across from India was a modern marvel while it was as wrong as using horse shit for a cooking additive!

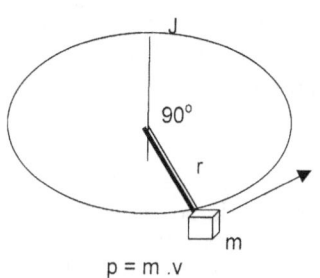

Newton said, "If a body is at rest it would remain at rest". That is it. That one sentence should silence my critics and vindicate everything I say about Newton. That shows the very backward thinking of Newton. Please, please show me one, just one, article or structure within what is thought of in terms of the Universe that is in the Universe that cannot or that does not move. Show me something in the Universe that could stand still. Atoms that move form everything and every atom is formed by sub-atomic particles that move! There are no such things that have the ability to stand still, notwithstanding even how remotely small or big they are, except and excluding any point forming time on the spot and in that case that spot is in time and not part of the Universe. What ever fill any point in the entire Universe moves within the Universe. Everything in the Universe is moving in relation to all things filling any position within the Universe. Time forms space by moving all there is in relation to time forming one spot by the way that time repositions all points being space in relation to any one point that has no space and therefore can't move. That point holds singularity and the rest represents what holds singularity, which altogether forms to become what singularity forms. That makes Newton's concepts as far as cosmology goes utter nonsense. For me saying this I am detested more than I am ignored and I am completely ignored. There is no one prepared to think about any of my ideas, just because I am not prepared to recite the greatness about Newton's thinking! When you detest what I say then find the reason for your detesting me is more about what truth is included about what I have to say. There is no force acting other than the force life conveys and life is not a valid force that is everywhere in the Universe. Life as a specific identity can't even form a presence all over the Earth but has to remain in very selected areas to be able to remain in life. Life is movement that can move or that cannot move in relation to what time moves. Life is that movement moving apart from the cosmos but life has the ability to move within the realms of the cosmos by using time and then extend the movement time produces. And even then when standing still life will move because life is the movement of a liquid in terms of a solid holding life. Life depends on movement to stay alive and therefore not even life can ever be still! Life will move, or become cosmic and a cadaver is what held life and when life left, then the cadaver became cosmic. I plan to extend t his four books later in and include another fifth book, which I plan to name **THE ABSOLUTE RELEVANCY OF SINGULARITY IN TERMS OF LIFE** in which I am going to explain that life is and never has been part of the cosmos and it is for that reason that life can enter the cosmos by influencing movement in the cosmos and afterwards leave by using a process we think of as dying. To return to matters in hand I maintain that everything moves and nothing could be in a state of rest. That makes Newton wrong!

Using the formula **p=m.v** or **F=ma** is total rubbish because nothing can stand still in the first place. The mountain that you may think is standing still is moving along with the Earth while circling around the Earth's axis while the Earth is circling around the Sun while the Sun is circling around the Milky Way while the Milky Way is circling…everything is moving eternally and never stopping! Then on top of all that movement comes the fact that everything moving is filled with atoms moving and the atoms are made up of sub-atomic particles all moving. If you can show the least particle that has the ability to not move I then will indorse Newton's **p=m.v** or **F=ma** but otherwise it is the rubbish I say it is. With that in mind the entirety of physics are based on $F=mv^2$ and later on I am going to elaborate on that.

However, I also know just because I am correct and I prove Newton incorrect that fact alone is enough grounds for you as a religious Newtonian to ditch this work even at this point. In modern standards Newton was as backwards as can be and yet, just because science has nothing better than Newton's incorrect and disproved ideas, they hail Newton as the greatest mind that ever lived. I researched the work of Kepler and found science doesn't even recognise his work while it is his formula that forms the basis of all physics. All persons in science hail Newton's work on the idea that **p=m.v,** while it is as wrong as thinking life could flourish on Mars. The difference in gravity in terms of the gravity the Earth has and where life evolved and then the sudden way life is transformed to Mars will not allow life to prevail on Mars and persons would be unable to think because people think by the gravity prevailing. Thoughts are the liquid called electro magnetic brainwaves while the brain forms the solid structure and the solid structure remains in form and intact in relation to gravity. Then there is the blood that forms the liquid with the brain matter forming the solid and this again forms the ability of thought where such ability depends on gravity. Gravity is movement and when forming mass the Earth takes control of the movement of the space only allowing the object filling the space to move in relation to what the Earth conducts. Thought is the control of the space in gravity surrounding the skull b y establishing movement of sorts in relation to the material within the skull performing by an electric charge within the matter. When gravity changes. Thought that controls life changes when gravity changes making life the product of thought and in that is the difference between species and races.

Thoughts are the liquid called electro magnetic brainwaves while the brain forms the solid structure and the solid structure remains in form and intact in relation to gravity. All Newtonians confuse mass and gravity because Newtonians wishes to create virtues about mass by giving mass inconceivable abilities and in that is what mass could never comply with. Gravity is movement and when forming mass the Earth takes control of the movement of the space only allowing the object filling the space to move in relation to what the Earth conducts.

When the object is in gravity the object is in free fall and when the object is in mass when standing on the solid ground, which prevents free movement, the object's free movement becomes restricted. By awarding the factor of mass the Earth then is preventing free movement. The object then has to turn within the range of movement the Earth provides. Being immobile and frustrated results in that the space that is filled with matter and then is forced into a position standing on Earth while only moving in terms of the Earth is forming mass just because it cannot move independently any longer. However gravity is the motion of space in time and mass is the restricting of such independent motion. Read my work and you will see! If professors are not prepared to read my work but always delegate the evaluation thereof to a person that is clearly not up to the task then what must I conclude from that…and that is what always happens because no professor could stomach my criticizing Newton? If professors do not even try to address the shortfall of Newtonian practises then what must I conclude? Then it becomes obvious that those academics being in the status and position that the professors are either detest what I say or do not believe what I say. If you as a professors do not believe what I say, take my challenge serious and prove to your mind why you accept mass has gravity as a result while you scrutinize my ideas. Put the accuracy of Newton's formulas stating at first to be $F = \dfrac{r^2}{M_1 M_2}$ then when it did not panned out as applicable became $F \propto \dfrac{M_1 M_2}{r^2}$ and afterwards by the intervention of magic became $F = G \dfrac{M_1 M_2}{r^2}$ to the test by exchanging the symbols with numerical values and then prove Newton accurate. When you detest what I say then find the reason for your detesting me including what I have to say.

Using the formula **p=m.v** is total rubbish because nothing can stand still. The mountain that you may think is standing still is moving along with the Earth while circling around the Earth's axis while the Earth is circling around the Sun while the Sun is circling around the Milky Way while the Milky Way is circling…everything is moving eternally and never stopping! If you can show the least particle that has the ability to not move I then will indorse Newton's **p=m.v** but otherwise it is the rubbish I say it is. With that in mind I say that the entirety of physics are based on **F=mv²** and later on I am going to elaborate on that. However, I also know just because I am correct and I prove Newton incorrect that fact alone is enough grounds for you to ditch this work even at this point. Whatever movement could be accomplished is by moving the space a^3 in a circle T^2 that also must link to movement in a straight-line **k**. That is physics applying in practise and that revokes all Newtonian forces. Whatever space moves will do so turning while going straight and that is the absolute crux of physics that puts Kepler at the basis of physics.

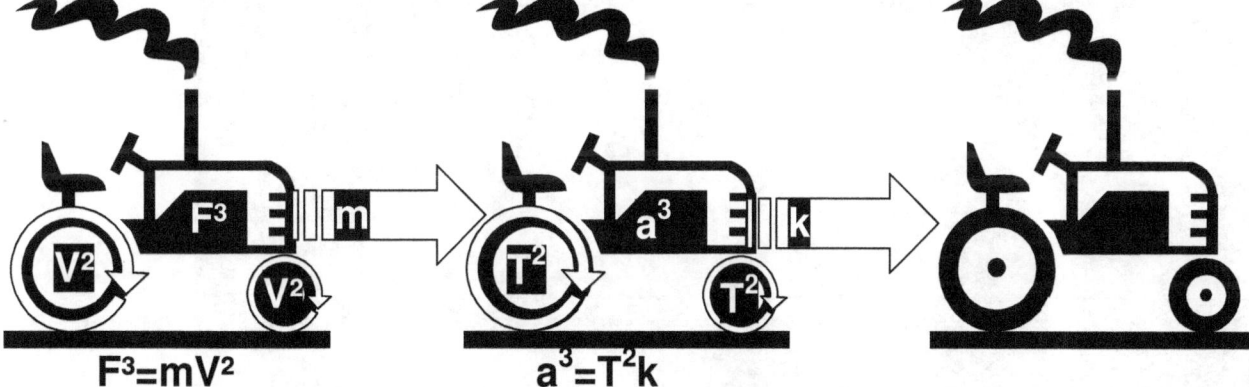

F³=mV²   a³=T²k

Everyone thinks that Kepler found planets rotating, with Newton being able to explain Kepler, which makes everyone more concerned about how Newton saw Kepler's work. The formula used in physics as a principle is **F=mV²**, which should be **F³=mV²**. **F³=mV²** is replicating Kepler's formula in detail as **a³=T²k**. By using Kepler's formula we have **F³=mV²** that is a precise repeat of **a³=T²k**. The duplication is so obvious that we have (F³ becoming **a³**) while (m is **k**) and (V² is **T²**). Einstein also only duplicated Kepler's formula by putting **E=mC²**, which also should read **E³=mC²**. Again that is precisely Kepler's formula **a³=T²k**. (E³ is **a³**), (m is **k**) and (C² is **T²**). In **E³=mC²** Einstein mimicked **a³=T²k**, Kepler's formula. (E³ is F³ is **a³**), (m is **k**) and (C² is V² is **T²**). So what is so brilliant about Einstein's formula if Kepler had it centuries before? E³=mC² is F³=mV² which is **a³=T²k**. Newton corrupted the formula when he added 4Π² to the formula and removed **k** that Kepler introduced while **a³=T²k** Newton ignored. Newton changed **a³=T²k** by using the symbols G (m + m_p) to replace **k** and then declared $a^3 = T^2$. I still wish to see the proof confirming Newton's changes as being correct notwithstanding that everyone thinks physics is entirely based on this conception. Whether the formula used is F³=mV² or is E³=mC², it still remains duplicating what Kepler introduced as **a³=T²k**. So I changed it back to Kepler's version of **a³=T²k** as to better the understanding of the foundation of astrophysics and mainstream physics. The entirety of physics is not based on Newton. It uses Kepler's findings to a precise duplication while science does not even recognise Kepler. Giving Kepler the credit due, the entire Universe becomes completely understandable…but then for my audacity to show mistakes in physics I am ignored flat! All I ever ask is prove the truthfulness of $F = G \frac{M_1 M_2}{r^2}$ because it is $F^3=mV^2$ that forms the basis of physics and that accuracy comes from Kepler's view of $a^3=T^2k$ that became Einstein's $E^3=mC^2$.

What I ask of readers is to beforehand forfeit the culture of Newtonian bias when reading this by paying attention to what I say and not about the degree in which I stray from mainstream science's thinking. This way the exercise will present many new ideas and explaining my new concept will become clear. There is so much to benefit from. Science has no idea what a Black Hole is while I can prove what a Black Hole is.

I challenge any Newtonian to prove that there is anything in the Universe that can and does move outside the will of time except when what that body hold forms movement by means of life. Only life and what forms the idea behind life has the ability to move above and beyond what time will allow. The top can move but the top moves as an extension of the manipulation of life. It is by the intervention of life that a motorcar moves, that an aircraft flies, that whatever we think of is in form does what it can do as an extension of life. A building is because of life and not because of some cosmic influence. The building will go to ruins when the cosmos takes charge and life does not uphold the maintenance of the building. To look at the building and see the cosmos and not life's intervention is totally corrupting physics. When we see a mountain then we see the cosmos. Look at a mountain and what one see is something standing very still as it moves around the axis that the Earth forms. The leaves of the tree moves but the leaves and the tree are filled with life.

Gravity is Π. In this book I set out to show that gravity is what Kepler said gravity is. Not what Newton concocted but I am going to prove what gravity is. Kepler proves that $a^3 = T^2k$ and I, the most unlikely person of all, found a way to interpret this correctly. I say correctly because the hogwash Newton made from this is not fit for serving pigs.

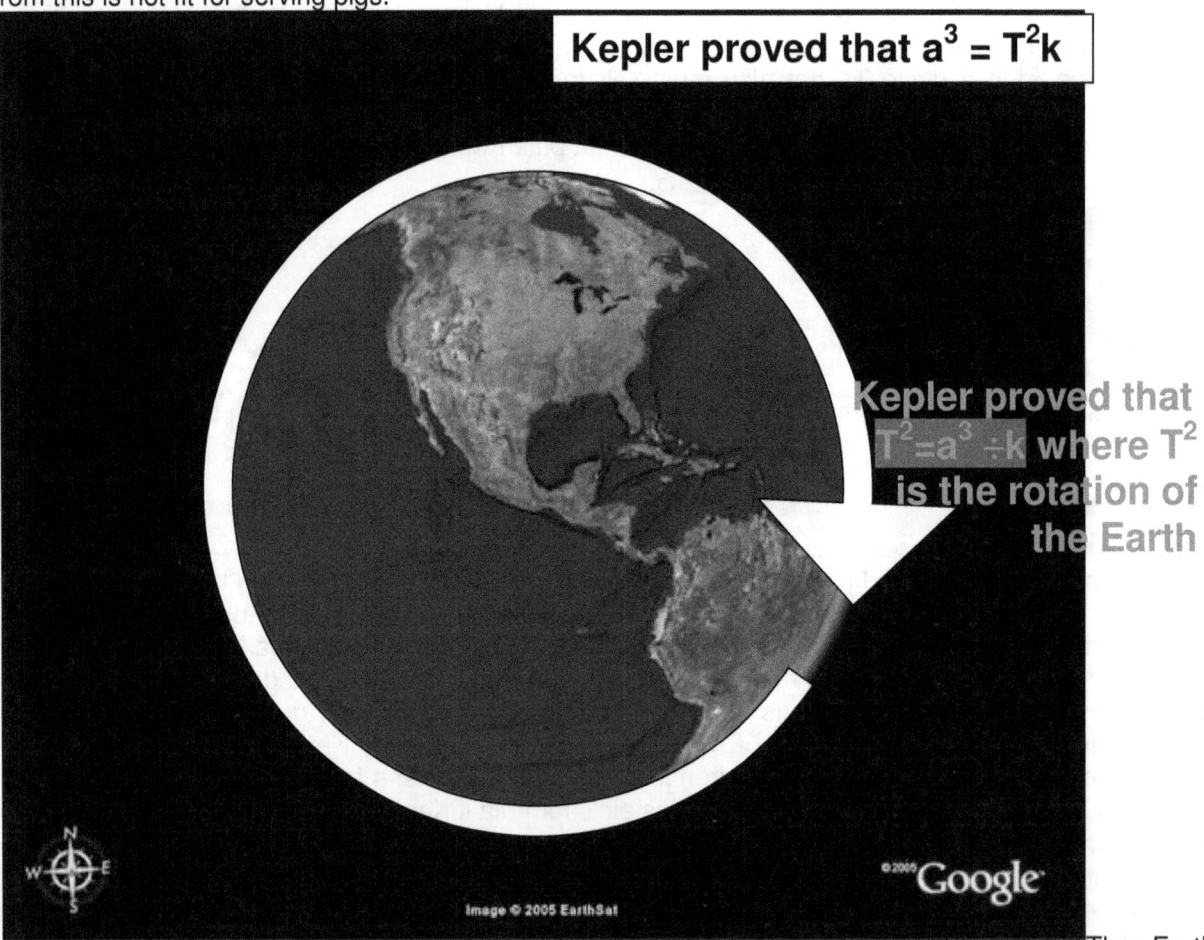

The Earth is just another sphere. As a result of the form the sphere holds there is a centre connecting the sides and the centre holds singularity. However by presenting a centre where all lines cross on a point that cannot distinguish sides since that point has no individual sides the centre holding singularity is inactive. Motion makes it active and the motion of space in time activates singularity to charge gravity that we find as a factor in the Coanda effect. That motion that establishes the purpose of space a^3 as a result of motion k

through time T^2 was what Kepler really presented as a formula. Gravity is $k^0 = a^3/T^2\ k$ and to install k^0 **the motion of space-time a^3/T^2 is required to complete the task.**

Another huge factor that favours the use of Π is that any expanding by any mathematically sympathising method will have to use Π since Π is the only route that a spot of no significance can develop into a dot with a Universe of development is when all possible sides progress on equal terms. The progress must be generated so that it can flow equally to all sides in all directions spontaneously where not one side will favour the growing process as such. Time in it's flow does form a bias but explaining that at this stage will involve to much other concepts which I would rather leave to the books.

There is only one way to permit such a flow and have a mathematically correct outcome and that would be the use of Π. The using of Π would ensure a dot. But the dot would have to form a sphere and a sphere is Π in relation to seven. This brings as back to the door of the Titius Bode concept where ten in relation to seven is the overbearing dominating issue.

When the cube, which is having six loosely connecting sides comes into contact with the sphere with six sides but also with a centre that charges gravity by its positioning of singularity the cube is destined to lose the one side that forms the contact with the sphere. The centre in the sphere removes the cube side that has the contact and the sphere allows that which is on the edge of the cube to become committed to the gravity of the sphere by descending to the centre. The centre will charge a reducing of space through implementing motion and the evidence of that we find in the Coanda effect which I shall explain later on. Only where such descending is no longer possible because material that form the edge of the sphere will not allow further moving by the descending object to the centre then at that point where motion becomes a tendency to move does the object receive a position of mass.

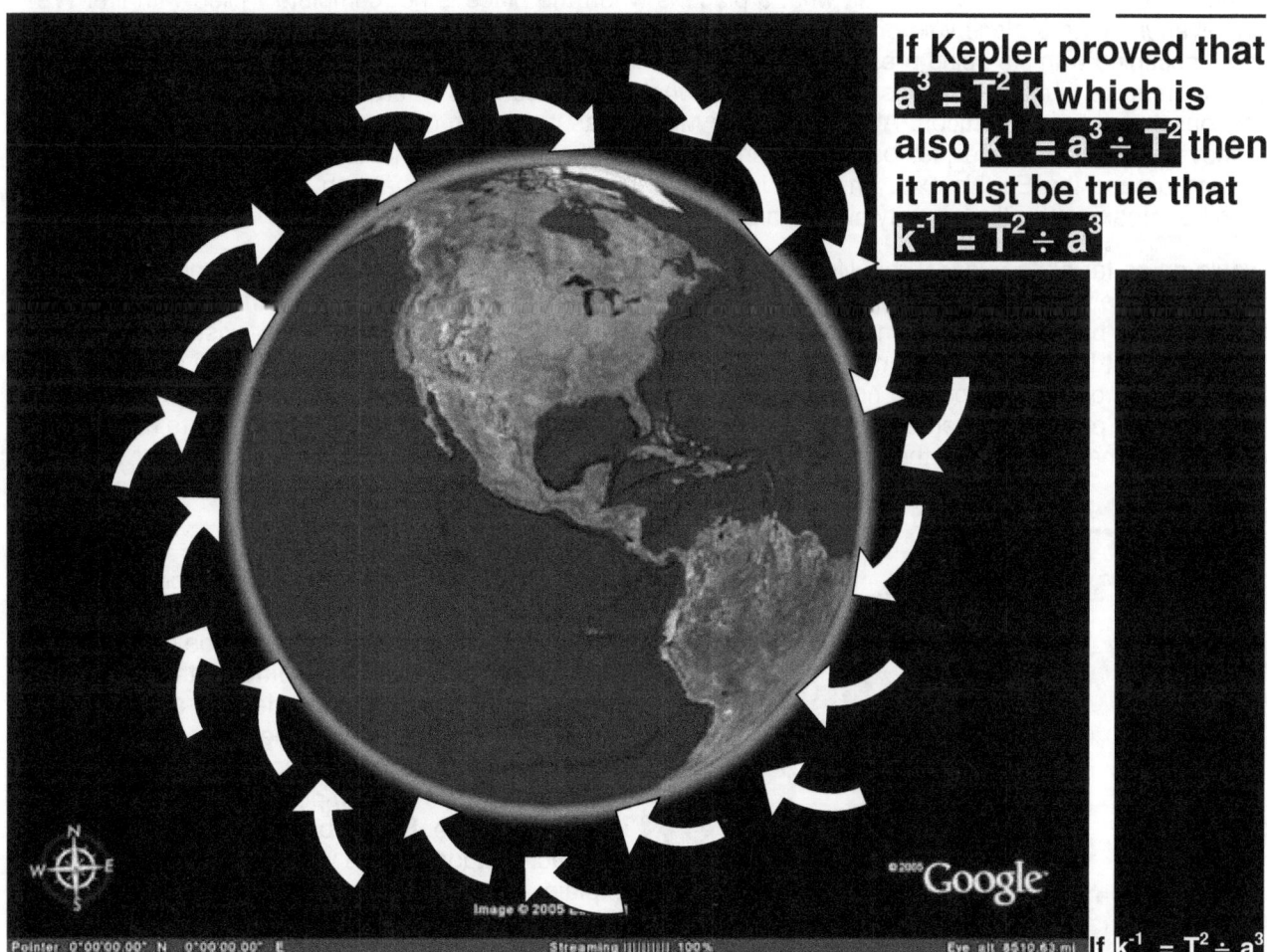

If Kepler proved that $a^3 = T^2\ k$ which is also $k^1 = a^3 \div T^2$ then it must be true that $k^{-1} = T^2 \div a^3$

If $k^{-1} = T^2 \div a^3$ then by mathematical interpretation it would mean that the spinning Earth is reducing space towards the Earth by the same principle applying in the centrifugal water pump. There is a movement of space flowing towards the Earth centre.

In the cosmos we find two forms available for use. The one is a six-sided cube that connects at the edges and can be found in whatever name one may attach to the angles that connect the sides but after all it still remains six sides with three sides facing three opposing sides. Then we have a sphere also with six sides with three in opposing to three others but in this case all the sides connect via a precise centre and the centre give the value of a seventh position that secures the precise location of all six edges and that keeps the sphere true. In the case of the sphere there are seven points where the six points representing the sides connect through the centre and the centre connecting is charging the form of the sphere to Π and with the centre forming singularity the centre is $\Pi^0 = 1^0$ that holds the form in charge and in truth.

But it only becomes mass after the motion or gravity that it had became restricted and the motion becomes frustrated and confined. The motion ends where mass confines the object to the motion that the larger object confirms. As the sphere spins it have seven positions in the sphere that is eternally part of the form the sphere has acting as one while there are six in the cube but with the demanding sturdiness of the sphere the sphere removes the one side of the cube where contact is made. The six sides in the cube then become five sides and five in the cube then has contact with the outside leaving one point in contact with seven points in the sphere. The centre will dominate because the centre charges seven points to gravity while the cube cannot secure the position of any particle that is in the cube. The cube with five lines that loosely connect and that don't form any firm position to secure material within the cube will be conformed by the spinning centre of the sphere, which generates gravity. The lack of the fact that the cube can confirm material within the form it secures while the sphere depletes any material to a position that relates to the centre concludes that the cube as form will lose one side and the sphere is going to capture one side from the cube and with five sides standing related to six sides and a seventh centre, which is commanded by singularity in the centre that is going to be affected by the motion of the six spinning around the seventh. In that it then is five plus five on the side of the loosely connected cube in form that stands in relation to seven in the sphere and that is why the Titius Bode holds a form of ten that stands in relation to five on the one side and another five which is part of the following direction. Space –time is built in such a manner. As the movement propels it will move from a cube of five to a cube of five and the total sides being affected will be ten on the side of space.

As we are on Earth we have no idea what a place is when the place is not completely filled with life. We can't even imagine a place void of life and yet everything except the Earth is totally void of life. Our minds are so occupied with life being present that removing the idea of being surrounded by life is unthinkable. In order to imagine what the cosmos should be and what fills the cosmos where no life is present we should draw our thoughts to the moon. All things on the Moon is as it was ten or twenty thousand years ago and no movement takes place just because there is no life that will bring movement other that the Moon spinning around the Earth's axis. The only ingredient within the cosmos that can move independently by any other means other than gravity playing a part is life.

I would go as far as to suggest that life is the movement of space using time but moving in addition to the movement time will allow. That makes life being movement bringing an addition to time moving space. Only life can move beyond the movement that time produces. Life is totally alien and non-existing within the cosmos except for some places on Earth…and those places are not even generally filled with all life forms. Life is not part of the cosmos and not being part of the cosmos gives life the ability to move and seemingly enter. Life can produce movement and in some cases influence movement and even manipulate movement but the movement is alien to the cosmos and in cosmic terms non-existent. Gravity on the other hand is movement in a specific order by the ratio of $a^3=T^2k$ and that Kepler proved with his and the meticulous research of Tycho Brahe.

It is commonly accepted that Physics demand respect because the general idea going around has the understanding that Physics only work with proven facts that cannot be in dispute or be disproved in any way. Well…I wish to bring to mind some of the facts that physics work with when academics as scientists only work with facts. Remember they are the ones boasting that if facts are not proven then it is fables and those very important academics don't waste time with fables because they only work with facts. Students, it is your liberty to ask them to explain what they say is such correctly proven facts. They maintain it is a fact that we have to have mass in order to produce gravity. Mass is responsible for gravity. If you don't have mass you're not going to have gravity.

Mass is equal to gravity and gravity is only where mass is. If mass is anywhere it should show its presence otherwise mass is absent. If a body falls it is the mass that allows the body to fall because the body receives gravity by ratio of mass and mass that produces gravity in relation to the mass available. It is mass that drags you down because the mass is in charge of the gravity and the gravity finds the value from the mass available. So what happens to balloons? Have they got anti mass or anti gravity? They are moving up when the air is heated. Mass pushes you down by the gravity it forces onto you. If mass drags you down then what are lifting you up in the balloon? If mass gives the gravity to drag you onto the Earth then why would the hot air lift you up? Is the hot air causing anti gravity or anti mass because gravity and mass drags you down or so Newtonians say. The balloon is lifting the passenger and all that is in the bag plus the bag plus the balloon into the air. So what is then pushing the lot up if it is mass that drags you down. Has the air not got mass because then the air can't have gravity and then the air must escape into the blackness of outer space because by going up it shows a resilience of either mass or gravity. We have seen that it is mass that pulls everything onto the ground.

Why would the air defy mass and allow the balloon to go anti whatever. We find mass being the equivalent of that which brings the object to the ground. The object has mass to produce gravity. Why then would hot air allow the balloon plus everything in the balloon to lift into the air? The balloon lifts in relation to the hot air that blows into the sack. The more hot air and the hotter the air is the more lift and the swifter the lift will be that the balloon provides. The issue sticking out is that the balloon then must not have mass because with anti gravity it is pulling up. Remember mass drags you down and mass can't pull you up and drag you down at the same time. Then what is pushing while mass is pulling or is mass pushing while what is pulling? The object is not going in the normal direction where it is dragged down by gravity and in all my life I have never heard one Academic mention anything about gravity lifting and that makes the lot very confusing. What is lifting up when the lot should be pushing down and why did everything connected to the balloon lose the mass and if it has mass why is it not dragging down the balloon? If you think this is a little confusing try what is to follow.

They teach you that it is mass that produces gravity and gravity makes you fall because while gravity makes you fall mass drags you down. It is because those mind controllers are lying through their teeth with a menace in which they are the experts, as they know just how to pull cotton wool over you eyes. Take a truck of 15 tons into an airplane. Put next to the truck a petite little dancer weighing 45 kilograms. Put next to her a frog weighing 150 grams. Then get this lot into the air by airplane and let them jump. Take note that you are told by the wise amongst us that it is mass that produces the gravity that pulls you down. We have just had a lovely debate on how it works and how mass drags you down and wondered if it then is anti mass or anti gravity that lifts you up with the hot air balloon, well take note of this as your airplane reaches 11 thousand meters which is eleven kilometres straight up into the air.

Now we drop the truck and the girl and the frog at the very same time from the airplane. The frog then pretends he drives the truck and the next scene he is dancing with the girl while the truck is falling as fast as a truck can fall. Who do you think is lying? Remember only one group can tell the truth and the other must be lying. Have you thought why one party is lying while the other party has to tell the truth?

The academic Brainy Bunch are telling students all over the world that mass is in charge of gravity and it is mass that's pulling you down. Then the mass is pulling the truck of 15 tons down since the mass produce the gravity and the gravity produces the fall which is three hundred and thirty three times more in a down direction than the mass of 45 kg is pulling the dancer down. The mass providing the gravity

that pulls the truck down is doing the pulling down of the truck one million times better than it is pulling down the frog. If the mass is doing the pulling by establishing the gravity the truck must fall 333 times faster than the girl and one million times faster than the frog. It is either that or the three has the same mass because they are falling at the same rate.

If the Brainy Bunch all too wise are correct the frog can fly to America and have a pizza in New York while the truck has a few micro seconds to get down if the girl is going to fall during the normal falling duration of a minute or so.

Everyone has seen skydivers jump out of airplanes next to cars and trucks and bags. Every one has seen they all fall at the same rate. The girl can do tap dancing around a jumping frog on top of the truck or below the truck and they can be inside the back of the truck galloping on fresh air inside the truck because the lot is falling at the exact same rate.

The academics wishes to brainwash you by mind control in accepting that it is the mass that the falling takes place and that mass is responsible for the gravity and by mass pulling you down it is gravity that makes you fall. Where is the proof of mass that according to them is that which is producing gravity. They tell you Galileo said all things fall equal and we can see from the TV monitors how all things fall equal. Where is the mass that makes the gravity to let you fall if all things fall equally? They tell you that the truck has a mass of 15 tons and that mass is making the gravity that is having the truck fall while the truck is falling at the same speed and distance than the frog does.

If you take that as proof then they got you. Then they brainwashed you into a zombie. Then if you don't repeat after them and echo every word test after test and exam after exam they will fail your papers and kick you from campus. That is mind control, better than what even the KGB is able to implement. You repeat after them and you live an academic life or you disagree and you go home to play with your toes. If mass is in the picture then mass must be represented by a factor of more than just one because if mass is not part of the overall picture then mass has a factor of one which proves that mass is not part of the equation since mass can't change the results. With all the objects falling equal mass has no role and if mass has no role then for my money academics in physics can't just go and put everything in as their hearts desire. If it is Galileo that is correct and if all things fall equal then mass has no part in gravity. If mass is the inspiration behind gravity the truck must fall a million times faster than the frog and in fact the frog should almost land in another country because that is how slow it falls.

The fact of the matter is that I don't wish to be near when any of this lot hits the ground because the truck will cause a quarry and the dancer will be a splash of red fluid while the frog might not be that worse for wear if the truck or the dancer doesn't land on the frog. But that is mass. The differentiation of having mass that would differentiate or having equality that would equalise all objects descending and then not having mass and between individual differences in mass by each component that enters the equation when the objects touch the ground. Then every one gets the mass it has. Only when they touch the ground and land on the soil is mass as a factor awarded. While they fall they all fall equal and there is no distinction between the falling at all. What then is gravity? The gravity is the falling. The gravity is the motion. While the object is in a state of mass it is not moving. The tendency to move and apply gravity is the part that the mass restrains. The mass is preventing the falling from continuing. It is the role of mass to prevent further falling and independent motion to continue. Some of then might even still honestly believe it is mass that produces gravity because they were taught that it is mass that produces gravity and never thought about the matter again afterwards.

They were brainwashed by their tutors as their tutors were brainwashed before them. You don't need the brainwashing because you now can find out what the answer is to gravity. You are the first generation that can receive the light of knowledge about what gravity really is, or you can be the last generation that will live in the lie. You are in a position where you can teach your tutors the truth about gravity if you read what is in the books. The truth is there and the truth is out and the truth will be because the truth is written for all that wishes to read. The academics on the other hand have ignored my work and my being on Earth for the past six years while I was writing them letters about gravity. They ignore me as if I am a rattlesnake because to them I am a rattlesnake. With what I say I will have them tumble down from their pedestals because by accepting my work they suddenly find their position equal to yours as students, and then they will have to learn my work in the same manner as you learn my work because to them everything is as new as it is to you. The Academics of the day have too much to lose to recognize my work and therefore have to protect their interest with all they can muster. For that reason if no other they will rather go on lying to you and cover their corrupt approach than face up to the truth and admit their work is lost.

The truth will be whether it is recognized by them and they can become the first to admit and repent or they will be the last of the laughing stock that those in the future will refer to as the bunch that couldn't see when things fall equal they cannot have mass and when things do not fall by mass then one can know mass has nothing to do with the falling and the gravity.

It is up to you as students to rattle their cages and make them admit they've been lied to as they are lying to you. Or you can be the last of the fools that couldn't see that when things fall equally they have no mass by which they fall. My book is written and those that read it first will know what gravity is. If you do not accept the role as being zombies that is brainwashed then confront these academics that treat you with disgust and betray your trust. They might tell you the mistake is not that serious and the damage is small but then how will they know how big or small the damage is if they don't even know what damage there is or what the damage is.

Science has stayed so far from the truth that they can't even see the truth any more. If you carry on you will learn about some of it and when you read my books I will entertain you with many more than you ever believed. My books will serve as the light switch that brings the light to you.

I charge your young minds to confront those academics about the truth. I wrote to them in the last letter where I informed them that they protect the criminality of their corrupt teachings because when the corruption is removed then nothing remains because they have lived a lie for too long. If you reach the need you may down load it because it is a fair bit of information.

If I come to you with a proposal about something I wish to share with you on condition that you pay me an amount to share with you what I know then I am an academic wishing to teach you. Have you a name for such a person that will force another person to pay him to be brainwashed and be mind controlled because the tutor has absolute control over the life and death of the academic future of the brainwashed being and therefore is willingly forcing this unfortunate creature in accepting what will never amount to the truth? I think they are called Physics professors and rule Universities as draconian authoritarian dictators bent on sadism.

Let's investigate the falling as such and see what happens during the fall. The truck falls at the same pace in which the girl falls, which is the same pace as that which the frog falls. If the truck falls at the same pace as the girl and as the frog there has to be a common denominator in this process and since the common denominator eliminates size form and shape we can eliminate mass. Mass brings distinction and the falling eliminates any form of distinction.

Empedocles Clepsydra of 450 BC

Connected pipe allowing filling of bowl by water

Round Container Filled with water

Water running from outlet at the bottom

A man by the name of Empedocles had established in 450 BC that when using a tool they called a clepsydra which was a common kitchen utensil at the time water would only run from the water sprouts at the bottom when the inlet pipe at the opening on top was free to let air in. From that science deducted that something fills the clepsydra or the water container because as soon as the pipe was shut by a finger preventing the "something" to fill the container, water would not release from the bottom. From that science concluded two thousand five hundred years ago that "something" and not emptiness flows down into the container and pushes the water out. It is not the mass of the water making the waterfall but the water is pushed out at the bottom by a transparent enveloping and unseen medium that fills the space the water vacates. The process has nothing to do with mass pulling because if it was mass pulling then the water that should be heavier than a human finger must rather try to pull the finger through the pipe as the water mass pulls to the Earth.

When I fall down a waterfall with a boat I travel the same pace, as does the boat. That could be because I am fixed to the boat by sitting in the boat. But my sitting in the boat has certain condition and one is that I can remain sitting because I fall the same pace as the boat is falling.

I fall down with the boat and the boat and me forming a distinctive unit falls at the same pace as the water that forms the waterfall falls. Should I at the time of my falling hold an empty mug in my hand and I wish to fill the mug with water, and then I will have to move the mug against the flow of water streaming down the waterfall. I will have to thrust my mug upwards at a faster pace than my descending is casting the mug down and therefore I accompanying the mug down the waterfall. My mug will not automatically fill with water or if there was water in the mug my mug will not automatically empty with water just because the emptiness filling the mug will be at a different pace than the content that is otherwise the filling of the mug.

The mug being empty falls as fast as the boat and I. The empty space in the mug is falling as fast as the mug will fall when the mug is filled to the brim with what ever can fill a mug to the brim. Notwithstanding the content within the mug or the content within the boat or the content within the water being within the waterfall, the very lot is falling at a similar pace. By lifting the cup while falling the cup will fill with water.

I am not putting the water into the cup but I am exchanging the space that the water holds with space that the empty cup holds and my action in truth has no bearing on the water filling the space, which I then transfer into the cup. I am filling the cup with space that at that point holds water but the holding of water has nothing to do with the transferring of space.

If I leaped from the boat and fell I would fall alongside the boat. The boat will be empty but will fall at the same pace and as the same space as I fall notwithstanding being empty. The mug being empty will fall at the same pace as the boat being empty which will fall at the same pace as the water in the waterfall and I would fall. The space in the boat, which is empty if I do not fill the space, will fall at the same pace as the empty space, which fills the mug, and the mug will fall at the same pace whether the space in the mug contains or doesn't contain whatever can fill a mug. The space filling the mug is falling the same as the water that would fill the space in the mug should the mug be filled with water.

The space in the boat is falling at the same pace as I would fall whether I am filling the vacant space in the boat or otherwise filling the vacant space next to the boat. It is the space that falls and not the object filling the space that are falling. It is the space that is filled or not filled that is dropping down because the space being filled is in decline. If it was not the space that fell the space within the mug would fill first as the mug and the boat fell because the empty space would first fill before it could take anything down. But since the boat falls as fast as whether it is being filled or not we can assume that the space which the boat fills or does not fill is falling as fast as it would fall whether it is holding the boat or I or the boat and I. The space not filled by mass also moves just as fast as space filled by mass.

When the object such as the mug or the boat or I connect with the Earth the Earth disallow the object free motion by taking any more space the object claims through to the centre of the Earth. The object now has to relent the space it claims and take on new space that the object claims to flow by contraction to the centre of the Earth. In forming a blocking it resists the flow or the gravity or space lining up with the centre of the Earth. The flowing of space by contraction is gravity but the object being in the space that flows becomes and obstacle through which the oncoming space must drag in order to flow to the centre of the Earth. It forms resisting of allowing space claimed to release to the normal flow when the object will not relent form in favour of gravity. This resisting such relenting of form and consequently forming a frustrating barrier that blocks the free flow of space towards the centre is time displacement of space and this relenting of space-time flowing freely becomes the mass factor. The density and the resistance that the particles show forms the mass that implicate the degree of the frustrating or preventing or disabling of such free flow of space through time and the displacement of space during time is space-time notwithstanding what ever irrational connection Newtonians wish to add too space-time. Allowing space to displace through time to form time is space-time and that is gravity.

It is not as if I wish to condemn and reject that which is in place without placing something of worth back into the process. All I ask is to read what I bring. Don't be a coward and stop reading as soon as you reach the point where I condemn what is in place! Just move past that to the point where I show what is wrong and how it can be corrected! Just judge me not for condemning what now is so apparently incorrect but for showing why I condemn what now is so apparently incorrect and what I bring to the table and offer as a remedy. See what I have to offer and not only what I am taking away. Don't set your sights on what there is to lose but take a view on what there is to gain! Do not reject me on merits you do not wish to instate because you have the fear you are going to lose what is instated. Do not judge me by using your double standards that is useless in the face of the truth. Rather look at the double standards you employ and do not judge me by using your double standards on me. Rather use your mind to detect what is double about your standards and then investigate with me what needs to change. Don't hide the truth. Don't hide from the truth and don't hide behind what you wish to portrait as the truth. Rather come out into the light for the first time in three hundred years and admit to the truth. Follow what I say and see for yourself what there is to gain by trying to detect what is wrong because we all know there is much wrong. The comet does not collide with the Sun and the Moon is not on its way to collide with the Earth in time to come.

Expand science and no the Universe for the Universe is the only aspect that has not the ability to expand. I challenge all of you Newtonians to prove $F = G \frac{M_1 M_2}{r^2}$ and not just to declare it proven because it is in use since the Dark ages. Expand your mind and double check the formula you all so vividly underwrite and support. Prove why do you support the formula in a modern and a scientific way. Explore the correctness that this formula $F = G \frac{M_1 M_2}{r^2}$ underwrites. Be a true exploring scientist and journey with me through the following pages while we venture on the quest to find and vindicate my incorrectness by proving the truth vested in the formula $F = G \frac{M_1 M_2}{r^2}$ that carries the entire physics everyone uses. Let us start where the lot should start and get two Masters together on one point of argument. Galileo said all things fall equal. That says all things fall alike. The first thing anyone brings in is the vacuum bit with the feather and the hammer and since we do not live in vacuum there is no chance of finding a feather that will fall as fast as a hammer. Since the feather does not fall as fat as the hammer we immediately jump to the conclusion that there are falling disparities because of the falling discrepancy we find between the hammer falling and the feather falling. Then what would give the feather the time to fall longer than the hammer does. Everyone concludes about mass coming into play and they are correct. But they are half correct while Newton still is completely incorrect by attaching mass to the entire idea of falling. Take away the resisting of the feather and replace it with something far less air resistant and one will come to a different conclusion.

We have to dissect what factor consists of gravity and what factor represents mass. Then we have to dissect which part does mass play and what part does gravity play. The falling object experienced no mass while falling therefore the falling or moving must be gravity's contribution. While objects are in motion those moving objects is experiencing gravity.

The object show mass when the object has a tendency to move but the motion towards the centre of the Earth no longer takes place. That means mass is the restraining of the motion or is that which prevents the motion or gravity taking place. On Earth, objects experiences mass by restricting gravity or motion with the Earth giving mass but taking away froo motion. By giving mass the Earth forces the object to become one with the Earth and move with the Earth as a pat of the Earth.

Persons falling will experiences weightless ness while falling and they have a weightless state while falling. One cannot then go on to declare that the factor, which prevents motion, is the factor that causes motion because that is totally contradictory. The motion takes place without the presence of mass because the frog and the truck are falling equally fast. When landing the motion of the truck and the motion of the frog ends. Then the two have very different mass values but neither shows the ability to break from mass and move further towards the centre of the Earth. Kepler said the space a^3 is equal = to the motion in a line k as well as a circle T^2.

While experiencing unrestricted gravitational motion a body a^3 is equal $=$ to the motion $T^2 k$ as Kepler said gravity is: ($a^3 = T^2 k$). When motion stops then afterwards when contact between an object and the ground is established, only then does weight or mass form as a result. While falling we find that gravity applies as individual separate space is moving and putting time in relation to the distance that the falling object travelled. That makes the falling factor the part that is the motion that confirms gravity. In the motion or movement we find the gravity because that even remains as a permanent attempt to move. Even when mass comes in as that which results in the ending of the gravity and in that gravity as a term is also forming the motion factor, still remains as an attempt to move. The while moving Galileo proved mass is not present because all things fall equal. Mass comes in when movement is retained and although the mass is present as a factor that factor that mass represents is what produces restriction of such a movement and not resulting in such a movement. The factor that mass represents is the containing of further downward movement. Looking at the factors separately it is obvious that mass as a factor cannot produce gravity. Mass is the restraining motion that leaves gravity as intending motion. Mass occurs only when motion is prevented and when mass prevents further motion resting objects leans against each other. When objects rest against each other they restricts individual gravity motion. Mass is a substituting factor, compensating for motion loss. When mass restricts motion gravity becomes the tendency of motion. Mass counters motion when the Earth restrains further motion of falling objects. When motion seizes, falling objects remains individual while still tending to move. The Earth resists further movement of falling bodies' movement restricting motion individuality. Having mass does not bring about gravity but it does restrict gravity's motion, which is what brings about mass.

Please note that in a web page such as this the information I am able to present is extremely limited. However, when purchasing the book, I am able to present so much more detail that science hide behind a veneer of centuries of brainwashing and mind control.

Those I accuse of practising mind control are themselves victims of this practise and so were their predecessors going back centuries.

I formulate mathematically what is the cosmos and what is in the cosmos without being within the cosmos. That is done because I prove what gravity is. By using the four cosmic phenomena, which is what the cosmos uses to form gravity, I show what "drives" the cosmos and what "drives" life being in the cosmos but not being part of the cosmos and I go much further than that. This I achieve just because I do not hail the greatness of Newton but I looked for other ways gravity could form. I show that gravity forms using the **Roche limit**, the **Lagrangian system**, the **Titius Bode law** and the **Coanda effect**. I uncover these principles by placing Π within the formulating of gravity and when using Π I bring clarity to the misunderstood cosmic principles. The list of the unknowns I can then explain is almost endless. I can even prove what life is and that is a first as far as I am aware!

Gravity forms by movement that establishes singularity initiating a circle in using Π.

I show why gravity is there, how gravity forms and what role stars play in forming gravity. There is no difference between how gravity and electricity forms and that I prove mathematically by decoding the cosmos. I prove mathematically when atoms spin they establish Π that forms the Universe. Whatever forms gravity has to link closely to Π since everything that has anything to do with gravity forms a circle that is Π by the value of the square radius. If mass has anything to do with generating gravity, then mass has to apply Π or otherwise mass has nothing to do with the forming of gravity. Everything using gravity forms a circle of sorts, which forms the curvature of space-time, which is Π and which curves light. The way the planets orbit the Sun and how stars spin has all to do with Π. In spinning in a circle, Π forms gravity as a centrifugal force that condenses space.

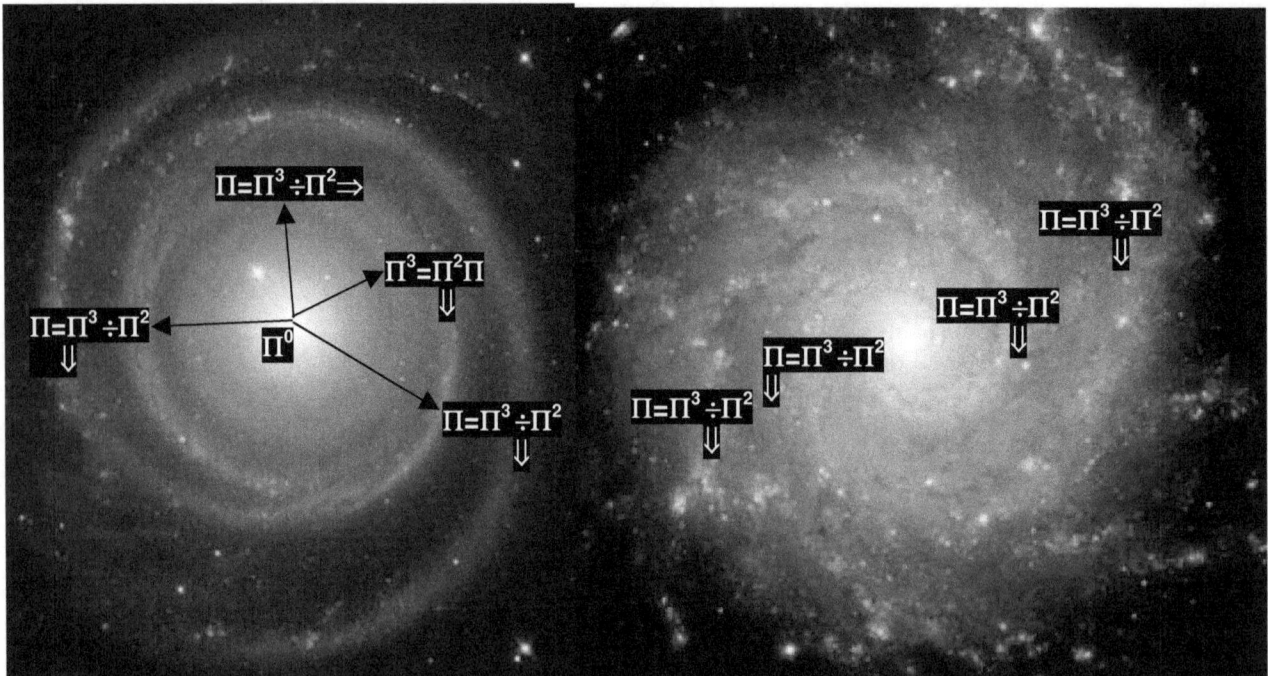

Every spot in relation to Π brings about a new space $=\Pi^3$ forming in relation to a circle Π^2 that spins in accordance with the centre controlling such spin that defines such space that puts such a relevancy in place.

Every circle Π^2 developed during a different era in time that represents another time forming another space.

All the gravity that is present in any and of all these images proves the gravity of Kepler:

$$\Pi^3 = \Pi^2\,\Pi$$

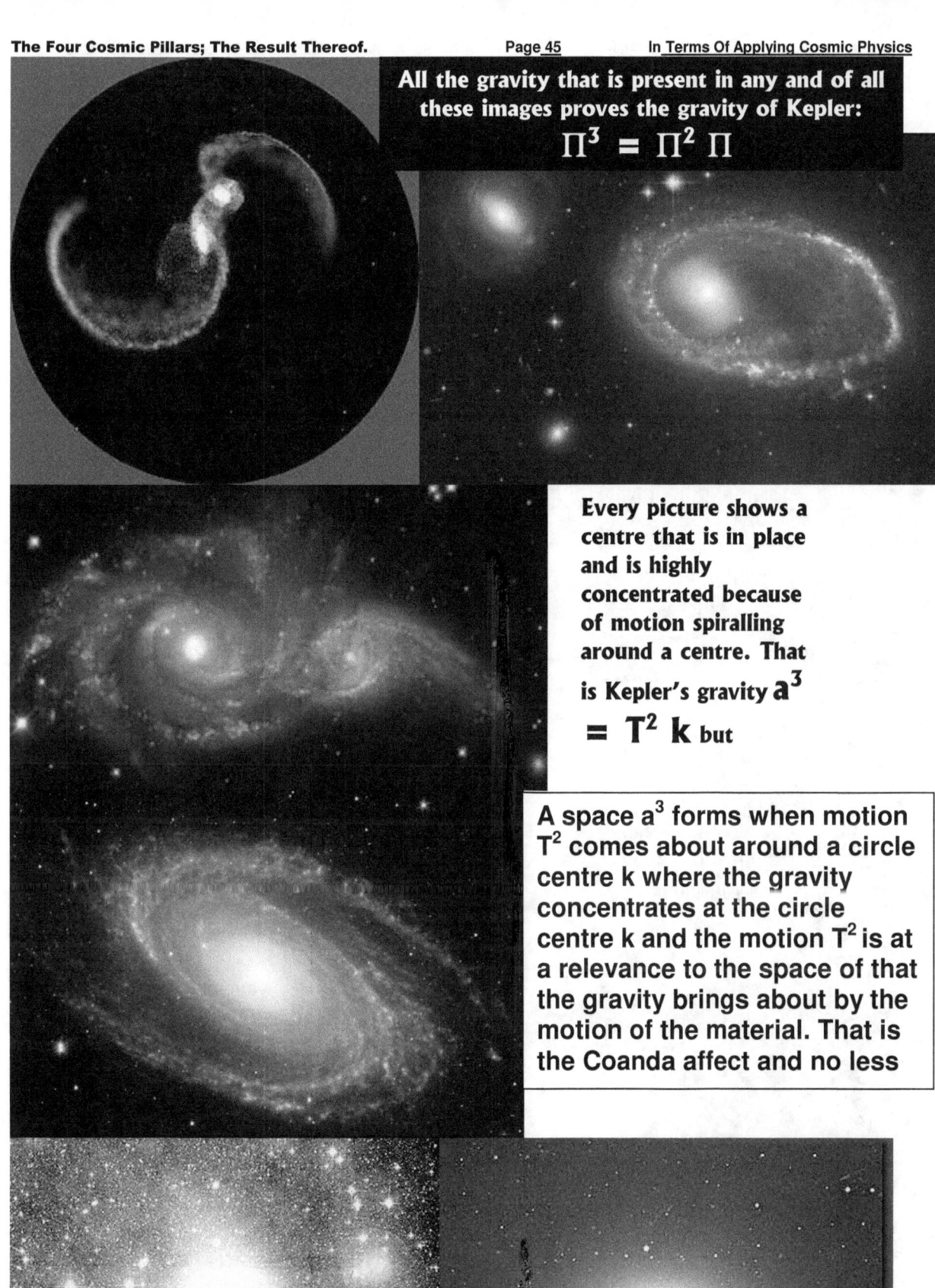

Every picture shows a centre that is in place and is highly concentrated because of motion spiralling around a centre. That is Kepler's gravity $a^3 = T^2\,k$ but

A space a^3 forms when motion T^2 comes about around a circle centre k where the gravity concentrates at the circle centre k and the motion T^2 is at a relevance to the space of that the gravity brings about by the motion of the material. That is the Coanda affect and no less

There are forever circles forming space being defined by relevancies using Π because Π forms gravity that forms space and space is time in history forming a representation of what was.

Space $a^3 = T^2 k$ is in motion

That forms gravity in the centre. That is the Coanda affect but it is too Kepler: $a^3 = T^2 k$

Have you ever wondered why a water drop so readily accepts the form of the sphere? When in orbit astronauts play with water and one see how the water drop takes on the shape of a sphere as a natural form. When gravity is outside the Earth in outer space and is able to choose form without the Earth choosing on its behalf, then the water drop formed by the gravity will take on the same form as the Earth and all other cosmic particles has. Ask yourself why the shape of the sphere comes about when the object can choose freely any shape there is to choose. In that answer is hidden the biggest "secret" the Universe has to offer.

Everyone generally accept that the true form the Universe would have will be in the shape of the sphere. Whenever there is a picture presenting the growth of the Universe, it is presented as a sphere growing larger. Why would our natural tendency be to put the image of the sphere fore ward when thinking about the entire Universe as a whole...but moreover why do we so easily accept without question that the Universe comes in the shape of a cosmic pre-cast form depicting the sphere. Well I know...because Kepler told me so.
He said gravity is the motion of space

Whilst recognising the work of Johannes Kepler, Mainstream science bluntly ignores the impact of his work, and in that they miss the full vastness of the wide influence of his work. Newton shrouded Kepler's work under a blanket of alterations which I show was most unwanted since Kepler's work needs no alterations or corrections and every one since then kept Kepler's work hostage under Newton's changes. It is therefore almost absolutely realistic to say that all information what you are about to read in this letter and article sent to you for your attention was never yet printed in the near or the distant past although Kepler's work has been with us for about four hundred years, during which time it went unnoticed.

It seems to me that any research predating Newton never came into use or into practise. My investigation of Kepler's work brought about a conclusion that no one yet arrived at concerning them with the findings of Kepler because no one scrutinised Kepler's formula before. Everyone is satisfied with Newton's version notwithstanding the incorrectness of it. The world seems satisfied with the idea that Kepler found planets rotating around a centre formed by the Sun and because of that Newton saw a circle.

Where Newton saw a mathematical circle and was unable to understand $a^3 = T^2k$, Newton added what he thought is mathematically required to indicate such a circle. Newton added a mathematical $4\Pi^2$ to the formula of Kepler and removed the distance symbolising measure that Kepler introduced using **k**. On the other side Newton changed the symbol of **k** by using the symbols G $(m + m_p)$. This is just a longer and probably a more detailed manner of indicating **k** and better defining of **k** but it symbolises precisely to the point what **k** stands for nonetheless. I wish to draw your attention to the matter of Johannes Kepler's findings that Mainstream science considers as resolved and closed for many a century while it is not. My investigating Kepler helped me too resolve other unresolved matters but it was only possible by using Kepler's work. This brought about the idea that the Universe is in a state of contracting towards a centre of sorts where mass will form this contracting. This was prevailing until a man by the name of E. P. Hubble came to the forefront.

Looking at the Solar system we find that all planets and objects not classified as planets and all things that is just simply forming solar debris has one thing in common…all apply the value of Π in the process where they orbit the Sun, which also uses the formation value of Π to construct the roundness the Sun has. Gravity has much more in common with Π than it will ever have with mass that produces gravity. Wherever singularity forms gravity, it involves Π which then results in gravity manifesting as some or other form holding Π as a major factor.

I have discovered that the Universe is not employing then concept of the Special Relevance of singularity, but there is a state of **_Absolute Relevancy of Singularity_** controlling the Universe. However notwithstanding by magnitude how significant this breakthrough poses to be, still I have no chance that what I state as my theory on **_The Absolute Relevancy of Singularity_** will be read, will be seriously considered and much less be accepted by those with the authority to change physics principles. The theory I introduce here and now would never be accepted in my lifetime because science in the Newtonian way is bent on believing in the marvellous, the outrageous and the magic of what can never be explained, although they claim to use facts as a basis. Science has no idea of what a Black Hole is and I can prove what a Black Hole is.

Yet, I feel I need to warn you whom are reading this letter that this work contained in this letter strays widely from mainstream science and for that there is a very good reason, but I should add that in the least it is thought provoking. I researched the work of a man that is most exceptional and even more prominent in the history of mankind. His role in the gathering of information furthering knowledge accumulating of the human species' efforts stands second to none while most of everyone is not even aware of the full implication of his work. While recognising his work Mainstream science bluntly ignores his work and in that they miss the full vastness of the wide influencing of his work. It is therefore almost absolutely realistic to say that what you are about to read in this open letter sent to you for your attention was never yet printed in the near or the far past although the work has been with us for about four hundred years during which time it went unnoticed. It seems to me that any research predating Newton never came into use or in practise.

My investigation of Kepler's work brought about a conclusion that no one yet arrived at concerning the findings of Kepler because no one scrutinised Kepler's formula. Kepler found planets rotating around a centre but Newton saw a circle and added what is mathematically required to indicate such a circle. Newton added a mathematical $4\Pi^2$ to the formula of Kepler and removed the distance symbolising measure that Kepler introduced using **k**. On the other side Newton changed the symbol of **k** by using the symbols G $(m + m_p)$. This is just a longer and probably a more detailed manner of indicating **k** and better defining of **k** but it symbolises precisely to the point what **k** stands for nonetheless.

Kepler said
$a^3 = T^2 k$ but that could also be $k = a^3/T^2$

Newton said a sphere is $a^3 = 4/3 \, \Pi \, r^3$, which is mathematically correct, however Kepler said the cosmos told him a cosmic sphere is $a^3 = k \, T^2$ There is the two distinct possibilities which Newton saw and which Kepler saw and both are most valid. Between the two concepts there is literally one Universal difference and the two can never be mistaken as promoting the same principles. It is true that Newton's method or formula of calculation $a^3 = 4/3 \, \Pi \, r^3$ when measuring the sphere is widely used, but Kepler received his code of calculation from a very high authority, which is none other than the Universe. It is the duty of the cosmologist not to reject Kepler's findings, or as Newton did, try to transform it into something that Newton could understand after such transforming, but to search for the meaning as Kepler received the formula from the cosmos.

'Ever try to answer facts about the Universe in as much as…what brings about the expanding?

Kepler said the Universe plus it entire content is expanding centuries before Edwin Hubble realised what he was seeing through his telescope.

We can test any of the following symbolic values in the mathematical expression and also test the principal behind the expression in which Kepler stated them. By such testing we will find that time after time there were never any corrections in the translations required since the translation thereof was never incorrectly presented and in that a case asked for no alterations to secure the correct reporting of the cosmic information being translated. By taking the formula on face value it can change as follows: **$a^3 = T^2 \, k$** can become **$k = a^3 / T^2$**

When translating Kepler's mathematical expression into English we can see what Kepler said also read as **$k = a^3 / T^2$** where **k** is one point from a centre point that is space **a^3** relating to time **T^2**. From a centre comes space-time. The centre **k** brings space **a^3** in ratio to time **T^2**, which are space / time **a^3 / T^2**. Reading this correctly cannot bring any dispute…yet it does…and it's been doing it for centuries on end!

Kepler was the very first person to mathematically introduce **space a^3** aligning a **centre k** and relating the resulting movement to **time T^2**. Not only did he introduce **space-time $a^3 / T^2 k$** but he also placed **space a^3** and **time T^2** in a relevancy **k** long before Einstein did and placed **gravity in space-time $a^3 / T^2 k$** even before Newton named gravity. He showed that space **k** is growing in the manse the Universe attend to space-time **$a^3 / T^2 = k^1$**. Kepler was the person who placed gravity as the ingredient in the universe that determines **space a^3** and **time $T^2 \, k$** and much more. Kepler was the first one that saw that gravity comprises of two factors being **k** or linear gravity and **circular gravity or T^2** as gravity keeps space in form while all is staying together.

. Kepler said **$a^3 = T^2 k$** and that correctly translates to a mathematical expression **$k^0 = a^3 / T^2 k$** which in the verbal statement in English translates that Kepler said that there is a **space a^3** which is **equal =** to the motion in **the time duration T^2** thereof between two specific points which holds a relation onto a centre **k^0** where from there forms **a straight line k** that is centred on the spot where space begins from **k^0 that produces k** as well as producing the circle therefore that spot **$k^0 = a^3 / T^2 k$** has hold **k^0** at a value of having the least space. The line **k** is centred onto a spot where space begins specifically at **k^0**. This point not only produces the line **k^0** but represents also the space that forms the eventual circle **T^2**. Therefore from the centre holding **k^0**, **k^0** leads to **k** that forms the roving space **a^3**, which is rotating at a distance **k** where **T^2** forms the outer limit of **k^0**. Mathematically **$a^3 = T^2 \, k$** will be **$k^0 = a^3 / (T^2 k)$** because **$k^0 = 1$**. But **$k^0 = 1$** also present the single dimension where all factors are a product of one. If one can locate **k^0** one will find singularity. That is where gravity is because gravity is strongest where space is least. Then that suggests that gravity is strongest at **k^0** because space is least. That is gravity because that is what keeps the orbiting objects in orbit but also that is what Newton completely missed when he changed Kepler's work. Newton failed to recognise gravity as the only ingredient in Kepler's formula. He admitted he missed this because he admitted he did not know what gravity is while Kepler explicitly showed what gravity is. Gravity is what keeps the orbiting objects orbiting. **$k = a^3 / T^2$** is **distance^1** = **space ^{3}/ time^2** forming from a pivoting centre **k^0**. That is a cycle and moreover it is a cycle formed **by space/time**. What Kepler said is that space is **a^3 in motion $T^2 \, k$.**

That says **space^3 (a^3/)** relates directly to **time^2** that uses the symbol **T^2**. This is also what I refer to when I say one has to read what Kepler did not say when one wishes to see what Kepler meant to say. Kepler introduced space^3 –time^2 long before Einstein's date of birth appeared on any calendar although Einstein is credited with the formulating of the concept of space-time and giving it a name. Going even further Kepler stated that the space **a^3** is on the move **T^2** around in a circle at a distance **k**. That is what that comet that we are discussing is doing. The space^3 (Comet) is circling the Sun using a radius **k** to establish the cyclic time^2 as a period of continuous motion and continuous motion is gravity. That reads much more correctly and closer to the truth than what Newton predicted what according to him (Newton) was happening in space. Remember in this statement I am separating cosmic principles applying from the way that gravitational principles apply on Earth. I distinguish that which is the rule in the cosmos from what we find ourselves trapped in on Earth. The two just don't mix. I am removing cosmic physics from normally accepted physics because the gravity concerned is not the same.

As Kepler said $k^0 = a^3 / k\,T^2$ **and we have to find k^0.** As a result of examining this proposition, I located two principle positions both holding singularity.

What is in the Universe is spinning. The entirety of everything forming the Universe is spinning inside the Universe and such spinning is always in the centre of one specific point, wherever such a point might be. In the **precise middle** of all **objects in rotation** is a precise centre where this pre-designated centre is dividing the object in rotation into sectors that will **start the spinning initiation** from that centre point. Thus, the spinning object **will have a middle point**, a very specific **centre point that does not spin** and only holds Π as a specific value because no radius can apply. But also the one value such a line **cannot have is zero** because the line **is there and holds contact** to the rest of the material bringing about that **zero does not start any** line and therefore the **value of the line must be infinite**, just as described in **accordance** and by **the definition of singularity.**

The condition for the presence if this centralised singularity is movement.

In considering the spinning motion in the fraction of time in the detailed instant every aspect of rotation will turn in every instant of change in time. Although the points had the same characteristics only one instant before, they oppose the characteristics it had just before and just after the very instant in which they are and to which they relate by similar points also in rotation. The fact of the graph proves my point in quarterly opposing dimensions and values,

As the rotating direction moves inwards, the rings will become smaller and smaller.
In dimensional terms, which I explain later on the value of **2k** relates to T^2. That relation extends to the next value where T^2 relates to **k**, which relates to T^2. The first space in the circle will then be T^2k. From the centre being in infinity one can realise by applying mental power the single dimension factor not seen but present all the same. Extending that into the 3D comes six **k** and any one of the six will further extend to form a seventh point as T^2 All this is a multiplying of $k^0 = a^3 / (T^2 k) = 7$.

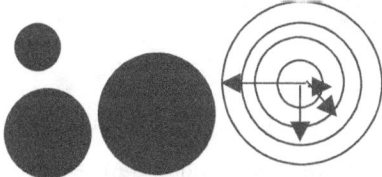

Looking at the affect of gravity it shows the precise quality of no distinctive point, as gravity never seems to end at a point but flows all over affecting all that holds a position in its sphere of influence. The gravity coming from China meets the gravity coming from America at no particular spot but intermingles without distinction.

Using the concept that gravity applies Π as the circle factor Π as well as Π^2 replacing r^2 the replacing by Π brings two values as Π and Π^2. That I found is the case with gravity and will be apparent when explaining the sound barrier as well as the Four Cosmic Pillars. In order to create a distinction I remained using r as the indicator of the cube or non-circle that has vacant space and by vacant space I refer to non-solid structures. In the solid structure I use Π as a value for reasons that will become apparent in due time.

This spot is the result of a most basic process of reduction as the Hubble constant is a most basic process of expanding during a matter of time. By reducing the line constantly the only value that will eventually remain without dispute from any party arguing about the facts is one followed by an exponential zero (1^0). By only having exponential zero instead of a numerical zero and a radius as one in the square (the radius effectively becomes one holding any and all sides on one point) such a point might become any value of any significant measure implicating anything but zero as the radius. By expanding the line, it will be an evenly spaced structure growing into the most perfect round dot ever possible anywhere at the point when it starts to grow.

The reducing of the line is one dimension in six and although such reducing is representative of two indicators all the other indicators must still be accounted for two. In mathematics there is a line being one quantity and the circle indicator Π being the next circle indicator. Reducing the line will erode the value of Π by ratio. That will eventually lead to having a circle ratio of Πr^2 and eventually lead to Πr^0 but that is not the point where the circle ends. That is where the ratio applying factor ends but it cannot exclude the circle. The circle as a concept can still reduce when it abolishes form to the single dimension. It is not the radius that is responsible for the circle but the figure value of pi and by abandoning π only then does all the aspects fall back into the single dimension.

The circle can reduce one step more when the circle eliminated r completely by returning r to a point of singularity r^0, but the elimination of r as the factor reduced the major factor to the single dimension in Π^0. That will not reduce the cosmos to zero, but it will only eliminate all potential lines r^0 to potential circles $\Pi^0 r^0$ and from there the circle Πr^0 will come about by manifesting as a line but that manifesting can firstly only establish a circle Πr^2.

The only value that singularity can have although the single dimension may host the entire Universe is Π^0. Pick a number and elevate it to the power of zero and in the process one may have established another point holding all points in singularity because that is the value of singularity. Only Π^0 or any other value holding one accompanied by zero as an exponential value can ever be the accurate value to singularity while singularity will then host the rest of all the possibilities in the Universe. This means that the entire Universe composes of and is made up of singularity... this much I am going to prove. Every point occupied or otherwise constitutes of singularity either under control by movement in a form we call atoms or being passive in a location we call outer space. I wish to repeat the position holding singularity because if I introduce anything new, then this centre singularity is the pivot of everything that I introduce to science, and also I refer to the top because Newton used the top as an example by which Isaac Newton missed everything that Kepler clarifies about cosmology.

In the sketch to the right above the circle to the right would come about from a straight line r growing influencing the appreciation of Π, but to influence Π would lead to a breakdown in r as Π and r are different entities. The circles to the left (black dots growing in size) shows a continuous growth by extending Π every time and since Π is the same part as the previous Π, only extending that billionth of a millimetre each time, the circle will be truly continuous without any signs of a break.

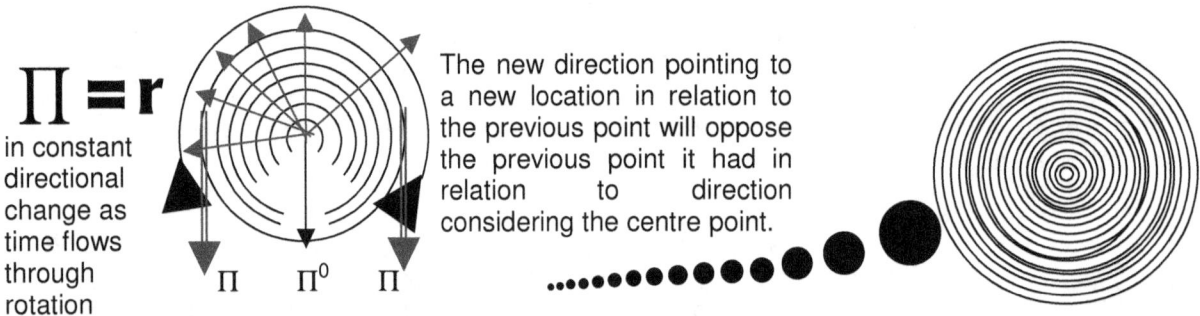

Let's go back once more and reduce the line by half every time. Then repeat the process until it can repeat no more. The reducing of the line by half every time will get to a point where all the ends land on the same position without any possibility if halving the two ends further. The points share one position and moving the points in any direction will lead too an increase of the line once more.

From this centre line that is only theoretical definable, but is still there all the same, an centralised line forms holding opposing values apart and parting the opposing values is what proves that this line forming has no status in space and yet controls all that holds space. The one side will turn left and by crossing this line holding no space, all the space will then turn right.

By moving any of the points, such moving by further decreasing at that point must then bring about an increase of space once more since the space at that point in the centre spot of the circle spinning has gone infinite. This also applies to the sphere that is a multitude of circles because the circle uses a line to indicate size running from a centre to an edge. By reducing the line and by reducing the circle the reducing will end up having the ends in the same position in the very centre of the circle. It is this fact of the moving in any direction of any point from that spot holding singularity that such motion will introduce space as the space exceeds the previous limits of singularity. What I am trying to say is by moving from the spot Π^0 to the dot Πr^0, such movement evokes Π^0 a spot to Πr^0 forming the unseen line and without the movement of the spot Π^0 to the dot Πr^0 to form the line Πr^0 the allocating or positioning of singularity Π^0 will not take place.

In determining this behaviour as part of a cosmic process where matter interact with matter in an laid down set of rules, once more we should be asking questions and this time it is whether the top will show the same behaviour in outer space as it does on Earth. With the reply of no it would not come as an admitting

Locating and finding Singularity

The entire Universe consists of lines running all over.

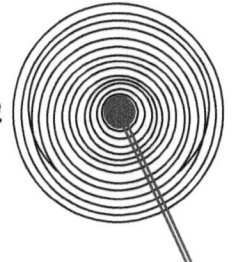

That point albeit hypothetical, is also as much a reality none the less and is placed where that point **must be standing still** because every line **running from that point** in **opposing directions** are also **in opposing directional spin the other or opposing side.**

In the **precise middle $k^0 = a^3 \div T^2k$** of all **objects in rotation** is a precise centre dividing the object in sectors that will **start the spinning initiation** from that centre point. The object has to rotate $T^2 = a^3 \div k$ in order to instate the space $a^3 = T^2k$. Thus, the spinning object **will have a middle point k^0**, a very specific **centre point $k^0 = a^3 \div T^2k$ that does not spin** and only holds Π as a specific value. One value such a line **cannot have is zero** because zero does not start any line and therefore the **value of the line must be infinite**, just as described in **accordance** and by **the definition of singularity**

In considering the spinning motion in the fraction of time in the detailed instant every aspect of rotation will turn in every instant of change in time. Although the points had the same characteristics only seconds before, they oppose the characteristics it had just before and just after the very second in which they are and to which they relate by similar points also in rotation. The fact of the graph proves my point in quarterly opposing dimensions and values,

The parting of directional opposing space will always form what becomes real and distinct when rotating, and by rotating around this line that is only theoretical the spinning is putting this line not rotating even in more distinct prominence. Because of the rotating that evokes the presence of the line not being there, the influence this line holds grows so much it covers all the matter from end to end, to a securing the spinning to a point in the centre that holds no spin value.

It is there as well as not being there by not spinning because it has no space to spin. Yet the line is there and therefore it has the most original value anything can have. When not rotating, the line disappears and only a diameter runs across the material with the diameter going as thick as the material will go.

When rotation begins, the line then forms according to a radius. While the line forms where the radius starts, it shrinks back to a hypothetical position claiming zero spin that through that it is not less distinct but more distinct because from that point every rotating becomes a piece of what ever forms part of what is then spinning. Because it spins the end of the line forming will clearly carry the singularity value of Π^0 to end at Π that then is implicating rotation by the value of Π^2.

When looking at the cosmos from whichever angle it indicates the fact that the cosmos is moving. Everything in the cosmos is moving and all things about the cosmos are moving in relation to everything else in the cosmos. Everything is forever spinning in relation to a point that could be any point that is not spinning and everything is all going towards as much where it is coming from.

Everything is on the move and always encircling something of making that centre point to seem to be of greater importance than what everything is that is spinning. A top can spin but the parameters of its spin are limiting the motion it can apply. By not spinning the top is still spinning as the Earth is doing the spinning on its behalf.

The spinning top that Newton dismissed as $\frac{dJ}{dt} = 0$ brings all the evidence any one needs in order to come to a conclusion that will bring any proof that the singularity governing the top connects too everything anyway. Placing singularity is fair and fine, but what will the evidence be in proving its activeness as part of the creation at large? The reason why we can be sure it is active is that when spinning it shows borders implicating restraining of further movements outside the set limits.

By going faster (past the upward border) the spin goes oblong where it actively tries to change the position the top holds to the Earth in relation to the surface of the Earth. By going too slow it once again shows identical characteristics. When going too fast it indicates an attempt to rise into the air, therefore relieve its singularity in an effort to part with the Earth's singularity. It shows unmistakable characteristics of trying to become airborne securing an independent position from the Earth, which holds it down.

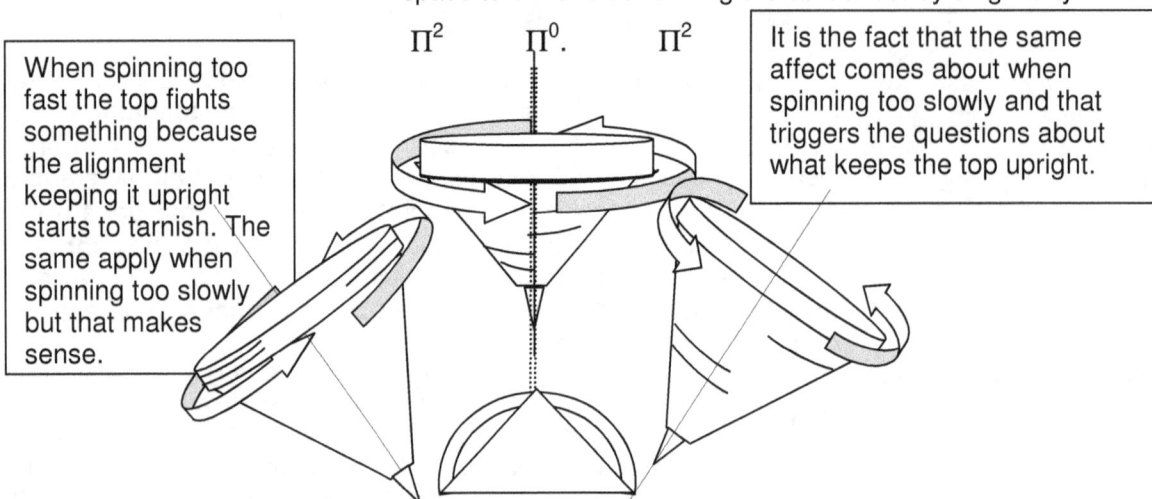

Pinpoint positioning of singularity Π^0 with Π positioning space to either side forming the border set by singularity

$\Pi^2 \quad \Pi^0 \quad \Pi^2$

When spinning too fast the top fights something because the alignment keeping it upright starts to tarnish. The same apply when spinning too slowly but that makes sense.

It is the fact that the same affect comes about when spinning too slowly and that triggers the questions about what keeps the top upright.

The rising above the position the Earth holds the top is clearly indicating the top is trying to generate more gravity as what the mass would be by which the Earth restricts the top and hold it in a position where it will form mass and then become part of the material forming the Earth.

When it is at the bottom we surmise correctly that it wishes to topple over and fall down, but something drives the top to put up a fight in order to stay upright. If the top falls down, this action of falling down will kill the centre line holding singularity and it is that centre line the top holds that is keeping the top erect. By destroying the line it will then enforce stopping the independent spinning of the top. Of course the bottoming out shows the same characteristics whereby we gauge that to be the normal process of falling down. If the bottoming is relative to the Earth's singularity and we recognise the process as normal, then the top of the limits should be just as recognisable normal.

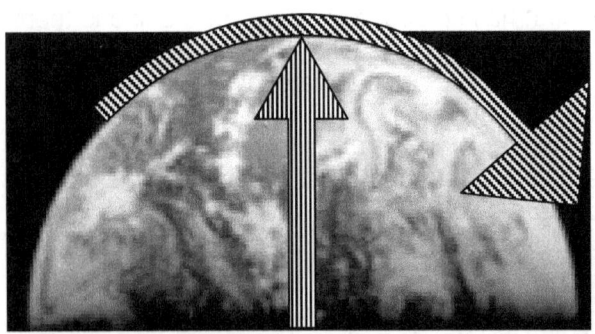

When any object is in a state of having mass the object has to be standing still and being having mass. The object has to be in a position of standing still in relation to the Earth while any object to form mass and it is where at that point secured in a position on the Earth at that point of of absolute rest while it is on the Earth. At a point excepting only the movement the Earth allows that the object with mass is resting while all the rotational movement is equal to the movement the Earth delivers where the Earth is rotating. Rotating at the speed the Earth dictates form the factor science call mass. When the object leaves the surface of the Earth such an object will have to move much faster than the Earth moves or have less density than is required to maintain a steady position on the Earth.

When any object is standing still being in a state of having mass on the surface of the Earth, an object has micro gravity because the individual gravity left to the object in mass is infinitive small and is left to become an indication of attempting further movement towards the centre of the Earth while the Earth's material blocks the micro gravity to move and hence apply mass in doing so. Mass is not something inherent of the object but is the annexing of the object given mass by the Earth to secure the position of the object to ensure the object becomes part of the Earth structure. Having micro mass (not micro gravity) is where the body in rotational movement extends beyond the limit at the point where the Earth surface would award a mass factor. The movement speed goes beyond the speed required by the Earth at the Earth limit where rotation velocity secures mass as a factor. By exceeding the rotational velocity at a higher rate, such movement would exceed the movement or gravity of the Earth that is required in order to grant a mass value.

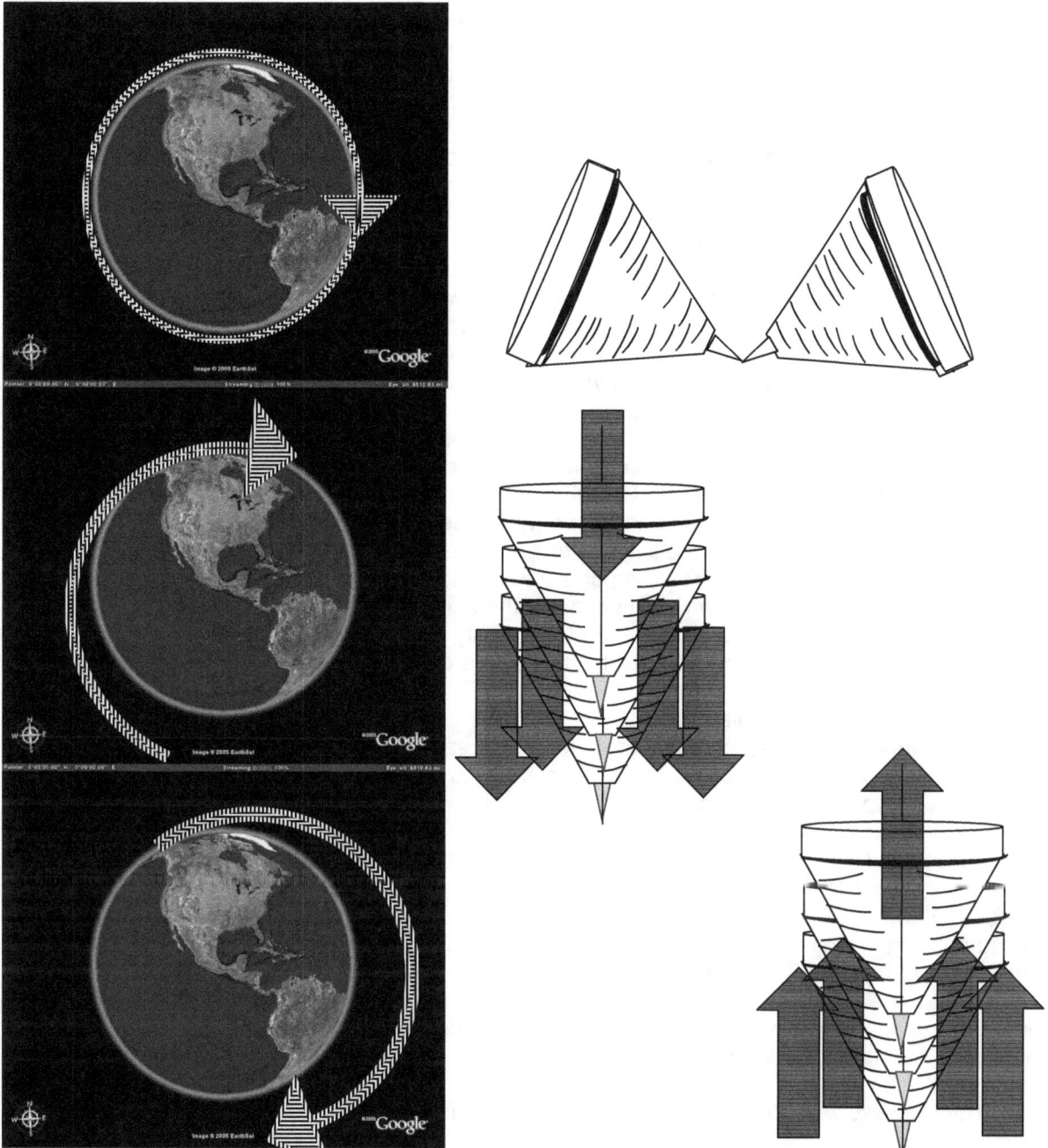

When the top is spinning it is this line that urges the top to excel from the limitation of the gravity of the Earth and extend up into the air and away from the ground. It is this centre line holding singularity that drives the top to lift up from the ground and fight the mass that the Earth inflicts as to retain the top with the limitation of enforcing the top into a state of mass where the mass holds the top onto the ground. The top is fighting the Earth's effort in restraining the top with mass by producing gravity that lifts the top into the air. It is the top's spinning that is producing anti gravity to fight off the gravity of the Earth. I have heard so many scientists refer to man discovering anti gravity as if such a discovery of a force of anti gravity will give humans the power only God can have. It is this mindset that I refer to as science wanting the marvellous, the magical and the unexplained. To the masterminds of science having anti gravity would come to the same as unlocking all the witches' forces and opening the Pandora's Box of forces while anti gravity is simply jumping in the air. If gravity is what is pulling you down, then anti gravity must be something lifting you up.

When the spinning has died down so much it will arrive at a point where the gravity of the Earth will reduce the spin of the top to lying still while the Earth secures the top with going into having a state of mass or having no independent movement which is having mass. It is at that point that the centre line within the top that is securing gravity by the spinning of the top will seize to be and the top will once more come to rest in a state of mass. The gravity of the Earth is fighting the gravity of the top which is equal to the singularity of the Earth is fighting to destroy the singularity of the top which is the movement of the Earth is fighting to destroy the movement of the top and all of these relevancies are all the same.

This issue is of cardinal importance and could deliberately be altered to hide the misinterpretation science wishes to connect to mass in order to hide the fact that mass does not bring on gravity but it is gravity that brings on mass. Mass is achieved when the object is resting motionless on the surface of the Earth while it is gravity that is still attempting to obtain movement as to try and move the object down to the centre of the Earth. This movement consists of two parts where one part is following the curve of the Earth while the Earth is rotating and the other part of the same movement is the thrusting of the object educing the object to move to the centre of the Earth. Mass is the result of gravity and not the other way around. Gravity brings on mass and mass depends on gravity to have any value or function.

I wish to draw your attention to the matter of Johannes Kepler's findings that Mainstream science considers as resolved and closed for many a century. Not one of the following principles was yet successfully proven but I believe I have accomplished that goal. I too am well aware that at the first glance you will immediately arrive at the opinion that the theme of the letter has to be considerably below the standard of an intellectual Master such as you must be being in the position you hold and therefore the normal research work you do. Not withstanding I hope that this writing may spark interest even at such a low academic level and grade in scientific sophistication development because I am about to prove that I discovered:
1) The location, the position and the value of **singularity** as a factor forming space-time
2) Finding **space-time** by dissecting Kepler's formula in relation to valuing singularity
3) Finding space-time, **proving space-time** and **aligning space-time** with gravity
4) The **working principals** behind and manifesting **of gravity** as a cosmic occurrence.
5) The **Roche limit**, and explaining the resulting of a law coming about from singularity.
6) The **Lagrangian system**, how and why that becomes the building form of the Universe.
7) The **Titius Bode law** and I show mathematically how gravity comes about from that
8) The **Coanda effect** and the producing of gravity through reproducing space-time.

Every one was sharing the Newtonian vision of a contracting Universe where the lot that formed the Universe we know would one day again come together and Creation will end where Creation started some time ago. Everyone was well accustomed to the idea that the Universe has mass that is pulling mass towards one another and we are in the centre of an ever shrinking Universe. That is what the lot of us can see… we are forming the centre of the ever contracting Universe and the bigger mass is pulling the smaller mass towards a centre This was what confirmed the cosmos where every Newtonian can vividly see. It is possible for every Newtonian to witness with his or her eyes while finding conformation through any telescope that all Newtonian-minded scientists are sharing the centre stage of the ever collapsing Universe. The one thing bonding Newtonians is the idea that the Universe is about to end where all mass contracts into one huge lump of material. That is why mass is pulling mass by force!

Then along came a man that had a good look at the Universe. He looked at the sky and came to a conclusion the lot was not shrinking but it was expanding. Any one that would look through his eyepiece could clearly see the lot was not shrinking. The lot was growing apart. In some cases he said the lot was racing apart. The Universe was growing by miles and not shrinking into nothing.

The main discover had a name and a position of seniority, which prevented others from pushing his opinion aside. The man was E.P. Hubble. Through his telescope any one could see that the Universe was expanding and the expansion was most rapid. This unleashed a problem the world had no name for. Everything known to science was at that point devastatingly unknown to science. The Universe was expanding and not contracting which made the Universe quite wrong in the eyes of Newton's laws. It is impossible to have any vision about Newton being wrong.

Newton could never be wrong because Newton was never wrong yet…so if the Universe is out of step with science, then science will correct such an abnormality by finding a way to defraud science and postpone the correcting that the Universe had to comply with since the Universe owed the Master Newton some apology. It has to be the Universe that stands to be corrected and not Newton! Did the Universe not know that he whom never can be wrong is in name Isaac Newton! Decisive action was needed. At this point I cannot believe that the most brilliant minds were so naïve and therefore I must suspect deliberate deception. Hubble was far too prominent to blow away and Newton was found wanting. At that point they put the onus of proof not on Newton but turned the focus away from Newton to what they then presented as to be the guilty party. The Universe had to comply to be corrected! When will the Universe confirm its incorrectness by affirming Newton's obvious correctness? Newton said the lot is contracting so how can the Universe have the audacity to expand? If they had to admit that Newton was wrong, the most intellectual science then had to admit they had nothing to show for all their minds brilliant work. Then all those the world thought about as being brilliant and that knew everything had to admit they knew nothing.

They put the onus of proof and converting onto the cosmos. There just had to be unaccounted mass hiding in some dark allies waiting for a chance to unleash the mass to form the required gravity! When will the cosmos come clean and prove Newton correct. When will the cosmos admit to a mistake and set its crooked ways straight by showing the hidden mass. When will it meet its diverting from Newton and reach a point where the Universe will finally come to comply with what Newton demands. It is the cosmos that is wrong therefore it is time to find out when the cosmos will correct its manner. To deal with such a task they needed a man with a bigger ego than he had an IQ. They needed a person that thought more of his abilities than his ability to grasp any complex situation. They needed a man that was presented as a genius without ever proving his genius. They had a man that filled the centre of the Universe, which then placed the man in a location from where the man could see the entire Universe. They had just such a man. He went by the name of Albert Einstein. For all the genius Einstein had, Einstein failed to see the most simplistic and tiniest mathematical rule. Einstein failed to realise that if there was insufficient mass at the beginning of the expanding Universe, the growth of the Universe will reduce the influence of such mass as a factor because as the radius grows, such growth will restrict the gravity by rendering the mass progressively more incompetent.

If the Universe is expanding as Hubble indicated, the growth of the radius will reduce the influence value of the mass as every second passes. The mass will become more and more wanting for such a task. Yet with this obvious shortsightedness of the genius Einstein, the genius saw him fit enough to calculate and measure something as overwhelming as the Universe. As in the case of Newton, Einstein was being an ego driven maniac that saw his abilities fit to measure and master the Universe and account for all the mass in the entire Universe while his mind was too simple to recognise the most basic principle of mathematics, the principle of relevancies or ratios. What a mathematical genius that turns out to be. While the radius enlarges with time expanding the cosmos at the same proportion does the influence of the mass factor reduce its influence on achieving any substantial pulling power and the mere fact that the radius increase as the Big bang developed the radius that should divide into the multiplication of the mass, this shows that at no stage further into the future going past the Big bang can the mass stem the growth of the radius because the radius overpowered the mass factor already at the very beginning when the relevancy favoured the mass at its most. Unless there is new material entering the Universe at a point, which is impossible, the entire concept is fraud.

The idea was never to admit wrongdoing on the part of Newton and Newtonian science but to post pone, delay and divert attention away from the truth. If there was not enough mass to start with, no dark matter can kick in later on and start secondary mass frenzy that at that stage will then be enough to bring about the required mass potential that will turn around the Universe from expanding to contracting. To establish a scenario that would hide all deception they got the man that has a bigger ego than an IQ, they tell the world this man is a genius while the fool does not recognise the least of mathematical principles because his Master Newton did not know the least of mathematical principles and they got him to measure all the mass within the entire Universe. While they did not even have any device (and will never have such a device) through which anyone would be able to see the entire Universe, they set of a scandalous misconception that this Einstein could calculate all the mass in the Universe. Surprisingly there was not enough mass so they thought up an idea of dark mass hiding from all prowling eyes and that dark mass would come to Newton's aid…but when would it come and aid Newton was the new question. Now a new task was in hand…to find dark matter that would hide mass and hide gravity.

If the dark matter did not develop enough contraction at this time, there is no chance in the future to develop enough gravity because the factor of what mass supposedly should have is tarnishing and tarnishing as the Universe expand. The bigger the radius becomes the less would the mass effect be. The community of astrophysics are trying to frame a picture where they set the stage in the way that if the Universe were stretched to a point the mass would not tolerate any more expanding. The mass will get frustrated in some way and show resistance to the increasingly elastic expanding. The gravity constant (I suppose) must prevent any further expanding. If ever there is a faculty ruled by absolute inconsistency and rubbish as the motto of logic it has to be astrophysics. Astrophysics holds the opinion that every measured kilometre represents nothing. The distance between the Sun and Pluto is more; therefore that which outer space is made of is more than in the case of Mercury and the Sun. Therefore Pluto has more nothing between the Sun and the planet than Mercury has between the planet and the Sun. Only astrophysics and all the geniuses guarding the principal of astrophysics can put a calculated value by measure on nothing. In fact Mercury has to have hundred times less nothing between the planet and the Sun than is the case with Pluto just because Mercury is a hundred times closer. Now make sense of that! Astrophysics can measure nothing in units of hundreds and still find such distance with nothing increasing. The figure containing nothing that puts Pluto at the edge of the solar system is one hundred times more nothing than what Mercury has where Mercury becomes the first planet in the solar system. That is astrophysics. The Universe comes from a point the size of a neutron. That makes the radius parting the Universe infinitely small. It just about removes the radius as a factor. At the very same implication it takes the pulling of the mass (if there are pulling forces converted by mass) to a level it will never again have. With mass more than it will ever be and the radius smaller than it could ever again be then the big Bang was the moment when all had to collapse, that is if Newton's formula is correct. As soon as the distance between the objects holding mass started to grow, the power and influence of the mass factor started to diminish in the same ratio. If the mass were incapable of contracting the Universe then, it will forever remain impotent in contracting the Universe. Then you may ask what is the story? Read on and you will learn how far Mainstream Physics stray from the truth and how big a cover up the paternity is protecting.

According to science the Universe started with singularity. Quoted directly from the Oxford dictionary of Astronomy the following:
The definition of singularity is as follows:

Singularity: a mathematical point at which certain physical quantities reach infinite values for example, according to the general relativity the curvature of space-time becomes infinite in a black hole. In the Big Bang theory the Universe was born from singularity in which the density and temperature of matter were infinite. The average daily temperature was "$10^{\alpha\beta}$ to 10^{34} K". This puts singularity at some centre stage that confirms the place this lot that formed the Universe began.

Being the onlooker the viewer has to maintain one position. From that position some particles would be circling a centre point, as the particles would be coming towards the onlooker. The other matter would be circling the centre point while rushing away from the onlooker.

At the very end the single dimension may come into the dynamics but where the single dimension comes in the factor of zero is removed. If there is space, there is a flow of light and a flow of light has to produce lines in relation to angles forming space between them. Something must be present to confirm space because there is an absolute difference between being in space and no space to be found. If there was a line that formed nothing that one line that forms nothing would completely destroy the other lines' chances of ever forming a triangle, let alone having all lines and they then have a total being zero. As shown in the example no line can form zero and therefore no mathematical equation as far as it extends to cosmology can ever bring about zero as a number. While there is space present there has to be three dimensions relating to each other by time and in three dimensions there has to be three lines in relevancy to each other by angles formed holding space in (at least) six opposing sides. Removing one line must bring about a flat Universe and that then will constitute nothing.

Cosmology is about light flowing by means of lines indicating space obeying the rules enforced by time in motion and light flowing dictates crossing space and across space light is using lines. The book: ***An open letter Announcing Gravity's Recipe*** is dealing with the subject finding singularity by removing the concept of nothing from outer space. By diminishing nothing one uncover singularity and the effort brings in a new perspective not yet introduced.

For your benefit I will shortly give a summary by which I hope to interest you in reading the manuscript: Compressing space produces heat. Releasing heat will bring expansion bringing about space. We call such a release of heat an explosion. In other words heat translate to space and space concentrates back to heat. The one is a product of the other where space forms expanded heat.

They are quick to show the time that was applying at the time being some thousandth of a second or the heat that was present being numbers we have no name for. The other side of the story they ignore. They ignore the other side of the story because in that respect it puts their promoting of Newton down to madness. If you reduce the radius applying at the present back to what it was at the time of the initiating of the Big Bang, you must also increase the influence gravity and mass had at that moment by the same number you are decreasing the radius. That is pure mathematics and the most basic physics of all concepts.

The shrinking radius will increase the effectiveness of the influence of the gravity that the mass can produce by the margin of the shrinking of the radius. If the Radius was infinite at that point, then that means the gravity was eternal. With the entire Universe being as big as a Neutron, the Universe was the size of an atom. If the Universe were the size of an atom and the mass within that Universal atom could not prevent the Universe exploding into immeasurable atoms, then it would not be able to retract all the atoms into one unit again. If there was not enough mass to start the contraction, there can be no contraction of mass that is producing the gravity at this stage. If the gravity is of such a nature that it allows a continuous growth of the radius, then the radius firstly cannot be zero as Newton suggested and the extending of the radius proves there is no contraction in the way Newton had everyone to believe. If Newton's mass contracting mass is true, then on the other hand it must have resulted in an implosion as that which can never repeat again. With Newton's formula of $F = \dfrac{M_1 M}{r^2} G$ forming gravity, then the Big Bang is just not possible because from that formula the Big Crunch must respond.

Prove that the definition science use about gravity being a force of mass pulling is correct. In other words prove that $F = G \dfrac{M_1 M_2}{r^2}$

Teaching ever since time began forms a pillar on which memory and remembering what you are taught is the most prevalent part of tutoring. The problem arises where the information studied is flawed and no one ever realised that. Fortunately this does not occur regularly, but if and when it does, notwithstanding the exceptional par it forms, ten becomes a major problem to deal with. When the student is taught, the student is expected to accept without argument. This mostly takes about all the time one lifetime presents. It has happened to physics and no one in physics yet realised it. Then the mistakes will carry on forming the past, carried over as flaws into the future for as many generations as it takes to realise the mistake and could continue indefinitely, if there is no clear minds working the recognise and correct what needs to be corrected. Judge what I present to you, for then you will realise with all my shortcomings, I present you with a truth that exposes short fallings in the basics of physics.

Show your academic worth and your educated dignity and accept the challenge I make to you and to all of your kind: I challenge one and all: **PROVE ME INCORRECT IN ANYTHING I SAY!**

First you should decide what belongs to the gravity factor and what forms part of mass. Newtonian science is of the opinion that when a body is floating up in outer space the body has micro gravity…that just can't be the case. Newtonian scientists confuse the factors being responsible for mass and for gravity because if not, then please explain which is gravity, the part that tries to move the body to the centre of the Earth, or the preventing thereof? We have to see that mass is created by the pushing of an object onto the Earth and from the pushing (not pulling) comes mass while gravity is what is doing the pushing. While resisting further movement mass comes into the picture and while moving towards the Earth or intending to move towards the centre of the Earth, that movement constitutes as gravity while stopping the movement leaves the object with having mass. Mainstream science loves to confuse the two issues because Mainstream science love to confuse everyone because Mainstream science is completely confused about the science they say they are the Masters of.

Is gravity that factor, which makes all bodies fall to the centre of the Earth, or is gravity that which prevents the further moving of bodies having gravity to fall further down to the centre of the Earth and then by restricting the movement, then forms weight or mass? By restricting movement towards the Earth a mass factor comes about which gives weight! It is presumed that the body has micro gravity because the body is weightless I outer space. This prompts me to ask the question underlying what has never been decided… what is gravity and what is mass. A body floating in outer space has maximum movement because when it moves slower, it starts to fall to the Earth.

At that point the mass (measured as weight / kg) is indefinably small while the movement that is applying is maximum in maintaining orbit. However, that is speed measured by distance (meters) travelled in time (seconds). Mass has a value, which is measured in the same currency in which weight is, and then mass is weighed as much as weight is and therefore, undeniably and in contrast to the logic of mainstream science's confusion and frenzy trying to confuse what can't confuse any further, mass and weight is connected as the same thing while gravity is movement notwithstanding mainstream science trying to put mass and weight far apart. If mass was equal to movement as gravity is, then mass must be measured in meters / second. Instead mass has the value which is the same as weight which is measured in grams.

When an object is having mass such an object is being secured in a position on the Earth at a point of having mass. At a point of standing still in relation to the Earth while excepting only the movement the Earth allows forms mass and is where at that point is where the rotational movement is equal to the of the Earth in rotation. Rotating at the speed the Earth dictates form the factor science call mass.

Having micro mass (not micro gravity) is where the body in rotational movement extends beyond the limit at the point where the Earth surface would award a mass factor. The movement speed goes beyond the speed required by the Earth at the Earth limit where rotation velocity secures mass as a factor. By exceeding the rotational velocity at a higher rate, such movement would exceed the movement or gravity of the Earth that is required in order to grant a mass value.

By orbiting at a specific distance, the distance from the Earth is determined by the rotational speed the object encounters. When the object reduces the orbital rotation (circular velocity), the gravity by slowing down will bring the object to start moving towards the Earth, which is falling and which is what everyone knows is to be gravity. One then must accept that mass is having an object being in a point of only moving with the Earth while gravity is the movement what is required to further move towards the centre of the Earth or the inclination of forming movement towards the centre of the Earth. That is the difference science never finally concluded…gravity is movement or inclining to move while gravity is that which comes about from standing still in reference to the Earth while having the Earth move.

Having gravity is forming movement that is inclined to move towards the centre of the Earth. That is called centrifugal force and therefore one could call gravity a centrifugal force. It is movement that is drawing space by concentrating the space, which is creating a flow around it towards the centre.

The space that is drawn and is forever becoming more concentrated and denser as it contracts towards the Earth we named as being the atmosphere.

As the Earth spins the spinning of the Earth is engulfed with space and the space surrounding of the Earth the Earth draws from the outside towards the centre. There are so many layers and each holds a name. The fact of why the layers form is left to gravity, which is left to mass, which is left to magic. The contraction of the space immediately around the Earth becomes dense and hot while the further t towards space the less particles the space holds and therefore the colder it gets. I put this incorrect view to the test in other work and show that what we think of, as being hot is in fact cold and what we think of as cold is extremely hot. However, that argument I leave for another opportunity because at this point we look at gravity in its most basic form.

There is one BIG centre pump pumping space-time towards the centre. There are nine smaller pumps; pumping space-time towards each one's individual centre and this is aligning according to the Titius Bode law of positioning the allocated position according to the specific requirements that the Titius Bode law prescribes.

The main factor required has to be Π since everything so far is reliant on the mathematical factor of Π. I will explain the Π connection in all four of the Phenomena call the Titius Bode law, the Lagrangian positions, the Roche limit and the Coanda effect and I will not use any force because forces belongs to witchcraft and in physics that is very absent.

Being a circle requires two factors and both those factors Newton dismissed in his search for gravity. More important is the fact that modern science are so well equipped wit the skills of mathematics and yet for hundred years after the fact not one in science came to a conclusion about Π having to be involved as well as having a diameter when dealing with a circle…any circle and the Earth is just another multi dimensional circle.

If r is the diameter, then the position science so feverishly award to mass has a point that actually holds Π as reference in the laws of mathematics.

Nothing stands still to anything else in the cosmos. When having "mass" the body is standing still in relation to the Earth. The Earth is taking on all the moving responsibilities and the body "with mass" is taking on the moving speed and density of the moving Earth. Everyone accepts gravity is taking a body "straight" to the centre but it is not. The Earth moves in two ways.

Modern science still supports the Neanderthal idea that a body fall straight towards the Earth as "mass" draws the body directly to the centre. This is as outdated as any view science may have on gravity having a pulling by force where such pulling aided by magical powers and forces.

Although being in a state of having a "mass-attack" "pulling" your body to the centre of the Earth in the state of having no motion your body is falling by 7° as it circles with the Earth around the axis of the Earth.

Not only is the Earth falling by 7° as it revolves around its axis, but also it is circling the centre of the Sun and by doing that the Earth is falling another 7° by rotation. This is pivotal in understanding gravity as a mathematical fact.

That puts the falling of the object completely in relation to the speed that the object holds and that places gravity by falling in direct relation to gravity by orbiting.

When a body falls there is no mass involved because all objects fall equal and this was accepted long before Newton started fantasizing about his mass involvement in gravity applying. The distance the object orbits measured from the centre of the Earth and the orbit circle holds a direct link to the speed or time in relation to space that the object rotates. If the speed in revolving declines, then the orbit circle declines and this reduces the distance the orbit circle is from the Earth centre or the diameter. The orbit circle is directly associated with the distance the orbit takes place measured from the centre of the Earth in a ratio of time taken versus space travelled through. This has to do with speed or movement and applies to all objects equally holding no specific relation to size or mass. It is a relation between the orbit circle (circumference) and the distance from the Earth centre (circle radius) and if that is the case, then gravity forms by Π having some sort of involvement and that throws any idea of mass playing a part in forming gravity out of the window where I hope it takes all of Newton's ideas of mass-forming-gravity with when going out the window. In forming gravity the centre line (diameter) holds a specific value to the orbit (circle) and with that being the case then we have to search for the part Π plays in the function gravity has and when doing that we can leave mass out of the frame because big or small, all things fall equal. Galileo was the one that proved that.

So you think that it is much simpler to maintain the argument that gravity is the force pushing the object onto the Earth only when the object moves at the same pace as the Earth rotates…and only then does the object finds mass or weight!

This is how I prove mathematically how gravity works. There is no pulling of mass or by mass or even that having mass plays a part in forming gravity. On the contrary, it is the forming of gravity that establishes mass when the space can no longer reduce and the reduced space locks whatever then has mass onto the solid surface of the Earth.

When the object moves while being in space or in contact (in relevance) with the spinning Earth, the object wishes to continue moving straight ahead while the Earth also moves straight ahead by turning 7°. Therefore, the Earth by spinning is falling away. That clears space or compresses space by the margin of 7° declining (compressing) of air / space.

The Earth is moving, constantly spinning and in this is contracting space by compression (we call this contracting of space in air the atmosphere) and while the air is getting more compact, it takes whatever is filling with space towards the Earth constantly at a rate of 7°. By the Earth rotating, it is compressing space and with space compressing it is moving objects in the direction of the Earth. That is why objects that is falling, has no mass and only the stupidity of the simple Newtonian mind will force scholars to accept that it is mass that is pulling gravity. There is nothing in the Universe that ever could remain still because everything cosmic that is filling the Universe is spinning while it is also at the same time moving in a straight line. The Earth is only moving straight ahead because the Sun is spinning and while the Sun is spinning, it is compressing space, which allows the Earth and all other rotating objects to spin around the Sun in a perfect synchronised fashion. This process is going on throughout the entire Universe.

Since the movement involves two equal phases that acts as one the double value of 7 in relation to forming ten becomes what forms Π. We have the movement of seven forming one direction standing in relation to singularity which is the square of 1…According to Pythagoras that will bring about fifty. Since singularity is equal a one and seven is combined with singularity, the equality of singularity brings about the seven uses the same attached 1 in the square making the fifty a combination of another fifty and from putting a double fifty in the square as Pythagoras demands we have ten as a result in relation to seven.

That is what gravity is. Gravity is space moving or changing position in time and when an object can retreat no further towards the Earth centre, it only then forms a solid that aligns with the spinning solid material and with that then receives mass… Gravity is the movement of space in regard to any one specific point…and that is also precisely what time is. Nothing is standing still in the entire Universe. There is not one fragment of a sub-atomic particle standing still in relation to any other particle through out the entire Universe that is standing still. Having mass is when one object is standing still in relation to the Earth forming a part of the Earth while the Earth does all the moving on behalf of the particle having mass as well as the Earth and only happens when through having mass the object becomes part of the rotating Earth.

This explaining flush down the toilet Newton's idea that gravity is being formed by mass that through some form of magical intervention is pulling on other mass and this is forming gravitational contraction, which is madness. This idea is going down the toilet and seeing it flow down the drain into the sewerage where it belongs. If one takes the formula **a³ = kT²** Kepler introduced, which Kepler received from no less than the cosmos at large, one find the **space a³** is equal to the movement of the defined space in a **straight line k** as well as a **circle T²**. In the cosmos no line can go straight without circling as well and no circle can go on without going straight at the same time.

In gravity there is a circle turning. Where there is a circle turning we have Π involved. Also we have a radius involved. The radius or distance from the centre Kepler called **k** and Newton classified this as zero ($\frac{dJ}{dt} = 0$), which is an insult to mathematical principles. Where a circle formed as a cube spins, three factors are involved, namely the radius, the spinning speed and most of all there is Π. Without Π the can be no circle and the Earth is a circle, even Newton should have been aware of this. Therefore there can be no gravity without having Π as a contributing factor.

Since gravity is tightly interconnected with a circle formed as a sphere and is spinning around an axis the main issue of research has to start with finding the factor Π. There is no connection with a circle and mass but for sure the circle will find in form an end serving in a measured value as Π. It is common knowledge that in calculating a circle the formula used is Π2r² or Πd². Would it not be mathematical plausible to start looking for mathematics in gravity and leave Newton's magical mass out of the picture, there might just be some common sense to be found in this. In science and in mathematics we have to see where true mathematics fit and what role has every factor have to play. We can't dump all findings on mass we since mass has no part in mathematics.

If we wish to award mass a value as a factor we then must see the role we give mass. Giving mass a value is issuing a body a value, as it would have being part of the Earth. We take a cube of water and we award the water a measured size. One meter by one meter by one meter of water would give a thousand litres of water which would leave a thousand kilograms of weight and that would leave a mass value of one metric ton or a thousand kilograms. The awarding of mass is giving the object a relevancy of being part of the Earth. In Newton's time the Earth was the Universe because people were getting used to thinking that the Universe was not spinning around the Earth but the Earth was only a small part of the Solar system and that was a small part of the Milky Way which was a small part of whatever was a small part of another small part of another small part of something getting into a bigger picture.

There is also a movement that should go straight but is in fact going in a circle and by never going straight but always circling the Universe becomes eternal on the one end and infinite at the other end. The infinite point I am going to explain in due time.

Science always awarded the position an object holds standing on the centre of the Earth with a measurable value of mass. This is not incorrect in normal physics but as far as astrophysics goes there is no mass factor present anywhere. The value of such a position should be Π in the relevancy of the movement that the Earth holds as Π²

This is the process whereby gravity forms. Mass has no influence on gravity except for resulting from gravity compressing the Earth.

Take the mass of the Earth (M_1). Multiply the Earth mass by your personal mass that any scale should indicate (M_2). After multiplying the two mass factors, then proceed to the following step by dividing the multiplied mass factors with the square of the radius there is between your feet and the Earth (r^2), which should not amount to more than a few billionth of a millimetre. If the answer in front of you is not 9.81 Nm/s² then there is something very wrong. The incorrectness has to be either one of two possibilities presented: The measured value of gravity is not 9.81 Nm/s² as science uses it, or Sir Isaac Newton's formula suggested as $F = G\frac{M_1 M_2}{r^2}$ is complete madness...Now which is it...you can decide...the force of gravity that the world of physics uses to do measurements is 9.81 Nm/s². If the answer you have in calculating your force of gravity is not 9.81 Nm/s², then it is either this measuring value of gravity that is wrong or it is Newton's formula that is wrong because by the calculation you did, the calculated answer you got could not possibly have delivered a measured value of 9.81 Nm/s². After all, science maintains it is the pulling of the combined mass in relation to the boosting that the radius would present to the force created that delivers the force of gravity! If by using the factors of mass and the radius does not accumulate to 9.81 Nm/s², then how can mass deliver gravity? Multiplying the mass of the Earth with the mass of a person and then bringing this answer in relation to the radius by dividing must be 9.81 Nm/s². If not, something is wrong with either the prescribed value science puts in place or Newton's suggestions.

To teach students that $F = G \frac{M_1 M_2}{r^2}$ are the measuring formula in determining gravity, while knowing very well it is not totalling gravity at 9.81 Nm/s², then doing that to students while enforcing a thinking pattern in the minds of a student is committing brainwashing because by forcing examinations on students, expecting them to confirm the falsified statements used that the tutors present as correct, is brainwashing, a way of enforcing mind control and it is manipulating the thinking process of students. If you can't prove that my manner of thinking is incorrect and you keep surmising that science is correct then recalculate the formula or start reading the rest of their shortsightedness.

Gravity is a constant of 9.81 Nm/s². This is used in all cases of scientific calculations. Mass is an individual factor that is different on anything on which it is applied as a measuring factor. How could something as different as mass that is never constant even on Earth form a constant such as the force of gravity and still be the same in all cases?

How does **M₁** connect to **m₂** forming $\frac{G}{r^2}$ when the connecting medium constitutes of nothing that then has the value of zero as the Cosmologists wish to value outer space?

In these following books forming a series of four and parts of this series about gravity there are the books entitled:

This Work disputes the correctness of the formula $F = G \frac{M_1 M_2}{r^2}$

Using the formula above as Newton did does not imply a suggestion or carry an idea across as a thought but must be seen to be acting as confirmation about a fact because one cannot suggest anything mathematically, one can only confirm a fact mathematically. There is no mere suggesting of any possible movement in a specific direction of any suspected behaviour by an object moving from and to a point as suggested, but this formula says the gravity of the Earth measured in mass at it's totality is colliding with the falling body's measured mass as the two factor's diminish the radius from both ends. This mathematical formula as it stands is no mere suggestion, but in its use it must back up or prove a fact!

This same space-time value ratio is also applicable where matter is moving through space-time as Einstein stated. This is also explained in detail and the difference that comes about when these two values are in place.

The value of the space-time revaluation is only applied to the occupied space-time belonging to the matter and not to space-time itself, but this leads to the concentration of unoccupied space-time, which in fact determines time in space.

You might think: **Very well, but what has all this has to do with the cosmos?**
<u>Here is but a very few changes that comes about.</u>

What is the Universe?

The Universe consists of two factors, which is space and time. These two factors are inseparable, undividable and one single unit. The gain to one factor is the loss to the other factor. There is occupied space-time and unoccupied space-time, but there is NEVER NOTHING. What is presumed to be empty space is unoccupied space-time! That is the value of the Universe, which I named geodesic space-time, and is the current value of time evenly distributed through space. The fact of "Nothing" can only exist in a person's understanding and perspective, but not in the Universe. Three factors rule the Universe:

1. Densified space-time (matter)
2. Occupied space-time (atom of elements)
3. Unoccupied space-time (The value of time in space)

There are only two energy forms in the Universe. The first is heat and the second is life. No other form of energy exists in the four dimensions that the Universe exists of. No force is to be found in the Universe, only balancing values.

There can be no such a thing as empty space. The Universe is time contained in space, which makes it space-time. Space has only one value, and this is to contain time and time provides space with a definite value.

Let us for one minute leave Newton's surmising about Kepler's failure out of the picture and concern us with what Kepler found long before Newton thought about what Kepler found.

Kepler said that the space a^3 is equal to the motion T^2 of the space a^3 distant from a specific centre k. That then is $a^3 = T^2 k$.

Reading this mathematically encrypted coded formula of the cosmos given to Kepler and keeping it removed from Newton it reads as the space a^3 is equal to = the motion T^2 of the space a^3 in ratio to a centre k.

What this proves is that gravity is the motion of space provided by time being the liquid.
Please allow me to explain. In the formula $a^3 = T^2 k$ the space forms as the space is in motion.
Newton suggested that $\frac{dJ}{dt} = 0$ where he stopped time to have the motion of the circle demolish the work that the circle does. That means he got time standing still or being T^1 and the motion $T = 0$. Let us ponder on that thought for a while, while remain with the formula Kepler suggested it will seem that according to Newton $a^3 = T^2 k$ and in that T^2 then becomes 1. Should that be the case then we have space going flat because $a^3 = T^2 k$ where $a^3 = T \times k =$ forming a square instead of a cube, and the Universe we have is a three dimensional cube in every aspect there is.

Newton's suggested G $(m + m_p)$ is just a longer and probably a more detailed manner of indicating k and better defining of k but it symbolises precisely to the point what k stands for nonetheless. I wish to draw your attention to the matter of Johannes Kepler's findings that Mainstream science considers as resolved and closed for many a century while it is not. My investigating Kepler helped me to resolve other unresolved matters but it was only possible by using Kepler's work.

I too am well aware that at first glance you will immediately arrive at the opinion that the theme of the letter has to be considerably below the standard of an intellectual Master such as you must be, due to the position you hold, and because of that, the normal research work you do. Nevertheless, I hope that this writing may spark interest even at such a low academic level and grade in scientific sophistication and development because I am about to prove that I discovered:

The first thing about gravity one must remember is that all things are spinning. There is no particle in the entire cosmos that is not spinning. The entirety of the cosmos is about everything spinning in relation to one specific point not spinning because that point not spinning at that point is excluded from space and is therefore not part of the Universe holding space.

In the centre of al things spinning a line forms coming from a dot that has no space. This is a dot that first forms and that initiates space that later forms. The line running from top to bottom of the spinning top comes from a dot that extends but just as the dot the line has no start and has no beginning. When reducing the radius it would eventually end at the line but crossing the line would land space in the very opposing quadrant. As soon as entering the line one has gone through the line. The line has no inside that can go even smaller yet we know that the line must have some ability to be able to go smaller.

When the top starts to spin and comes erect, the dot as well as the line that forms can't have a value of zero since the line is there for all to witness. If the line was absent only then could it be valued as zero. Realising the value of this line holds the entire principle of gravity in perspective. The line is there so it has a place but it has no space. The line is formed from the dot and since there is no space it has to be singularity. Since singularity could have one value and that is having 1 as a value this dot and line forms singularity. In the line there is no space because as soon as anything enters the line it moves right through to the other side and the line by itself is there without ever forming space to confirm its presence.

What evokes the line is the movement of a confined space that is filled with material spinning around a centre that holds no space. The line comes from a point that can never start but for material evoked by movement and in space it could never have a beginning except for the movement of material placing the spot without space in place to form a line which never could have space. The spot as well as the line could never go smaller.

The line forming a circle confirms Π as the product of space forming. The movement of space forming is then measured as Π moving from one point to another point and that movement takes Π into a square value as Π^2. However without the movement the line does not form and therefore without moving the line forms no gravity and therefore gravity is all about confirming the line coming from the dot in the centre.

Let's find $a^3 = T^2 k$ and see where it is hidden.

The sphere is a circle in many facets and therefore we will approach the sphere as one multi dimensional circle, however the sphere as such remains one circle to the power of many. When investigating a circle one would draw a line from one edge running through a centre all the way to the other edge. In doing that we would find the measure of the diameter, which is most important when trying to establish the worth of the sphere. The circle has Π to indicate form and then uses r^2 to establish the worth of such a circle.

Using the radius symbolised as r is drawing a straight line.

In any circle or sphere the size only depend on the fluctuation of r in the square as a component to the circle or sphere but that does not affect the form by indication of Π in any way there may be. The conclusion from this is that no line can start at zero because that will be a mathematical impossibility. This statement by itself excludes zero and with zero excluded one then begin to appreciate all the rest of the concepts governing corrected cosmology. A line or spot starting at zero would therefore be shorter than the shortest line possible. For obvious reasons can no line, or any line grow or extend from zero because such a line must then quit zero and become something, thus abandon its original value

In any circle or sphere the size only depend on the fluctuation of r in the square as a component to the circle or sphere but that does not affect the form by indication of Π in any way there may be. The conclusion from this is that no line can start at zero because that will be a mathematical impossibility. If a line started with zero, that would nullify Π and that would leave the form without having any form because $\Pi \times 0 = 0$. This statement by itself excludes zero and with zero excluded one then begin to appreciate all the rest of the concepts governing corrected cosmology. A line or spot starting at zero would therefore be shorter than the shortest line possible. For obvious reasons can no line, or any line grow or extend from zero because such a line must then quit zero and become something, thus abandon its original value, should such a line wish to progressively become more of what it was before.

When a line starts off with zero that would mean the start of the line has a different value to the end and a line holds conformity through out…no line can end by applying zero as the concluding worth. When any line is starting from point zero it can never leave zero because of the influence of being zero disqualifies any possibility of growth, or even being present at any point to grow. If the line then had to grow in all directions at the same pace the line must therefore be a circle or being three-dimensional, a sphere. Flowing from this fact is that in the Universe there can be no zero point or unfilled space. In the case of the growing sphere the value of the circle is Π, and that is where creation started. That gave me the clue where to start looking for singularity. One would find singularity in the value Π and the value Π will be in all things rotating in a circle. You might wonder how does that apply to the cosmos and moreover to gravity? Mainstream science promotes the idea that outer space as far as outer space goes, is filled and even overflowing with nothing. The nothing is overflowing because the nothing is expanding it is growing! You cannot fit nothing into outer space because it just will not fit; there is just too much space to harbour nothing. If any of the factors in Kepler's formulae represent nothing that is what you will get, it would be nothing.

The Universe in its total entirety will be nothing. **$a^3 = 0$ $T^2 =0$ k=0, which then mathematically could only translate to $a^3 \div T^2 k = 0$.** If the argument seems ridiculous it is not my mentioning such a fact that is ridiculous but the mere fact of the reasoning that also became an accepting of the valid ness of recognising that it is nothing that fills the cosmos that is the silly part. The basis of such an argument and the fact it could be accepted by science is what is making it as such ridiculous. It is the fact that one must argue about such a ridiculous matter about an idea that nothing is overflowing while it is filling the Universe, that allows the ridiculous part to enter the conversation because the trend reminds of arguing about fairies and little people existing or not and such argument is nonsense. If space is nothing then nothing has a distance having an indicating number to use. The distance would be indicating just that value being zero or the capitol O indicating zero while every planet has a precise distance it is located in terms of the Sun as well as in relation to each other. Try and indicate what is measured and calculated in value in outer space in measured space, while having that going in kilometres or astronomical units and then finding it is nothing in multitude filling that distance. The distance between the Sun and Pluto **is Pluto is 5900 X 10^6** kilometres of space, but in that statement we take it that the one as a factor used in determining whatever constitutes to form the measure of a kilometre. By adding one **5900 X 10^6** times is present in such a multiplication because adding 0 5900 X 10⁶ will still amount to 0. The one constitutes the presence of a fact being a statement of a value compiling to present the measure of space as it is in a distance. By saying the distance constitutes of nothing we have to substitute the one factor with a factor of zero. Then the calculation must read **Pluto is 5900 X 10⁶ X 0 = 0.** Including nothing as to state the presence of that part contained by the calculation delivers the total of zero. It seems as if science has ignored this issue by simply not thinking about the fact and therefore simply ignoring that what is measured forming the sole value of space has a practical worth, but put the value of nothing as part of the distance in calculation because that is what is measured and then see how one can multiply by using zero in mathematics and reach a distance holding a value other than zero when multiplying by zero.

I agree that what is filling outer space is invisible, but also it is there, it is present and being present and there while being invisible disqualifies whatever is there from being zero which will mean it is not there. Then what will be there, will be invisibly small, still be a line because every aspect of the Universe forms lines while also will have the potential to fill space and can still form a measurable unit. That then must be 1 because while 1 x 1 = 1, 1 + 1 = 2 and that qualifies that invisible thing to be present (1 + 1 = 2) but at the same time be completely invisible (1^3 = 1). I knew what has to be true about that which I was looking for had to be singularity because singularity can only have one value and that is 1.

To find the invisible I had to locate singularity. I realised that my effort to locate the point holding singularity enabled me to backtrack the exploding Universe to its origins. The Universe is a sphere because it is filled with spheres filling the void spaces (not the nothings) and in that I first had to investigate the visible.

Newton's mathematics says a sphere is $a^3 = 4/3\Pi r^3$ while Kepler said a sphere is $a^3 = T^2 k$, and both are equally correct because the cosmos gave numbers to support its statement.

With Kepler $a^3 = T^2 k$ and with mathematics the volumetric size of space must either be according to the measure of normal mathematics if it is a cube then three sides form a^3 = L x B X H and in the case of a sphere the measure will be $a^3 = 4/3\Pi r^3$. This was like comparing a triangle in relation to the half circle and the line. It predates mathematics where the numerical use of determining a value was not yet established and only form was in use. It is equal to a time when we find in the half circle standing 180° related to the triangle (180°) and both still are equal to the 180o of the straight line notwithstanding the obvious differences used in form. However the starting point of these forms has to be equal and also has to be not zero to have the end be equal and result in all being equal in value in the end.

I too am well aware that at first glance you will immediately arrive at the opinion that the theme of the book has to be considerably below the standard of an intellectual Master such as you must have, due to the position you hold, and because of that, the normal research work you do. I realise it is dealing with a subject school children learn but in that comes the issue that goes unnoticed. Nevertheless, I hope that this writing may spark interest even at such a low academic level and grade in scientific sophistication and development because I am about to prove that I discovered:

Lines mathematically cannot start at zero because there is no evidence of zero as a factor in mathematics. Should you disagree with my statement the question in need of answering is this: **What will the length of the shortest hypothetical line imaginable be and moreover, what would the total overall length be in that case?**

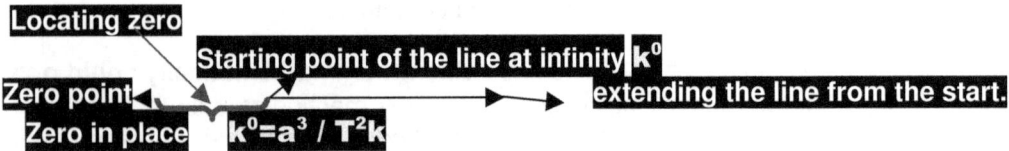

Lines mathematically cannot start at zero because there is no evidence of zero as a factor in mathematics. Should you disagree with my statement the question in need of answering is this: **What will the length of the shortest hypothetical line imaginable be and moreover, what would the total overall length be in that case?**

Kepler said a sphere is $a^3 = T^2 k$.

In honesty we have to realise that we cannot dismiss the whole formula that Kepler produced just because it doesn't match the scenario set to determine volumetric size as does the Newtonian version does. Kepler's version holds a foundation based on movement and it is in the movement we find the measure and not in the size as Newton's mathematical formula does. In Kepler's formula the entire formula is formulating a circle being motion. However with the correct interpretation we find so much more that just motion. It is $a^3 = k / T^2$: That is what Kepler brought into civilization for all time to come. He saw space a^3 being in isolation due to the time it uses to move T^2 claiming such space forming independence according to the lines k indicate. Let us look at the factors in more detail before we proceed with the rest of the book.

a^3 symbolises in a mathematical interpretation of implicating the three-dimensional space.

T^2 is representing the period or time that Kepler suggested we should use to calculate time that holds the orbiting planet in direct contact with the space in relation to a very specific centre moving from point T_1 to T_2 in relation to a precisely placed centre k^0.

k is the space taken from the centre to the end of the line k from which the planets must have grown if one accepts the Big Bang growth of particles and the affect of the Hubble constant on all cosmos material. The specific value about the centre is most important because from the specific centre gravity always apply the strongest influence.

The turning T^2 of any circle holding space a^3 is valid only if in reference k to a centre k^0.

Space a^3 will always be circling around as T^2 is in a position referring k to the centre k. That is what Kepler said when he said $a^3 = T^2 k$. Kepler indicated space a^3 will forever fight for independence and show separate individuality in remaining apart as identifiable cosmic components by means of motion. Every space will cling to independence indicated by k through fighting off the integrating of another coverall unifying unit by applying the motion of T^2! The problem we have to solve is what will the cosmos use to secure such independence between all particles? What sets space apart from the rest of space? First we have to admit that Kepler was the one that introduced the following.

Kepler gave us the answer to the following but no one ever took notice!
Kepler was the one that discovered **space / time** as **space** a^3 = **time** $T^2 k$
Kepler was the one that discovered **singularity** as $k^0 = a^3/T^2 k$
Kepler was the one that discovered **gravity** is holding **space-time** relative by the measure of distancing k as $k = a^3/T^2$ and $k^{-1} = T^2/a^3$

Kepler said gravity in space is about the area a^3 that would always keep equilibrium with the time T^2 it takes to travel the distance of the full circle position placed by the indicator k, therefore adjusting k as the need arrives. With k shifting in length a^3 will have to readjust and therefore T^2 will find a new relating value each time. This was the finding of Kepler and came after his intense study of orbiting planets.

Translating Kepler's mathematical expression $a^3 = T^2 k$ correctly to the verbal statement in English Kepler said that there is a **space** a^3 which is **equal** = to the motion in the **time duration** T^2 thereof between two specific points which is a straight line k that holds a relation from a centre k^0 to an end k where the two ends run from the beginning of k^0 to connect at the end of k. I might not be the smartest boy on the block but I'm not that stupid either. I know how to translate mathematics into English… and I translate as follows:

a^3 must have a volumetric interpretation because the third dimension is sure evidence of multiple conjunctions of dimensions put together in three sides opposing three sides having the third dimension in place. The fact that any symbol uses a value to the **third power** a^3 indicates **space** or a volumetric established and separate unit. Using a cube by three dimensions symbolises a cube, a room, a space to be filled, a unit able to hold other ingredients on the inside when empty or partly filled. It is space because it is volume using the third dimension.

T^2 is an indication of something having a cubic nature other than the square forming motion that is provided by the motion the square indicates, which is where the moving object is representing a third dimensional object that is moving from point to point and it is this point to point that multiplies into the square. The space is moving as a unit from one point to another point and the moving between the points are represented by a flat square or following a flat distance between two points. The cubic space was in one instant in one place and then the second instant in the other and because time can never stand still or become single dimensional (this I am about to prove) insisting that time must always support the motion it consist of or space as well as time in time cannot be. It is motion that is taking time, which is motion in the second dimension moving the space in the cube.

k^1 is the symbol used to indicate a straight line between two points with a definite beginning and a specific end position. It is the location where the form in question is holding space and where the space was and where the space are going to be in very next split instant that follows. This indicates points of representing **k** in different time positions to which the points will then be multiplying indicating form as result of the square. The movement indicating not a square surface but movement by the square indicates the time the journey took to move the space from one point where **k** is indicating the location of the space to where the next indication of **k**. T^2 will shift **k** where **k** indicates the space that formed as a result of the movement T^2 of being the space a^3 indicated the point end of **k**. However, since time represents the square T^2 and with **k** being the distance that prove that the **k** represents the distance the space a^3 representing the form it is obvious that T^2 represent the time that represents the space a^3 in the square T^2 through the motion. It is the distance moving space in the cube to complete time in duration in the square of motion; therefore **k** is permitted to be in the single dimension.

Reading this mathematically encrypted coded formula of the cosmos given to Kepler and keeping it removed from Newton it reads as the space a^3 is equal to = the motion T^2 of the space a^3 in ratio to a centre, which is relevant to the positioning of **k**. If we bring in the full equation it will be $k^0 = a^3 \div (T^2 k)$ which means half of space is solid and half of space is liquid where liquid is moving. However, it is also true that everything through movement defines a value in relation to one point holding singularity k^0 and that is what the formula $k^0 = a^3 \div (T^2 k)$ underwrites. What this proves is that gravity is the motion of space provided by time being the liquid. Please allow me to explain. In the formula $a^3 = T^2 k$ the space forms as the space is in motion.

Newton suggested that $\frac{dJ}{dt} = 0$ where he stopped time to have the motion of the circle demolish the work that the circle does. That means he got time standing still or being T^1 and the motion $T = 0$. Let us ponder on that thought for a while, while remain with the formula Kepler suggested then it will seem that according to Newton $a^3 = T^2 k$ and in that T^2 then becomes **0**. Should that be the case then we have space going flat because $a^3 = T^2 k$ where $a^3 = T^2 \times k =$ forming a square instead of a cube, and the Universe we have is a three dimensional cube in every aspect there is.

I am of a very different opinion about Newton's point of view where he declared that forming a circle moving $\frac{dJ}{dt} = 0$, and by doing such movement removes Kepler's relevancy factor. This places a value of empty space in which a top would spin and Newton missed the difference there is between a top spinning and a top laying on its side on the Earth. There can be no such a thing as empty space. The fact that space is valid removes an empty connection because space can be anything there is in space. The Universe is time contained in space, which makes it space-time. Space has only one value, and this is to contain time and time provides space with a definite value.

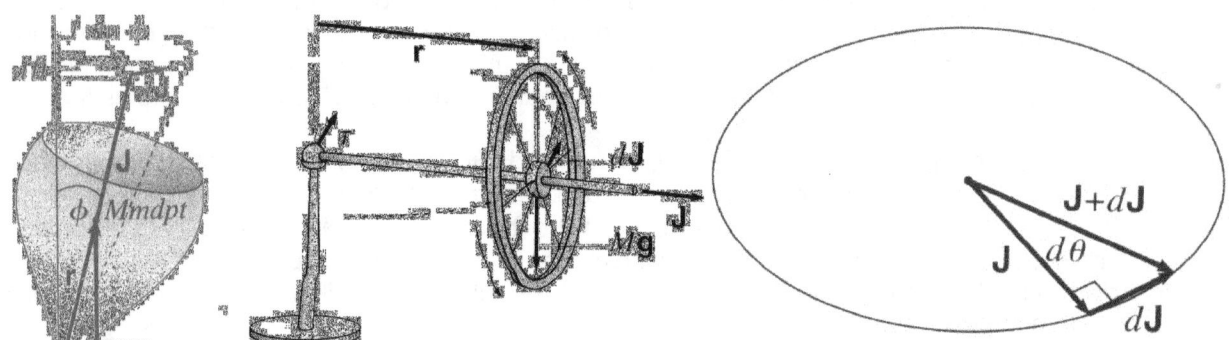

I do not disagree for one instant with Newton's calculations and therefore I am not going into repeating the entire calculating process. All of the calculations Newton made are very correct except the eventual and final conclusion Newton came to. Newton never understood the mathematical concept of time playing a part in physics. In the time of Newton singularity and the relevance thereof had no feasibility in any concept regarding physics. Newton had the concept that time could stand still and that is impossible in physics or any other place. Time can never stand still because time is forever moving by establishing space.

Being the mathematical genius as Newton is so often portrayed as; Newton had very little insight into mathematical possibilities, because when he suggested that $\frac{dJ}{dt} = 0$ he made one huge mathematical blunder. Newton or no other person may place any two objects in a direct relation and have an outcome of

Zero. Much surprising is that not one mathematical genius that came after Newton drew the correct conclusion that forming $\frac{dJ}{dt} = 0$ is mathematically not acceptable. Newton saw that dividing something into something else could bring about zero and that is impossible. In concluding that $\frac{dJ}{dt} = 0$ bringing in zero as a legitimate value Newton found a way to replace Kepler's symbolic relevancy value of **k** with using the symbols G (m + m_p).

In doing that Newton painted a picture that has no real meaning except where Newton tried and succeeded to put mass into an argument that has no true validity in cosmic principles. This is just a longer and probably a more detailed manner of indicating **k** and better defining of **k** but it symbolises precisely to the point what **k** stands for nonetheless. I wish to draw your attention to the matter of Johannes Kepler's findings that Mainstream science considers as resolved and closed for many a century while it is not. My investigating Kepler helped me to resolve other unresolved matters but it was only possible by using Kepler's work.

Newton did not think the situation through when he contemplated about gravity. Newton should have thought about factors keeping the gyroscope upright while the gyroscope is spinning. The gyroscope will fall on its side when not spinning and in that position the "Earth's mass" could play a part since the gyroscope fell on its side. However, as soon as the gyroscope started spinning, the balance shifted in favour of a position wherein the gyroscope stood upright. What then came about had the ability of keeping the gyroscope upright. This is rotational movement and I explained how rotational by the square of the double seven forms Π and Π is forming the curvature of space-time and in that bending of space-time we call the atmosphere that keeps the gyroscope square with the Earth and through that the gyroscope stays upright. That is evoking singularity that is establishing gravity in relation to the Earth evoking gravity through also spinning.

Newton found mathematically that the movement of the top by spin removed the value of the radius $\frac{dJ}{dt} = 0$ where quite the opposite applies. The spin of the top $T^2 = a^3 \div k$ positions the relevancy that **k** as a factor produces by initiating singularity k^0 on both sides of the relevancy $k^0 = a^3 \div T^2 k$ as well as placing singularity in relation to the spinning top $\frac{dJ}{dt} = 1^0$ because that is the correct mathematical principle coming from the equation.

The spin of the top does not eliminate the relevance of **k** but institutionalise the measure of **k**.

Trying to find a measured value for **k** is showing no understanding about what **k** is. The value of **k** is finding the space that **k** indicates in terms of what moves. The indicator **k** identifies the space a^3 that the top claims in terms of singularity k^0 that the movement T^2 isolates from the rest of singularity $\frac{dJ}{dt} = 1^0$. The value of **k** is dictated by T^2 as the movement isolate the space a^3. The measure of **k** is the relevance **k** claiming on behalf of the space **k**, which uses the relevance of **k** to limit the space spinning in accordance with T^2.

There was a Universe forming in a time that everything present at this moment was so small that only form was in place and this was when the triangle, the half circle as well as the straight line was equal 180°, with no numerical values in place yet. At that point the line must have been so small it had reached a point not yet mathematically dividable in any way. The dot that formed was so small during that time that if at present any further dividing that took place, such dividing would have brought growth because there then would form space between the sides going in the opposite direction. However it is important to realise that anywhere we might locate Π^0 we also locate 1^0 because Π^0 is 1^0. The dividing brings all there is having all sides moving literally on the precise same spot, and I have located singularity in just such a spot.

I came to the conclusion that the spot I found had to be singularity purely on the grounds that that spot holds only one side to serve as a start to the starting point of all directions possible. In that side is only one spot where there is only one side applicable and one dimension present. In that spot space ended. That point is serving as a position for all possible points and cannot allow further dividing as it is in the smallest line or spot there may ever be. In the very centre of any and all circles spinning we find this point holding no space and therefore forming Π^0 that is 1, which is singularity.

Again I indicate the precise location of such a point. What is in the Universe, is spinning and therefore what I am referring to, applies to everything holding a place in the Universe and therefore this which I mention directly links everything holding any space whatsoever in the entire Universe. In the **precise middle** of all **objects in rotation** is a precise centre dividing the object in sectors that will **start the spinning initiation** from that centre point. Thus, the spinning object **will have a middle point**, a very specific **centre point that does not spin** and only holds Π as a specific value because no radius can apply. But also the one value such a line **cannot have is zero** because the line **is there and holds contact** to the rest of the material bringing about that **zero does not start any** line and therefore the **value of the line must be infinite**, just as described in **accordance** and by **the definition of singularity.** As I am introducing a very new idea, I wish to explain in better detail what I try to convey. While the toy top is spinning one will find singularity by moving the rotating line or radius progressively to the middle by reducing the length the line has from the edge to the middle. At one point all further reducing must end but the ending cannot include zero or nothing because the rest of the line still attach the rest of the top. As the rotating direction moves inwards, the rings will become smaller and smaller.

That point albeit hypothetical, is also as much a reality none the less and is placed where that point **must be standing still** because every line **running from that point** in **opposing directions** is also **in opposing directional spin the other or opposing side.** In considering the spinning motion in the fraction of time in the detailed instant every aspect of rotation will turn in every instant of change in time. Although the points had the same characteristics only one instant before, they oppose the characteristics it had just before and just after the very instant in which they are and to which they relate by similar points also in rotation. The fact of the graph proves my point in quarterly opposing dimensions and values.

All stars hold a centre point in singularity where that centre point has the value of being the equivalent of all the atoms where each atom holds a centre has an atom's worth that combines to form an equal to the value equal that the star's singularity is worth.

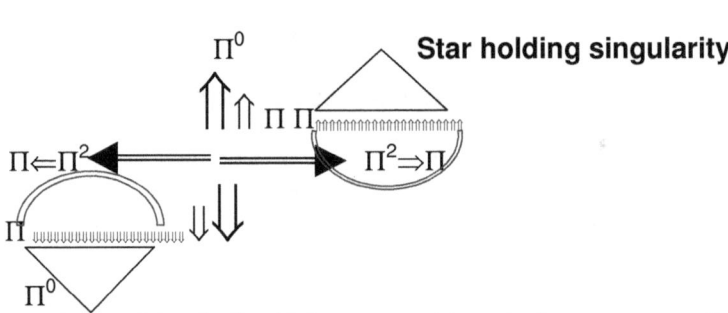

The point is in everything that rotates and everything in the Universe combines to form a connection that connects everything to all other things.

Applying Newton's second law F=ma one arrive at the formula GMm / r^2 = m ($\omega^2 r$) claiming a zero

Not Applying Newton's second law F=ma but applying Kepler's formula on motion as $a^3 = T^2 k$ or $k^0 = a^3 \div T^2$ $k = 1^0$

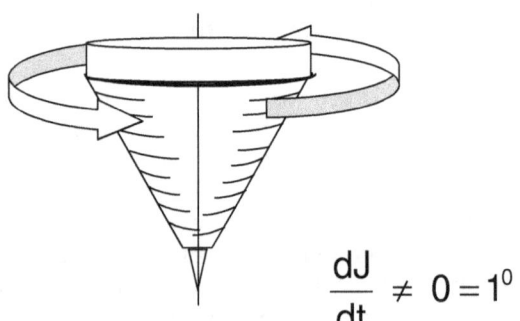

$$\frac{dJ}{dt} \neq 0 = 1^0$$

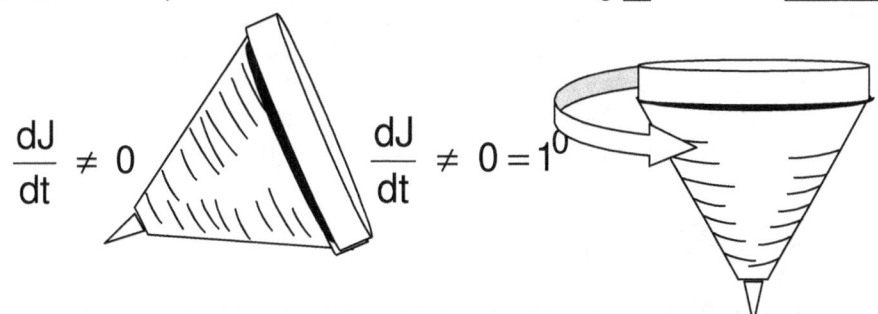

Let us have a look at the bicycle. It is said that the bicycle works on a balance and by science mentioning that the rider of the bicycle is applying a balance, in that the entire problem is solved. That is so typical of Newtonian simplicity about a very complex issue. It is the same as putting gravity down to mass pulling by some small particle called the graviton without ever showing any ability to look more intensely to find a solution for a very complex problem. Creating a graviton or creating dark matter to look for solves all the unsolved issues.

Saying the riding of a bicycle is due to balance is the same as putting everything in the Universe down to mass taking charge of particles. As the wheels spin (T^2) the relevance of **k** leaves the bicycle firmly attached to the ground and in doing that it confirms the space in location (a^3) in terms of singularity k^0. Newtonians would call this having mass or whatever. Then having the bicycle moving forward in terms of individual cycling gives a relevance of (**k**) to the peddling power and the movement (T^2) then is about having momentum in relation to the Earth spinning.

Getting back to my first argument about a line and that no line can start at zero but has to use singularity as a starting point, this is all the proof I require. The line **k** coming from the centre (singularity k^0) forms by forming an initial dot Π^0. However, I went on to say that whatever the line used to start with has to continue in order to repeat the same that began the line. Therefore the line started with Π^0 and it has to continue with Π^0 until such a point, as it must end with Π. Whether the line is Π^0 or is r^0, or uses 1^0 the outcome all refers to singularity being used.

What Newton suggested while never realising he did suggest the following, is that the movement puts singularity $\frac{dJ}{dt} = 1^0$ in position on the outside of the moving circle. However, by using $\frac{dJ}{dt} = 1^0$ Newton placed emphasis on the turning movement of the circle. That Kepler also found without ever realising what he found. Kepler said $a^3 = T^2 k$ which is $k^0 = a^3 \div T^2 k$ which is $T^2 = a^3 \div k$ which is the circular T^2 movement validates the space a^3 in relation **k** to a centre k^0 which is exactly and precisely what Newton said when Newton said $\frac{dJ}{dt} = 0$ that actually should read $\frac{dJ}{dt} = 1^0$. The location where Newton placed singularity as being singularity established by the movement of space $\frac{dJ}{dt} = 1^0$ I named **eternity** because there nothing can ever go bigger or become more. Whatever was and is and will ever be is locked in that space I named **eternity**.

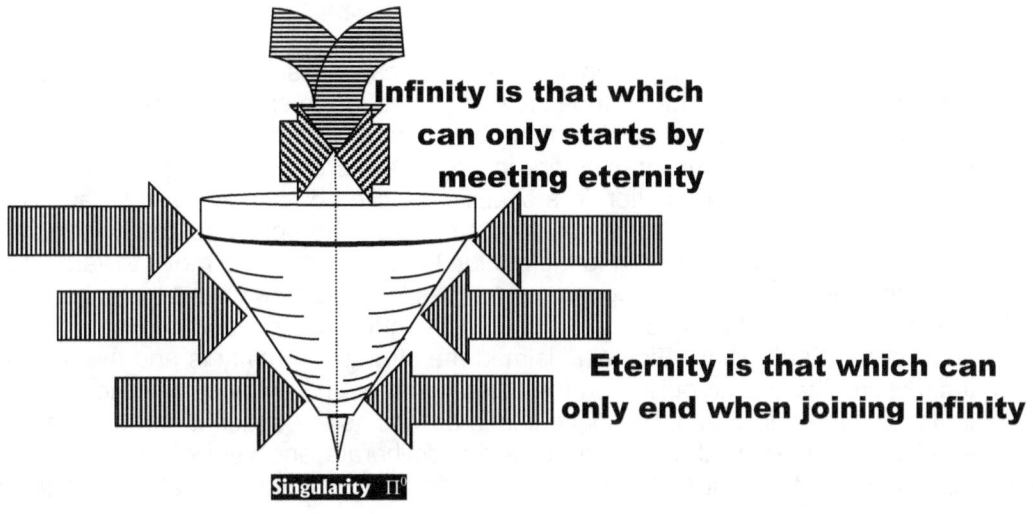

Getting back to my first argument about a line and that no line can start at zero but has to use singularity as a starting point, this is all the proof I require. The line **k** coming from the centre (singularity k^0) forms by forming an initial dot Π^0. However, I went on to say that whatever the line used to start with has to continue in order to repeat the same that began the line. Therefore the line started with Π^0 and it has to continue with Π^0 until such a point, as it must end with Π. Whether the line is Π^0 or is r^0, or uses 1^0 the outcome all refers to singularity being used.

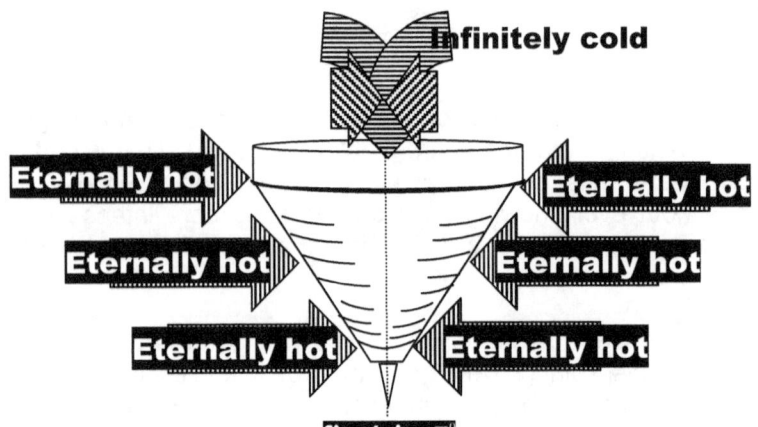

By reducing the line we come to the end of the mathematical equation of the circle but the circle does not end there. That is what Newton did not recognise from the figures the cosmos represented to Kepler. The circle only secures the final cosmic figure and the value to singularity where all things have equal value. The movement of the circle splits singularity in two sectors. By forming Π the circle has to form Π^2 due to the movement coming about in securing the space Π^3.

Kepler chose to use different symbols too those being valid, but the concept remains the same. Kepler said that $a^3 = T^2 k$ while I show that $\Pi^3 = \Pi^2 \Pi$. It still confirms that movement Π^2 = is the forming Π^3 in relation with Π singularity Π^0.

At that point the half circle and the triangle and the line must start since all three having many different forms have equal value at 180^0. Only after that point does mathematics begin where all factors in 1 have the value of 1 being 1^0.

In that conclusion one realises something must separate singularity from all other factors because singularity hosts all other factors but is by own initiative Π. That will be the spot of origin. That will hold the eternal spot...the smallest spot ever because all spots that ever can be was secured in a position in the centre of that spot. Because of the progress singularity follows from the single dimension singularity only allow mathematics a start at Π^0 progressing further too $\Pi\Pi^0$ and from there the line is born as $\Pi\Pi^0\Pi^0$ or $\Pi^2\Pi^0$ $\Pi 3\Pi^0$ $\Pi 4\Pi^0$ $\Pi 5\Pi^0$ where Π^0 then may form the concept and value of r. But the line starts at $\Pi^0 = r^0$. Because cosmology is singularity based and the value is $\Pi\Pi^0$. This escaped the attention of the greatest mathematician about the work of the greatest cosmologist ever because Newton incorrectly introduced $4\Pi^2$. The introduction of $4\Pi^2$ exaggerated the value of time and removed space / time from the concept. Mathematics in cosmology does not apply pi, pi is the root value of all concepts in cosmology. The factor pi impersonates as much as it represents singularity. This is my argument with which I support my claim that I made

The fact of form proves that the sphere captured all sides that can possibly influence the sphere. The sphere therefore holds $k^0 = a^3 / T^2 k$ within the boundaries designated to the sphere. When a body is placed in a location on the outside of such spherical borders that object seems to float in any direction. There is no control one can establish which will secure movement in any specific direction of preference except by releasing heat to counter act the required motion in a specific direction of choice. We all have seen what happens to any object that comes into the border area of a sphere. The object suddenly is motivated by motion to follow a specific designated direction and the motion leads the object to move towards the centre of the sphere. It is as if the support of the six opposing sides has lost one side where the sphere took over the control and movement starts in the direction of the Earth centre. The support of one side is literally removed by the centre of the Earth where Einstein claimed the strongest gravity is and the motion of the object starts in that direction. There is no pulling on the object but there is removing of space by the centre of that specific point leading the object and the space it is in as well as the space it carries to move to the centre spot. In the sphere the borders the sphere holds are deliberate and very distinctly placed edges forming a specific distance from the centre. The centre is also proven beyond any debating. The centre of any sphere has to be at the very point where space completely falls away. That will put that space at that point in the single dimension and centre is the single dimension.

The claim becomes obvious when observing the connection between the half circle, the straight line and the triangle, which could also promote all the qualities lurking behind the pyramid. Consider the connection between $180°$ sharing three different forms all part of mathematics where each is different in form, but equal in value and then one may realise in considering the very basic in mathematics being the Law of Pythagoras on which all mathematics are focused. The triangle stands in for one factor represented by one at a value of $180°$. So does the straight line become a factor of one and the half circle also becomes one where the factor of one equals all $180°$. All three are most seriously part of shapes in the cosmos. Revalue any one form to zero and the rest too must follow and share the same value.

The line dips into infinity every time it passes infinity when it cuts through infinity. The line going into infinity comes natural as the line progress because all lines are infinite dots linking from one point to another point. That brings about that coming from infinity might change in angle bit that directs the route and not the form. The form is all the same but the relevancy changes between infinity (going flat) and eternity (staying 3 dimensional) and in this one finds Einstein's flat Universe not going flat. It is not the 3 dimensional going flat but the governing singularity 0.991 relating to 1 and then again the governing singularity relating to Π.

Singularity is not only part of the spinning top or the spinning atom or wherever things seem to be obvious.

When you walk outside and look at the vastness of the blue sky or at night at the black night sky, you are physically part of singularity in the part of $1°$, the part that moved away from 1^1. You are within the part $1°$ that has no end because it has only one side, which is the inside. It is $1°$ going nowhere. It is the part that I named the spot that had the dot 1^1 moved away from. For having no better initiative I called the one point that formed the spot and the other point that formed the dot and I explain that the spot always grows into the dot where the one being 0.991 increases to 1. This increase in singularity is what forms the Hubble constant where Newtonians visualize that the Universe that can never grow then according to them grows.

This is the dot that has no start. It is 1^1, the part that released from the spot $1°$ when motion parted singularity. It came apart when motion unleashed the dot 1^1 that has no start from the spot $1°$ that has no end. It is the Universe born from motion that was driven by heat. It is still there because once anything is part of the Universe and forms a principle within the Universe it has nowhere to go but to remain within the Universe. Walk outside at any time and you are a witness to the result.

In the centre a spot forms a dot. Then spot holds a value of 0.991 whereas the dot is one. These values I explain somewhere else but it all relates to the values Π forms. By forming a spot $\Pi°$ the next action is that singularity forms a dot that extends to forming the edge Π. This is a nock-on effect and the one point serving singularity cannot go without the next point holding singularity. That is the reason why something in the air falling to the ground has no mass while something on the surface of the Earth receives a value such as mass. Then object touching the ground forms a relevancy relating to Π that connects to the centre of the Earth that holds singularity. If Π forms that immediately charges three equal values as $\Pi^1 = \Pi^2 \div \Pi^3$, $\Pi^2 = \Pi^3 \div \Pi$ and $\Pi^3 = \Pi^2 \Pi$. This is the result of singularity deploying space-time

In the spinning top we find that singularity $\Pi°$ can be generated by motion. But singularity $\Pi°$ has no motion within the dimensions we find allocated to the Universe in which we live. Since the singularity found in the centre of the spinning top is in truth just a mathematical point, which means in mathematical terms the point with no sides cannot even be calculated as a factor since the measure thereof goes beyond what mathematics ever can calculate. Mathematics has a use within the 3D Universe but singularity that keeps the spinning top attached to singularity governing the gravity of the Earth, that singularity is truly single dimensional and beyond mathematical measure. It is singularity $\Pi°$.

As a result of examining this proposition, I located two principle positions both holding singularity. The cosmos is made up of one type (1^0) that is in two categories where one type moves and the other type does not move. The one is a liquid and the other is a solid. There is life forming a principle time position being other than comic time but in addition to cosmic time that because it is time has to have a position of 1^0 and then every cell forming the body forms that part holding a position in the cosmos because the part is structurally part of the cosmos and that connects to Π.

The condition for the presence of this singularity that forms everything, controls everything and is everything is the centralised to a centre singularity holding relevancy $k^0 = a^3 / (T^2 k)$ that forms by movement $T^2 = a^3 / k$ of space $a^3 = k T^2$ placed in relevancy $k = a^3 / T^2$ that is centrifugally going both ways $k^{-1} = T^2 / a^3$ thereof (**Newton's 3rd law**).

In the line that forms when the governing singularity charges a controlling value on space that forms we have the governing singularity 1° that relates to the controlling singularity 1^1 but then 1^1 becomes 1° where it then relates to 1^1, which then becomes 1° that in turn relates to 1° that becomes 1° that relates to 1^1. This is eternity moving across infinity and in that creates the relation of singularity 1° 1^1. This is the spot stretching to become more than the dot and becomes a line. This is 1° that relates to 1^1 that becomes 1° that relates to 1^1 is time moving forward and time forming space is what keeps the top erect and spinning.

On the inside of space forming the outside of time there forms a line that can never end because the factor time forms the line and time has no outside. Everything that ever will be, will be because it is inside this line formed between that which can hold everything formed in space and the line that can hold no space and that there is no inside too. This space ends where this that has no end meets that what there can be no start too. This forms a Universe that has everything on the inside just because there is no outside to what this is. Why would I not use names…it is because it is Biblically named? This is eternity1^1. This is infinity1^0.

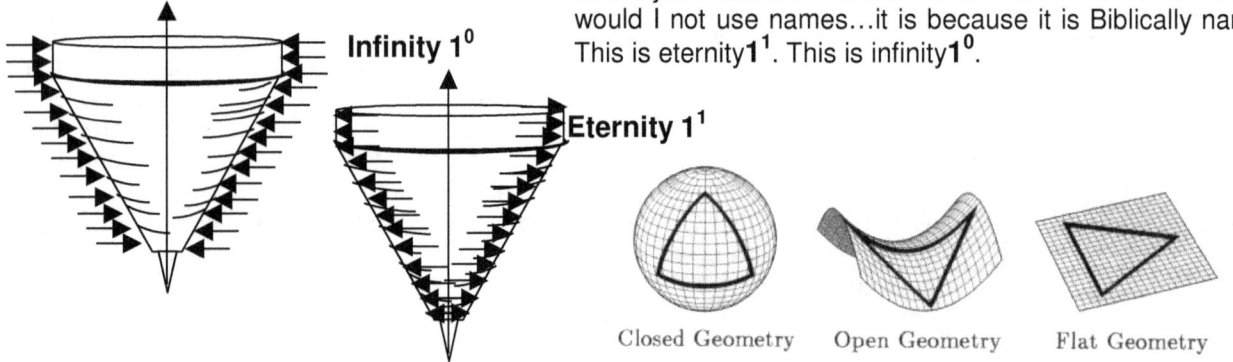

Closed Geometry   Open Geometry   Flat Geometry

This 1^0 can never start or begin because this is the beginning of everything 1^1 there could ever be. Then we have the Newtonian thinking about the Universe forming a possible saddle, a possible flat or a possible sphere. The way Newtonians look at the matter is by approaching the Universe as if they are not part of the Universe but are positioned as godly entities looking at the concept of the Universe from a dizzy height. The picture of the three possible Universe forms portrait just that demeanour thy have. The truth is that we are in the cosmos and the cosmos is expanding away from us as much as it is shrinking away from us. We can't look at the growth as if we are detached positioned at a far-off distance. We are part of whatever is there that is forming the Universe. We have the ends of the Universe unite by material forming the Coanda effect, which is the interaction between infinity and eternity. We have no place to look at the cosmos develop in a saddle or as a flat shape because there is no dimensions other than the dimensions ruling us being part of the cosmos.

One cannot observe the Universe as if one is studying an oil painting. One cannot even look at the picture as if one forms part of one of the oil painting molecules because even that would bring an incorrect realisation of what applies. One has a peculiar position in the Universe where everything is moving away from you outwards as well as moving away inwards because everything is moving in regard to the position the observer has. The Universe has no outside but only has an inside going outwards and an inside going inwards from the vantage point of the human onlooker. If science does not recognise man's position in relation to the entirety the Universe holds then every study done in this regard would be meaningless. The Universe from our perspective is as much expanding as it is contracting.

Looking at the picture science gives the Universe an edge, be it curling like a saddle or flat or in the forms of a sphere, whichever form can't apply in any way. It does not matter how we wish to use a form because from where we are the Universe is formless. The big mistake is their giving the Universe an edge where eternity ends. Where there is an edge there has to be something on the other side of such an edge and therefore there is nothing on the other side but a continuing of what is in the Universe going on endlessly.

There is no limit to the Universe except for the contact made by that, which can never reduce connecting to that which can never stop expanding, and that limit is at the border of every atom forming the Universe. It is well documented that science is under the impression that they have the ability to look at the Universe at a point they could observe the edge where the Universe ends. Science must realise that the limit they find when searching the Universe is their equipment being fallible. The cosmic researchers are using equipment that is limited and the equipment fails to go further than the distance they find a limit of.

The Universe is limitlessly going outward in every conceivable direction and the time limitation of the light they can explore comes from the limited ability of the equipment with which they observe and the lack of possibilities the equipment offers. The Universe just can't go into a saddle formation, it can't go into a flat formation and it is a sphere at the point where singularity forms space by time forming space, but this form is at the beginning end and this continues then endlessly and eternally as long as space applies. From where we are that sphere is forming on the inside and on the outside space has that form eternally. On the outside the Universe is a formless sphere going on eternally. We have to acknowledge that the Universe is an inside going on and on to the inside as well as the outside.

We are like a water molecule in the sea with immeasurable many water drops going in all direction as far as it is limitless. The approach science has is that they wish to look at the Universe from their calculating perspective and with a mathematical formula they then find a position of being unattached from reality while from their lenses they look at something that does not exist and try to be opinionated in such a regard as if they see a painting forming the Universe. There is a huge limitation placed on man's ability that science does not care to take responsibility for forming that limitation of man, not in mans smallness and not in man's ill equipment. Man is handicapped by his own smallness and limited abilities but since Newtonians put their position in regard to be godly entities they would much rather transfer such inabilities to the Universe and therefore handicap the Universe by giving the Universe the edge and the limit and the age.

They can put a limit of 12 billion light years in the distance that the light has travelled, but instead of declaring that such is all ability their equipment is able to observe being limited in vision by imperfection on the part of the equipment, they would much rather have the light coming from the end of the Universe and limit the space and age the Universe have. This approach is ridiculous and is testimony of the small mindedness of the Newtonian mind. It is the limitation of man manufacturing such limited equipment and only having the ability to go that far back 12,5 billion light years. The Universe goes eternally much further back than that but man has only the ability to see 12,5 billion years back into the past. The Universe starts and ends where singularity serves up time as Π^o to form Π, we also find space forming $\Pi^o = \Pi^3 / \Pi\Pi^2$.

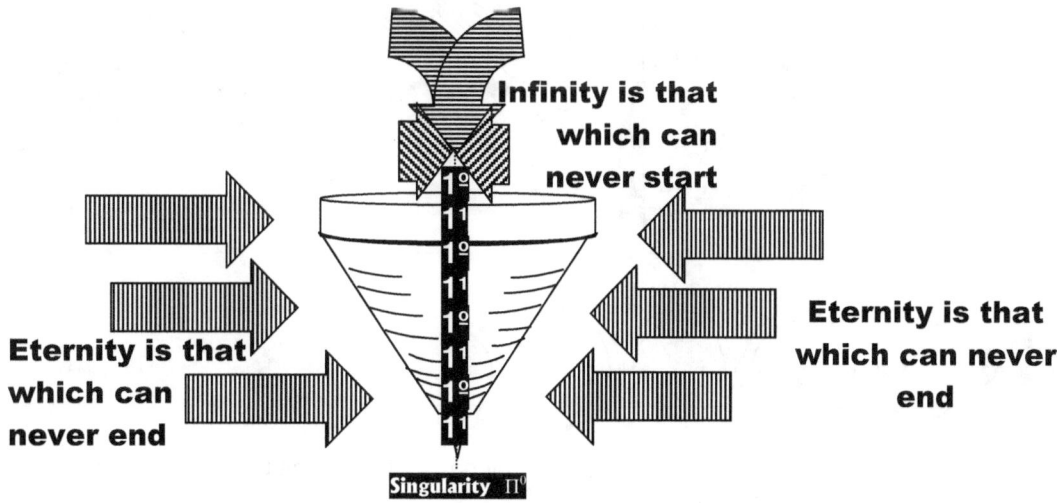

1^0 to 1^1 change 1^0 to 1^1 change 1^0 to 1^1 change 1^0 to

1° = 0.991 going onto 1^1 going back to 1° = 0.991 going onto 1^1 going back to 1° = 0.991

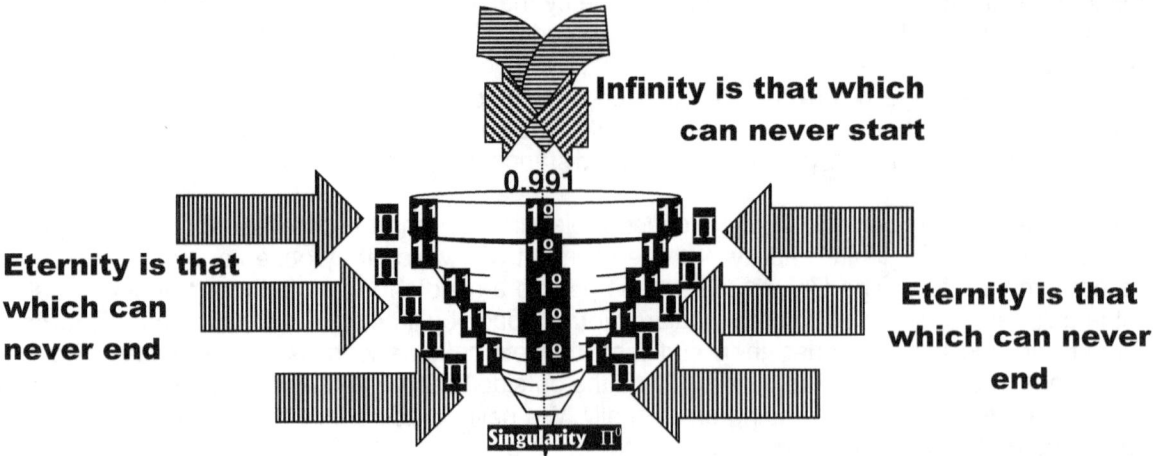

Gravity is time, not only is it time related but gravity is time. Whether the movement is electricity or wind blowing or water flowing or just mirages on sand dunes imitating water, it remains time discrepancies applying. Time following the line which I refer to as the spot 1° forming the dot 1^1 is according to all accounts delivered by Π the turning of (1^1 = (7)+1°=(7)+1^1 =(7)) = 21. Then the growth of space ads by .991

1^1 Time coming from the past
+ 1° Time in the present
+ 1^1 Time going onto the future
= 3 Value of time moving

1^1 + 1° + 1^1 = 3 and 3 x 7 = 21, which leaves a value of 0.1416, unaccounted for.

Π = 3.1416 − 3
= 0.1416
    Movement of material through space is (7 + 7) / 10 = 1.4
1.4 / 0.1416 = 9.86 = $Π^2$ and that proves that time moves by replacing on e position in the future that then grows to form one to become three and to result in one more future point.

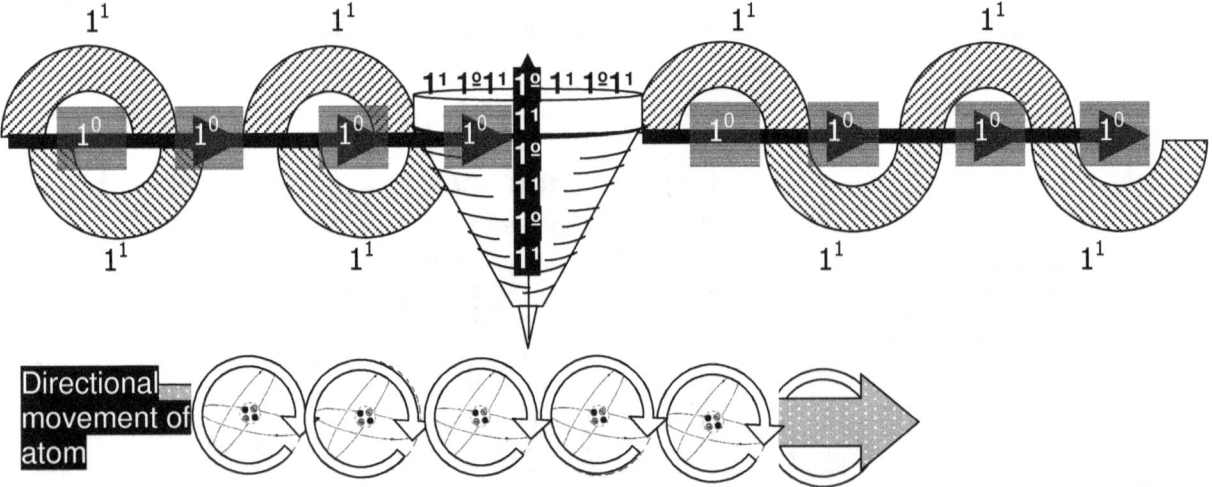

The way the top spins is exactly the same manner, in which the atom spins because the top begins to mimic the atom by having independence.

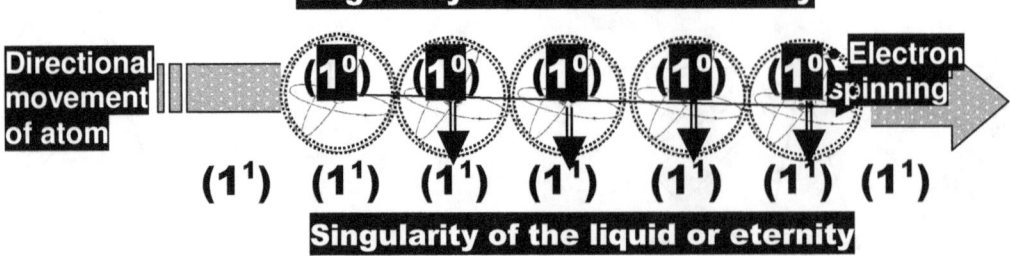

We also find in the atom the relevancy form between 1^0 and 1^1 or $Π^o$ to form Π, we also find space forming $Π^o = Π^3 / ΠΠ^2$.

The Four Cosmic Pillars; The Result Thereof.

Page 75 — In Terms Of Applying Cosmic Physics

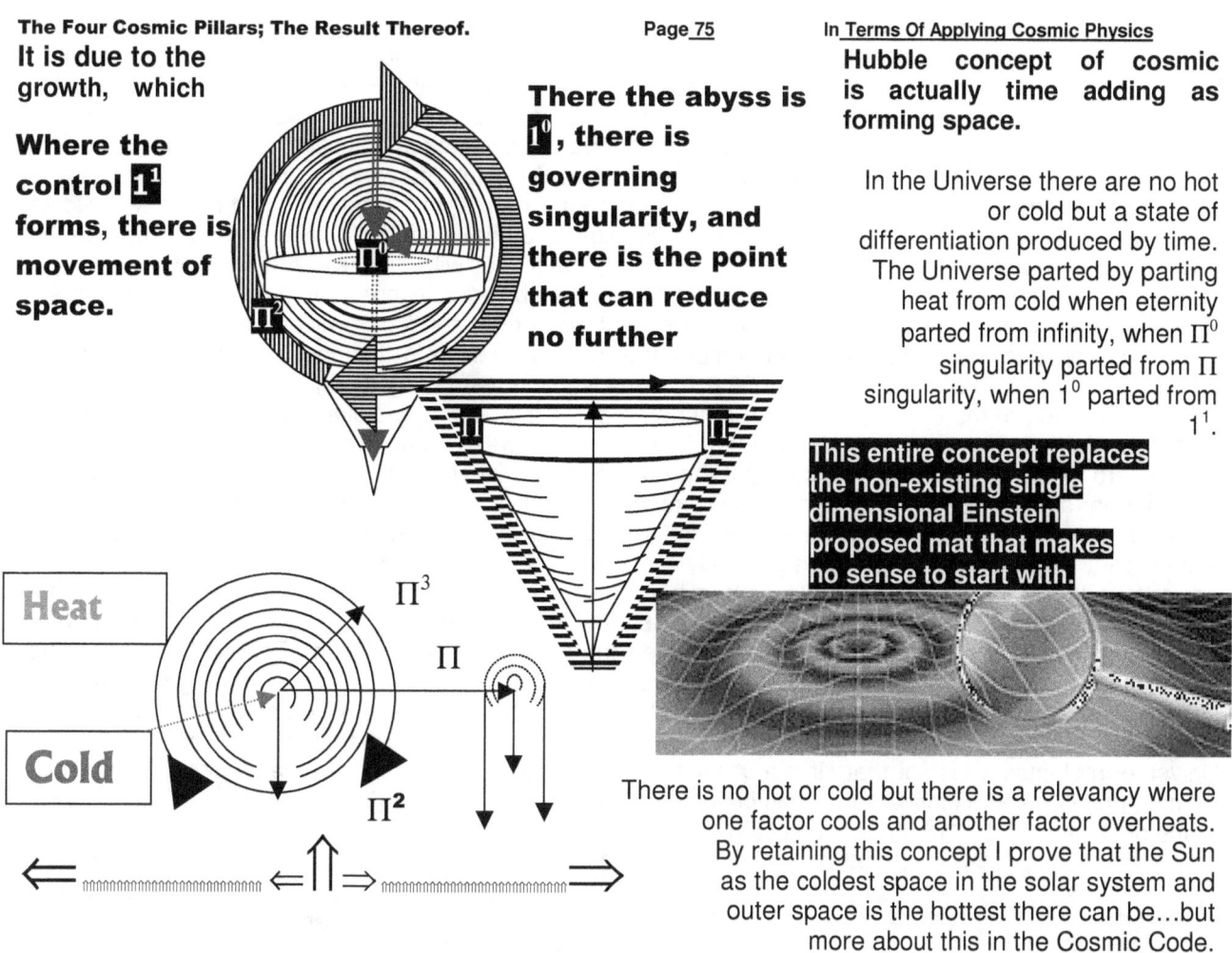

It is due to the growth, which

Where the control 1^1 forms, there is movement of space.

There the abyss is 1^0, there is governing singularity, and there is the point that can reduce no further

Hubble concept of cosmic is actually time adding as forming space.

In the Universe there are no hot or cold but a state of differentiation produced by time. The Universe parted by parting heat from cold when eternity parted from infinity, when Π^0 singularity parted from Π singularity, when 1^0 parted from 1^1.

This entire concept replaces the non-existing single dimensional Einstein proposed mat that makes no sense to start with.

There is no hot or cold but there is a relevancy where one factor cools and another factor overheats. By retaining this concept I prove that the Sun as the coldest space in the solar system and outer space is the hottest there can be…but more about this in the Cosmic Code.

As a result of examining this proposition, I located two principle positions both holding singularity. The cosmos is made up of one type (1^0) that is in two categories where one type moves (1^1) and the other type does not move (1^0). The one is a liquid and the other is a solid. There is life forming a principle time position being other and very different from that which forms cosmic time but life is in addition to cosmic time because it is time has to have a position of 1^0 and then every cell forming the body forms that part holding a position in the cosmos because the part is structurally part of the cosmos and that connects to Π.

The condition for the presence of this singularity that forms everything, controls everything and is that everything is the centralised to a centre singularity holding relevancy $k^0 = a^3 / (T^2 k)$ that forms by movement $T^2 = a^3 / k$ of space $a^3 = k T^2$ placed in relevancy $k = a^3 / T^2$ that is centrifugally going both ways $k^1 = T^2 / a^3$ thereof **(Newton's 3rd law).** In the line that forms when the governing singularity charges a controlling value on space that forms we have the governing singularity 1° that relates to the controlling singularity 1^1 but then 1^1 becomes 1° where it then relates to 1^1, which then becomes 1° that in turn relates to 1° that becomes 1° that relates to 1^1. This is eternity moving across infinity and in that creates the relation of singularity $1^o 1^1$. This is the spot stretching to become more than the dot and become a line. This 1° that relates to 1° that becomes 1° that relates to 1^1 is time moving forward.

From since the time that man discovered intelligence (if he ever did) man has been with the presumption that the Sun is the hottest centre in the solar system. Later on in the present time, it came to someone's attention that the Sun also holds the solar system in gravity. The Earth by its standard and dominating its sphere of which it can control with influence is the coldest centre in the space of its domain being surrounded by the hottest space and it holds the moon centred to the Earth by freezing the moon in a ring to the Earth. The gas planets are the hottest centres in relation with the most heat and they all hold their satellites captured by a hot centre. All space structures hold in every centre there is that is confirming their independence at that point of securing independence the centralizing of the most heat it is able to concentrate and from that centre holds all material captured or controlled in the domain of what that forms the independence of the structure. By moving the structure moving cools in relation to outer space overheating and therefore expanding. In brief, that is gravity where movement cools and standing still brings about overheating and with it expanding. I can go on and on but heat in the centre couples gravity to space-time, just as if Kepler said before he was spoken for on his behalf and without his permission or his agreeing to it.

It is very easy to say the Universe holds anything there is contains all and occupies the lot of everything, but with a human mentality the practical implications are far and wide. However, understanding this concept proves to be something a little more advanced! Every poster or picture science show their vision, about the Universe indicates a concept suggesting boundaries. We look at galactica in the same way as we see city limits and we think of them as independent states. We think of every galactica as a Universe and we do not see the Universe starts and ends with the atom. The rest of formation is just clusters of atoms that could hold much or could hold little where size makes no difference. Newtonians are grappling with the notion that one day they would be able to build a computer that will run the Earth without the need of human or life in particular and that shows how little understanding science-wisdom has for the Universe. The Universe is growing every instant and the growth of the current conductor will bring a rise in the flow of resistance lowering the amps, which will render the voltage useless in the end. Everything is made to destruct with time and so would the Earth and so does life self-destruct by aging the body in using cosmic growth. Thinking of the cosmos in term of a hundred percent efficiency yielding machine is as Newtonian as Newton is and is of course so Newtonian it is as much hogwash as all the other Newtonian ideas they put forward as scientific truth. The idea that the cosmos stays what it was being what it is, is completely Neanderthal and laughable. Every point time provides to form space changes by space growing as time confirms new points that will provide space with a changing dynamic lasting forever. Every instant time moves on it resembles the cosmos in precisely the same way as it was before except for one difference and that is that nothing is the same as it was before because time moves the cosmos by changing the cosmos just one point that has no space but forms space by controlling space. This is where Einstein's flat Universe falls flat. As soon as exploration released the world from a flat Earth, Einstein brilliantly came along and gave us the flat Universe, which is not true. Is accepting that idea not expanding the Universe into much wider misconceptions than there was before…? To marry our logic with Einstein's calculations of singularity brought about some bizarre ideas. To incorporate a three-dimensional Universe into a flat Universe in singularity is not that simple. After I concluded this I then realised I was looking for Kepler's formula that detailed all the factors I was in search of. The cosmos is $a^3 = (T^2 k) = a^{3+2+1} = 6$ Putting this in terms of Π it is $\Pi^{3+2+1} = \Pi^6$ and that I prove to be true.

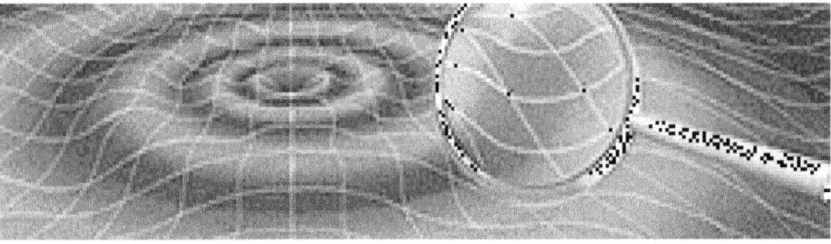

Let's take a view of how the Newtonian see the single dimension unfolding in the grid that is sporting a mat forming waves. How they could ever think that a picture such as this gritted wavy mat could represent singularity goes beyond my understanding…but then again I am always accused that I am unable to understand the essence of Newtonian brilliance! With the mat in the picture being a square it holds space and having a grid it holds space, so the single dimension concept flies out the window. Where there is space there just can't be singularity. We see it has one topside because we view the picture from the topside and having the topside on which we can look down upon, and then that topside has to have a bottom side that is underneath our view. Where the topside ends we are left with what is between the topside that we see and the bottom that we can't see. That is space that is between the topside that we can see and the bottom side that we are unable to see. If there is visible top then there is a visible bottom that should be visible when holding the other side upside-down. Then having a top and a bottom excludes the idea the Newtonians try to portray about this grid forming mat where this mat is being single dimensional. But let's give them some slack and allow the Newtonian brilliant mind some space to hang him or her further. The grid has a hole in the centre and if there is a hole that hole then has to lead somewhere because all holes are measured in depth going deeper into the space being three dimensional and by being single sided the mat are unable to form a hole leading deeper as it is going nowhere. From what I can see this picture is as multi dimensional as the Universe is. The list with compelling theories and proof can and does fill many libraries of mathematicians with brains more than hair. According to the theory on the Big Bang, there has to be a Big Crunch in order to form a Universal beginning of an end. According to present facts presented, the Universe has an age ranging from between 10 and 25 thousand million years, which puts a limit on the endless and limitless. The overwhelming favourite age is placed at $13,5 \times 10^9$ years and is the most widely accepted number in years or Earth circles around the Sun, which by its own accord is madness because where was the Earth and the Sun to form years when the Big Bang erupted? The first scientific formulated theory that was accepted, implied that the Universe was static, going no where as it stayed the same which meant we had a nicely organized Universe that maintained itself uninterrupted and existed in a state of regularity where matter was evenly destroyed as it was created. This meant that the Universe would be precisely the same, unconditional to changing the position that matter was distributed through out the Universe. This gave the Newtonians an unbelievable chance to play God with their mathematics and to redesign what they thought God got wrong and correct about the flawed imperfections in the Universe! This had to be the Newtonians' favourite because it was filled with constants every Newtonian could calculate to their hearts desire! Then someone came and showed a Universe made of everything that always changes and this brought a lot of anger because now the Universe is not filled with constants to use in calculations. As time progressed and a lot of research was done, no concluding evidence was found to substantiate this fact of a static Universe; because no proof could be found at any place where matter originated spontaneously. There was no white hole in relation to a Black Hole where material came streaming out. This implied that as matter changed, there was no proof that it was ever renewed. The list of theories born from that confusion is as many as they are bizarre. To make sense of everything I placed the Universe into dimensions and after implementing the dimensions a lot began to make sense. I placed gravity where there is only a line, moving from somewhere and going somewhere, but never having a defined beginning or end.

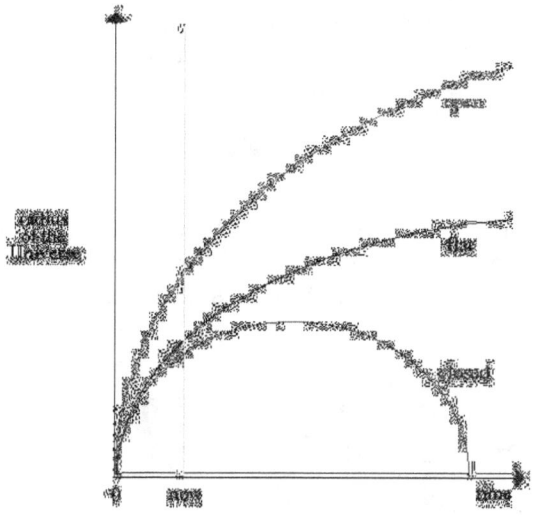

Looking at the Universe from an overall perspective as would be seen through the lens of a telescope is looking at something that never existed. The telescope draws the Universe flat to the instant the onlooker views the Universe. One can see in the same instant what happened 4.5 thousand light years ago in one area of the Universe while also seeing what happened 4.5 million years ago in another area of the Universe and again see in the same glance what happened 4.5 billion years ago in the third area and then come to a conclusion as if the picture portrays the very same instant. The instant is here and now drawing the light the observer see flat, but what the observer sees happened many instances apart. This is the same method horoscope culture wishes to bring to reality by placing light coming from different eras in relation to the second the applying to the onlooker and from that tell the future. When what happened 4.5 billion years ago happened, that which happened 4.5 million years ago and that which happened 4.5 thousand years ago was not yet in place and by the time what happened when 4.5 million years ago happened, everything at 4.5 billion years ago changed unrecognisably but even today those changes did not arrive at our station in time. That puts what happened at 4.5 billion years ago a non event when 4.5 million years ago took place and so did what happened 4.5 thousand years ago also at that point still was somewhere in the future and for all purposes a non event that did not yet take place.

Therefore no one can draw any sustainable conclusion about development of the Universe from what could be observed as the Universe and this makes diagrams as the one next to this writing a joke if not completely dispensable as science fiction.

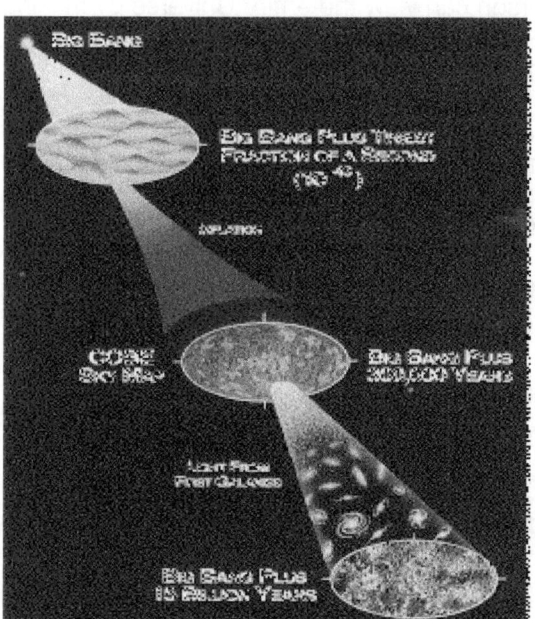

Every person alive knows that there is a Universe that is not flat and this Universe is chained to time. One can draw all the lines you wish, but as soon as you bring in time forming space, the Universe becomes a four dimensional, moving "something" that cannot turn back and cannot skip forward. It is on tracks going in the direction, wherever that may be, through the movement of time. However this is true, it is also true that every instant in the Universe is totally independent from every other instant in the Universe and this makes the Universe no oil painting in which you can see things in the present being in the same frame as things on the horizon of time because the space might be space, but it represents time that applied when other space was not yet applying or the other space did not apply any longer.

Any view of the Universe should not be seen in context of other things in the Universe but completely independent of all other things in the Universe because what is viewed applies in context of all other things not applying yet, or not applying any longer. All space $\Pi^3=\Pi^2\Pi$ that is seen stand independent $\Pi=\Pi^3\div\Pi^2$ from all other space seen as is identified only by the space $\Pi^2=\Pi^3\div\Pi$ moving in relation to what the time table indicates as time differentiation between what can't move Π^0 and what can moves $\Pi^{-1}=\Pi^2\div\Pi^3$. It is the speed by which the space moves in context to the space not moving that determine the duration of time applying and the emphasis should fall on the movement of the space and not the size of the space moving.

The Four Cosmic Pillars; The Result Thereof. Page 78 In **Terms Of Applying Cosmic Physics**

I have previously shown how much incorrect was the opinion of science by having the approach to cosmic concepts that everyone including Einstein gave eternity borders and an ending. There were always an edge forming space in either a flat or a saddle or a sphere but whatever Newtonians envisaged, they saw a border ending eternity. There is no ending to eternity because as much as the line goes straight, the line curves to conclude space.... but there is even more ... by growing straight a line forms a circle at the same moment and with the line circling the line will end where the line began and in that lies the eternity factor of space. That is Π and that is gravity. However, I have to repeat what I said about Einstein's thinking of a flat Universe.... as soon as the ending of superstition released the world from a flat Earth, Einstein came along and gave us the flat Universe. To marry our logic with Einstein's calculations of singularity brought about some bizarre ideas. To incorporate a three-dimensional Universe into a flat Universe in singularity is not that simple.

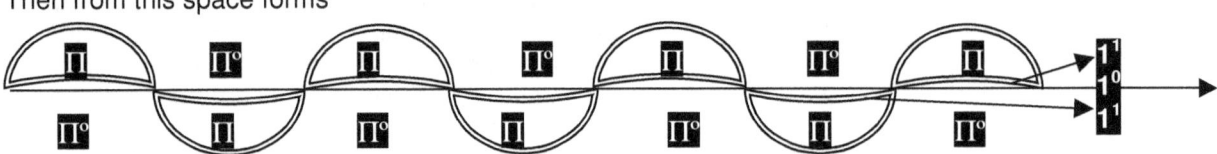

Singularity forms in the relevancy between what is hot (1^1) and what is cold (1^0) or between that which forms infinity (1^0) and that which forms eternity (1^1). When there is movement, (1^0) evokes (1^1). This is the instant in which the onlooker views the Universe which is time and which is flat. Space is not flat but time is.

Then from this space forms

When there is movement, singularity in infinity parts from (1^0) and this evokes eternity (1^1) which establishes $\Pi^0\Pi$ where this carries singularity forming time and time becomes space and at the very same instant in time forming singularity $\Pi^0\Pi$ then from singularity Π that forms space $\Pi=\Pi^3\div\Pi^2$ where space contracts $\Pi^{-1}=\Pi^2\div\Pi^3$ as much as space expands $\Pi=\Pi^3\div\Pi^2$ because forming space is equal to the movement thereof $\Pi^3=\Pi^2\Pi$. In the Universe singularity stitches everything into a woven concept we call the Universe. There is the premier singularity that serves as a beacon to everything carrying singularity and that singularity forms part of every singularity charging space-time. Everything is growing but not according to human perceptions because that is growing in alliance with the cosmos. I am not going into much more detail but to say that everything there is (1^0) connects to everything there is (1^1) as $(\Pi^0\Pi)$. The network linking what there is to what there is, is not linking on this side where there is no reality, but is linking where reality meets infinity by uniting with eternity. The fact that a person can view the entire Universe by making contact holds what there is (1^0) in view of what there is (1^1) and this takes place on the side where space is of no consequence because time in eternity meets time in infinity.

On the outside of time there is space that can never end because then time has no outside. Everything that ever will be, will be because it is inside this that can hold everything but nothing can hold a place in this space that there is no outside too. This space ends where this that has no end meets that what there can be no start too. This has everything on the inside just because there is no outside to what this is. Why would I not use names...it is because it is Biblically named? This is eternity. If there is eternity, then there is infinity.

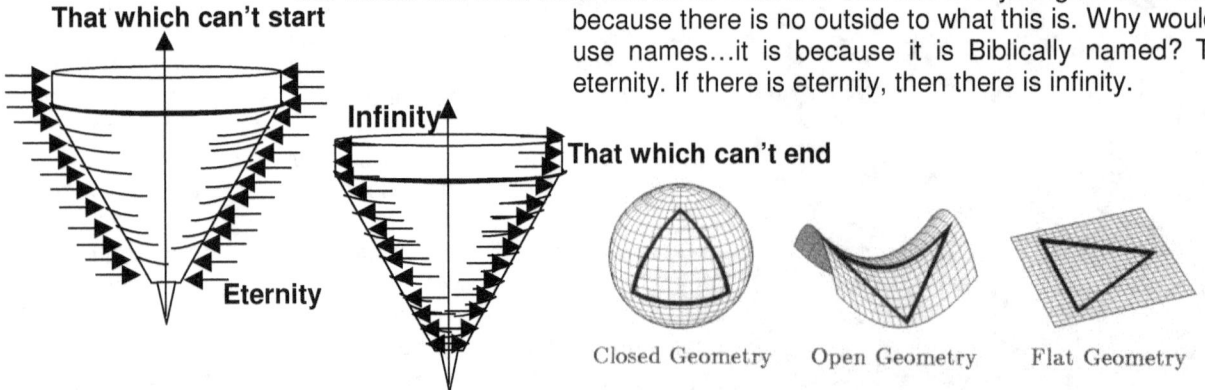

This is infinity. This is infinity and this is eternity. This can never start or begin because this is the beginning of everything there could ever be. The entire approach to previous cosmic concepts was that everyone including Einstein gave eternity borders and an ending. There were always an edge forming either a flat or a saddle or a sphere but whatever Newtonians envisaged, they saw a border ending eternity. There is no ending to eternity because as much as the line goes straight, the line curves to conclude space.

That is the main issue derived from Kepler's formula $a^3 = T^2 k$. There is no outside to eternity because eternity forms the entire outside that will ever be available. There is no inside because this is the inside to whatever could be on the outside of this that is representing all the inside there will ever be. The Universe is what is inside that which can have no outside and what is on the outside of that which can never have an inside and reality is where these two factors representing that which can never end unite with that which can never start and moreover, the unification forms reality found only on the other side of where time produces space. This is where eternity becomes connected to infinity and this is where (1^0) unifies with (1^1). This is where (1^0) connects with (1^1)...and that is where we living in the Universe are not because where we are, there is no connection of what has to connect in order to allocate us to where we are.

These proposed Universe formation concepts that Mainstream science proposes prove to be examples that are a sure indication that without the use of mass and without the use of pressures then their thinking of science in the Newtonian convention boils down to nothing. The problem these suggested Universe concepts have is that it places borders in the Universe and allows the observer to look at the Universe from the perspective God will have. They haven't got the insight to see that the viewer observing the Universe on the outside belongs inside the Universe because to the Universe there can be no ending or for that matter a start to the inside. The fact that this places a picture of the Universe from afar and not of within the Universe as it truly must be, and in that incorrect approach, al the incorrect surmising nullifies the entire idea. Here is a far better explaining of the Universe starting at its starting point and also at the ending point.

In the centre of al things spinning a line comes from where a dot first was. The dot came from where a spot once was. The line comes from a dot that extends but just as the dot extended from the spot it once was, the line has no start and has no beginning in the sphere we could observe. When trying to locate the line we have to reduce the circle forming inwards going all the way to a point where no further reducing to the centre of the rotating circle is possible. As soon as entering the point where the line is, one has gone through the point where the line is. When being on the entry side, the circle direction would be the very opposite of the direction of travel on the other side of the point leaving the spot where such a line is situated and because the line forms the divide of the two directions of travel, therefore the line could never form space. The line has no inside that can go even smaller and yet we know that the line must have some ability to be able to go smaller since our understanding of the concept insist on this reality by the value of Π. The line forms the control of the directional movement without ever participating in the directional travel.

That point without space is where the Universe starts because everything that can ever be, starts at the point that could never start and which can have no beginning because it is where everything is beginning and is without limits because it holds no space. That point is in the centre of all material, which puts that point in the centre of the Universe holding material or holding the ability to hold material. That point shows where eternity parts from infinity. However, to get eternity to part from infinity, eternity needs to move because infinity can never move and therefore will forever stand still and the point forms the line we call the axis only when moving, thus having infinity part a position from eternity.

There was Π^0, which was α^0 or if you would rather have it Ω^0 or it maybe was 1^0, but more correctly it was all the above and the beyond because multiplying what ever constitute the mentioned will bring about what is mentioned to a precise equality. It was a spot that was not. It was a line that ran eternal but because it ran eternal and kept repeating exactly what was before to the precise what came afterwards the line was there and was eternally running, while never changing in the least or growing by any measure. Infinity and eternity was one because when it was one, what was one repeated as one and the process cycled back to before one and before one could be reached to form one again. It was such a continuing of the monotony, no change ever once occurred and therefore never did the running produce progress because the progress was in the perfect repeat of what was before continuing as the same going forever. The duplication brought contraction to the smallest detail and the detail was so small it repeated one as one. Then Creation came about by heat differentiating from cold and the perfect became imperfect forming two. I have written a book about how this process came about but that is most probably my most complicated book I was ever able to write. This is precisely as Geneses one verse one describes how it happened and changing even one word according to the Bible of the process applying translates everything incorrectly.

That is where our atheists get one hiccup. Everything that I show is as real as the Universe can be and yet not one point is part of the Universe we see. At the start before the start when eternity met infinity nothing changed and after the start of the start, nothing since then remained the same and as eternity repeated the past it met infinity at the next point holding the future. Before creation started, eternity ended on infinity every time but the instant Creation came about eternity shifted one point away from infinity to establish space. Before Creation came in place, the repeat brought eternity and the repeat was so perfect that the repeat continued. The repeat still is with us as much as we are within the repeat, but for one position away from the repeat. There was something beyond the Universe that instigated breaking the perfect cycle, which change the institute. There was something that brought a difference and we are within that difference.

That difference was space that time left behind as a thought and that space is what time leaves behind and is what we move through as much as what we see at night. Oh, how stupid and how thoughtless the minds of atheist and other atheistic animals are. Baboons do not recognize these factors revealing the position allocated to infinity as well as eternity because they cannot think and are therefore atheists. Animals are not able to realise that the true value of every point securing singularity albeit (1^0) or (1^1) is without space and is therefore not in this Universe but that makes them dedicated atheists just because they can't appreciate the truth. Atheists are unable to see what is there is true but is beyond what is observed. The location is where space is not and the only value such points have, is the trajectory of time bringing on movement. The points hold relevance and that holds space as a result of time leaving footprints but the true essence is in a place that has no space at all.

By finding where it all began is equal to finding where the line began and therefore we have to trace the line in order to trace the development of the entire Universe. As could be seen in how far we have to go back in finding the first development it is clear that it went beyond where mathematics may take us. At first the Universe did not become more but only focussed better on detail. What is at present was present because nothing can be new to the Universe that started out to be what it presently is. Time forms space as the legacy of what time was.

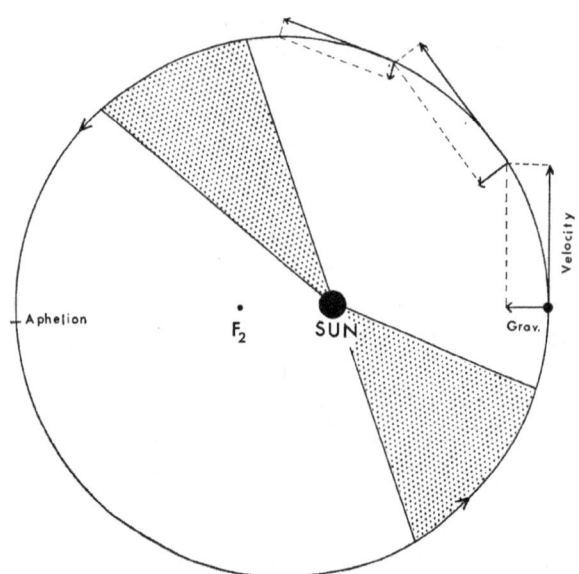

An object can rotate in outer space as long as it can maintain a speed that will keep the object rotating in that orbit in relation to the centre around which such an object is rotating. Newtonians see everything going straight because that would be the obvious way to look at the situation, but that would be backwards looking at the cosmos. No line can go straight because part of all lines in the cosmos is the curve. There is always some relation between the factor of how much k influences or how much T^2 influences and the combined unit determines a^3. The formula Kepler **received from the cosmos says that any specific space a^3** is defined by two factors comprising of the line coming from the centre k that determines the circle forming T^2 Another huge idea this leaves is that space is the result of what time leaves behind as time moves onto and into the future. That reason too has its footing in the Titius Bode law because 7/10 is half of 10/7 and by going seven the factor also has to go 10. The compliment of this is gravity at Π^2

E. P. Hubble (1889-1953) confirmed an expansion through out the Universe, which contradicted all we thought we know about our Universe. According to the accepted the Universe is in a normal state of contraction because that is what $F = G \dfrac{M_1 M_2}{r^2}$ imply. Every person is very aware of the idea that the universal expansion would not last for ever, but has to start with some contracting effort at some point.

These bodies will collide and destruct, without a doubt. When $F = G \dfrac{M_1 M_2}{r^2}$ apply, there should not be any force, which is able to keep them apart. Known for almost a century and a half science has failed to give any explanation about this cosmic phenomenon. This phenomenon should not occur with Academic's laws about gravity. If the Universe is on a contracting, we have to first find proof about the location to where such contraction is pointing. In order to locate the contracting we have to locate the centre of the Universe, which means we have to locate singularity. With singularity eternal small, holding the place where the Universe started, we first have too differentiate between singularity and zero, should we wish to find singularity In modern science the phenomenon we know as the Roche lobe comes more and more to the foreground, indicating an undeniable interaction between orbiting structures sharing a common axis. That axis science at present do not recognise, not withstanding the reality and un-deniable proof there is behind all evidence. As apparent as it is to me, I went about divorcing $F = G \frac{M_1 M_2}{r^2}$ from outer space and applying it to specific positions where one may locate singularity, which we have to locate.

The information given in the next few pages is representing the line of thought I tried to get introduced to the world at large via my work as contained in the books I hope to get published someday and this is the most basic I can give about the information that I tried to get published. When you have read all four books in e-format that what you have read forms a mere framework about the information I present in my books I wish to have published. I sent many, many manuscripts on many, many previous occasions to even a lot more publishing houses and publishing universities without having any joy from my actions but then on the last occasion of me trying to get work published, I sent three manuscripts forming the collection of books to about thirty publishers worldwide in an attempt, yet again, to find a publisher.

I sent this introduction letter accompanying three manuscripts to more or less thirty publishers:

PO Box
Town
Province
Republic of South Africa
Tel
E-mail

Dear Sir / Madam,
I am Peet Schutte, a South African resident residing at the address given above.

I present my books to you for your attention in view of a possible publishing venture.

That which you are about to read is proving the physics phenomenon called gravity.

Gravity has never been understood but ever since Newton introduced it as a concept, it has always only been accepted to be what Newton envisaged gravity was. However, as widely accepted as the concept might have been all that time, this was only based on accepting and proof to the fact became about as merely culture passed on from generation to generation without any person bringing substantial proof about Newton's suggested hypotheses. However, notwithstanding the overall ongoing accepting of the culture as a truth that then became a religiosity all the while it served science, still Newton's vision as the accepted theory remained void of proof and remained in place without ever finding any backing to secure the reliability of the facts. In three hundred and fifty years there was no shred of evidence coming from comic evidence to prove the attraction Newton claimed that was going on between cosmic objects and without that, it is nothing other than just being a theoretical vision that is serving as the baseless and unproven factual concept, which it is.

I disprove the hypothetical presumption that is been accepted concerning the principle called gravity where such presumptions stood as the truth and this is ongoing ever since Newton made presumptions as to why the phenomenon he named gravity exists. I show that there is no attraction by force between cosmic objects and that the Universe is not in decline or in any state of collapsing. I prove that the Critical Density theory is a criminal supposition aimed to divert public attention away from Newton's failings to accurately substantiate his theory of attraction. It is a criminal farce they put in place as they called this the Critical Density and by which science claimed this density is going to "correct" the cosmos. It is a hoax never seen to the extent it is during any time in the entire history of the world. The question is if the substance producing gravity is there either by being dark and invisible or bright and visible, the fact that it is there must bring about gravity that will bring about attraction. Why the attraction by critical dark mass is postponed and when is it going to become activated and what is it then going to activate the mass of the suggested dark matter into realizing the force that is coming into attracting the Universe. The concept is totally void from reality. If the matter is there then the mass is there. If the mass is there and mass is what produces gravity and gravity brings on attraction of other mass then the attraction must collapse the Universe? It can't be suspended and come into action somewhere in the future.

For the first time since Newton gravity as a cosmic phenomenon and the physics principle is explained and understood. If you care to read my manuscript you will see why and how gravity applies. I remove the insanity and introduce reality.

In the past I tried to convey my explanation to academics in physics without success. By addressing academics on the level their standards of conversing requires, such level of conversation in the past with previous books might have been well above the understanding ability of students, but I rectified that mistake in this manuscript.

With this book I turn my explaining to students and I tell students how they are brainwashed into accepting and believing Newton without ever being offered proof to substantiate Newton's mythology of gravity attracting objects.

This is as simple as I can present the subject and the content should be clear in the form it is in this manuscript as I present it in this book to even school going students. I am not able to simplify the facts any more than I do it in this presentation.

Reading the information will allow students to see to what degree and level of intensity they as students in physics are being brainwashed into believing in Newton without ever finding hard evidence to prove Newton as being realistic and truthful. This process of brainwashing students into submission is ongoing for generations and every generation is the victim of the previous one as much as they victimize the following generation of students, forcing the following generation into believing in Newton while the theory put forward by Newton never became fact but became culture and folk law.

As I said before, in the past I tried to converse with academics on a higher level than I do it in this book and I was shot down without given any response to why they reject my work except that they are not willing to go against the grain of Newton. In this book my aim is to show clearly to students that Newton and his mythological concepts have never been proven and I show how they can question Professors tutoring in physics to either prove Newton for the first time in almost four centuries or come clean and admit Newtonian science is a fraud.

I supply students with question to uncover the fraud upholding the fallacy called Newton.

Every student studying physics and even studying physics at whatever level should obtain the information I give in this book in order to start to question those that claim to be experts on the subject of gravity. It will show how much expertise in Newton's gravity is truly available.

By questioning the academics the world would learn that Newton's mythology has never been proven and I give the reasons why there is gravity as well as what gravity truly is.

By addressing the questions the students then are able to ascertain that Newton is fraud covered up by a conspiracy so large it has never before been equalled.

There is no way in a billion years that the use of this formula $F = G \frac{M_1 M_2}{r^2}$ would apply by proving that the Universe is shrinking. There is no attraction between objects with mass and anyone that says that must prove that by using the formula correctly and show that it does apply. Academics build an entire physics philosophy on the formula $F = G \frac{M_1 M_2}{r^2}$ that can never work. Let any student insist that those academics tutoring them in physics prove by example how much the solar system shrunk in distance since the days of Kepler. In the era we live in, the calculations can be made instantly with using new technical devices and the figures Kepler had is still available. Let them publicly compare the figures and show how much did the solar system reduce in space, if indeed it did reduce as Newton claimed it has to do. Give students the chance to force those academics in physics to admit that the expanding of the cosmos and even the solar system is on going even between the Moon and the Earth. It is either they prove Newton's formula $F = G \frac{M_1 M_2}{r^2}$ is reliable or they admit that expanding is in place and that Newton's formula $F = G \frac{M_1 M_2}{r^2}$ is as accurate as the Cinderella story. At present Mainstream Physics is hiding behind a lie they named the Critical Density Theory. They use the excuse of trying to detect so called dark matter that will later on produce the gravity that will allow the cosmos to start shrinking and therefore then vindicate Newton's failing to show that the Universe is indeed shrinking. Everyone knows we live in an expanding Universe and with an expanding Universe there is no attraction in place as Newton claimed it has to be with his formula $F = G \frac{M_1 M_2}{r^2}$, This Critical Density Theory is explicitly criminal and has the motive to criminally deceit the public openly in order to cover the failing of Mainstream Physics understanding of physics. They protect their incompetence in understanding physics by protecting the incompetence they use as a pillar on which physics are formed.

Those that bring in the so-called Critical Density Theory do so with the purpose of hiding Newton's failure behind a complete hoax in order to detour from the truth. Let those claiming that the Critical Density Theory is going to enable the Universe to contract show how they use Newton's formula $F = G \frac{M_1 M_2}{r^2}$ to prove it and show therefore when this mass will kick in and why the mass is going to form gravity because in an expanding Universe it is not doing it at present! Let them with $F = G \frac{M_1 M_2}{r^2}$ show when the mass is going to form the gravity that will start the contracting again because Hubble proved it is not doing it at present.

After Einstein failed to calculate that enough mass will be present in the entire Universe to bring about contraction as Newton claimed it will do, then Mainstream academics in physics started to promote the so called dark matter theory where unseen material is present all over in the cosmos that will supposedly bring about the change of direction in cosmic development.

Let the students ask why will the dark matter that they all pretend to search for at present not generate the gravity at present in the here and the now in this minute, if it is there and it is indeed able to generate the gravity by way of the mass it then has to have at this minute and why that force being there through the mass being there does not do the contracting at this minute.

If the mass is presently not visible or not luminous, then why is the force not contracting the expanding Universe because being luminous or not luminous, the fact being luminous has no input on the generating of gravity and if it is present then the mass has to produce gravity. If the gravity is produced, be it from a luminous object or a dark object, then why is it not at present charging the gravity if indeed it is mass that produces gravity that brings about contraction. If the mass is present this minute, then the question to answer will be when will the mass start to produce the gravity because taking the nature of the formula $F = G \frac{M_1 M_2}{r^2}$ into question, we find that the attraction between mass must reduce in the growth of the distance coming about from the Hubble expanding. The nature of the formula puts in relation a reducing factor in mass when the distance r^2 is increasing. By increasing the factor that divides then the factor on top will reduce the influence it should have, if the factor dividing is increasing the influence of the divided is going to diminish. The time that increases the distance in the expanding Universe reduces the influence of the mass if the mass indeed will promote contraction by attraction. The more the expanding goes, the less is the chances that $F = G \frac{M_1 M_2}{r^2}$ can ever apply.

The so-called Critical Density Theory is the biggest swindle ever devised by man. Let those academics that the so called Critical Density Theory is going to bring about Newton's vindication then say why the mass, be it luminous or not, is allowing the expanding to take place at present because being the it has to have mass that has to produce gravity that has to produce attraction if this rule indeed applies and in that also say why it is not enforcing the cosmos to compress space and when the mass will start to generate the gravity that will bring about the turning in cosmic development direction and where at that point will the turn about take place. The truth is that Newton's contracting and attracting is a hoax and there is such a cover up going on in the presence of mainstream physics to protect Newton that covering Newton became criminal deception bigger than anything that came before it. Again I say there is no attraction or depleting of space in the Universe taking place at all and every time any one promotes this idea, they are busy covering up fraud with more fraud. Let those that do so bring the figures to show that this attraction is taking place and that the cosmos is indeed declining in distances between objects in space.

If you give this work to any academic in physics for assessment, they are bound to condemn my work, as they did on numerous occasions in the past and therefore when you present it to academics in physics, then be sure that I tell you they will again advise you negatively about the nature of my work. Be sure that this is going to happen as I predict it will beforehand and that they will condemn this book because if this is accepted as the truth, then everything previously written on cosmology and the human understanding of the cosmos is going to go to the fireplace as fables. It puts egg on the faces of the most famous and most brilliant group of academics there can be!

On the other hand, when you listen to their advice as it has been happening with all my previous attempts then you will lose a lot of money just because you chose to cover their deceit. Therefore I am telling you that if you do publish this book you are going to make a shipload of money, as this is truly controversial in every aspect as much as this is blowing the truth about the foundation of science to the surface in every aspect. This is truly controversial because with the Da Vinci code as much as it is bullshit, it still is controversial without supplying substantiated evidence where any person can gauge this controversy with substantiating evidence and the information to use in such gauging is in every science handbook in every school. The subject touches everyone that ever studied physics and that even includes those that went no further than at a school level.

But while knowing your publishing process and the normally followed procedure is that you require the conformation of academics in physics about my work, I am not naïve believing that you will jump for joy and embrace me as if I am a gift from god, given to your publishing house. If you do present my work for scrutiny to any academic in physics, then in doing that the rejection of my views is part of their work to protect the science being in place but when receiving their rejecting of my work, be sure to just ask them those questions I tell students to ask their tutors. One question you should ask them is to use Newton's gravitational formula to indicate how much closer did the Moon come to the Earth in the past fifty years and by using Newton's formula then let them show you how they calculate the time when the Earth and the Moon will collide. If Newton's $F = G \dfrac{M_1 M_2}{r^2}$ is feasible, credible and as accurate as physics claim it is, then it should be no problem to ascertain the time the Earth will end with having one massive collision with the Moon! Test their honesty with this simple question and see who is bluffing and who is honest.

Tell those academics you charge with assessing my work to explain the Hubble shift in terms of Newton's attracting and where there is indisputable proof that the Universe is expanding in accordance with the Big Bang concept, let them show how the Big Bang could ever be a fact of the past if the Universe is contracting according to Newton's mythology.

If you want a true assessment of the book then just read the book yourself because knowing academics they will shoot my work down in flames as they did hundreds of times before. If they will agree to the accuracy that my work present, then they also admit to my work's accuracy and in doing that they then condemn their personal work. Every academic going into agreement with my standpoint condemns their personal work by agreeing to mine. That, no academic will ever do because they are not in virtue of seeking the truth as they all wish to claim but is hunting personal profits and professional acceptance.

Lastly, be sure about one thing…I am not holding my breath wile I am waiting on your positive decision and I am taking this book and my message to the Internet but I realize in doing that it also is the slowest possible way to get my message across to students. That way has failure written all over but at least I might begin a new craze of which your firm then has no part. If you decide not to publish because they denounce my work, then make sure that you thank those academics that are also deceiving you and that they helped you by advising you to losing out on a fortune in the making with this book. Please read the condensed version first for your benefit. The truth is if I gauge reactions coming from the past presentations I submitted, I am not expecting much and I am not sweating in anticipation while waiting anxiously on a response, but be sure to know that eventually this time you have the most to lose. I know I have nailed gravity and I prove gravity for the first time ever and even then if it takes another decade through the internet, I will eventually publish what I discovered because it is me that discovered what I wish to have published while then your institution will eventually lose out because you are not the publisher. This book is going to shake and vibrate many mountains and make money sometime somewhere in the future and if you are not on the wagon at such a time, then you fell off the wagon somewhere with a wrong decision and are not coming along with the money-making ride into the future.

Yours truly
Petrus Stephanus Jacobus Schutte.

The Four Cosmic Pillars; The Result Thereof.

From about thirty publishers I contacted by sending them manuscripts (this time alone) I received one positive sincere reply. More important is the fact that this is the only positive reply I have received in eight years sending my work to countless Universities and publishers. I will forever be filled with gratitude for the help I received and I publish the letter I received as I have received the helping letter.

Dear Mr. Schutte

Having browsed through the cover letter and the attached files, my colleague Alexander Grossman, who is the Editorial Director for our Physics and Maths publishing list, would like to suggest that you submit your work to a journal rather than further considering the idea of a book. This is the usual way in science for new theories or concepts and he is confident you will receive constructive feedback from the reviewers.

He suggests that you submit your work as an original paper to our physics journal 'Annalen der Physik' (www.ann-phys.org)? You will then receive feedback from the Editor in Chief or the reviewers.
I trust this is helpful.
With best wishes
Lesley

Lesley Smith
PA to Mike Davis, Managing Director
Wiley-Blackwell Life Sciences
John Wiley & Sons Ltd
The Atrium
Southern Gate
Chichester
West Sussex
PO19 8SQ

Wiley Bicentennial: Knowledge for Generations
1807 - 2007
Tel: +44 (0) 1243 770110

John Wiley & Sons Limited is a private limited company registered in
England with registered number 641132.

Registered office address:

The Atrium, Southern Gate, Chichester, West Sussex, PO19 8SQ.

This writing that I received was the first ever response (positive or negative) I have ever had in eight years of contacting publishers as well as many, many Universities and for that I mention the person and the name of the publisher to show my utmost gratitude.

From this I contacted ***Annalen der Physic*** and sent a copy of **The Veracity Of Gravity** as well as a letter in which I tried to entice the Editor in Chief Prof. Ulrich Eckern to read my work. In stark contrast to the about (believe it or leave it) 1500 + Academics I contacted during the past many years I did receive a reply from Prof. Ulrich Eckern. I supply the letter that I sent to Prof. Ulrich Eckern be it in a much-abbreviated form and I also include the reply I received from Prof. Ulrich Eckern.

PO Box,
Somewhere in the bushes,
0000
Limpopo Province,
South Africa.
Telephone
E-mail

Prof. Ulrich Eckern,
I have been conducting my personal research on cosmology since round about 1977 or thereabouts and after coming to conclude my personal view on gravity and how gravity affects nature, which clashes with Mainstream Science head on, I started writing books on the matter in order to convey my perspective on the subject. Then for eight years I have been writing various books on cosmology and all my attempts were aimed to bring across my findings. To my dismay I have not found one academic promoter or publisher willing to endorse my views. I have been rejected as far as I have introduced my work notwithstanding the fact that no one even once could prove any inconsistency or one mistake I made in all my declarations about cosmology. At the end of last year (2007) I was given advice by a British Publisher to contact your institution. This is what I now am doing. I am suggesting a new understanding on gravity and moreover a new standing on how cosmology functions.

I DISCOVERED A MISTAKE IN SCIENCE.

I cannot be more aware that my mentioning the mistake would bother you even in the least, but still I have to try.

I am well aware that I have no hope that my effort in sending you a copy of this manuscript would urge a response from you, but without my trying, I thereby then forfeit even the least chance of finding a positive response that I might evoke from your institution, however slim the chance might be and with that I then deny myself the possibility of having any chance I might ever have of even having doubt about my effort in sending you this manuscript. I am sending a manuscript and not a brief theory proposal jotted in an essay format to be published in a journal because what information this book holds, will change every aspect currently understood by science, and when I say that it is not a madman boasting, for I challenge any person that are able to read this work, to show that my statement of changes that has to come to physics are not true. However, my claim about this work having the ability to bring change to science cannot come from brief communication about a few topics or a number of thoughts to be studied. When one says a piece can change science that would entail a massive undertaking covering a comprehensive background of detail in a study of gigantic proportions. This is what this is but also this is just a small fragment of what detail the entirety of my wok holds that is consisting of more then twenty other books covering the total view.

There is no chance in hell that the content containing the fundamental changes required to correct physics as a hole can be dealt with by simply reducing the information just this book presents as it is, to an article presented in a journal. If it is thought that the information presented in the other twenty or more books will also be added to the information in this book, the entire thought of confining the concept to an article in a journal becomes bizarre. The information that just this book holds as it is presented in this book form alone, will change the understanding of mathematical physics completely, and yet as I say that, the madness that accompanies the words I utter cannot escape even me and even I see how ludicrous it sounds and the obscenity in such a statement rings through the mind of any person reading the statement, however true such a statement might be.

It sounds like utter madness when I declare that I am the only person on Earth that is correct about science and I can see how any one would find me the joke of the century by saying this, and yet I dare any person to prove I am incorrect when I charge these claims. This could only become true and sound in fundamentals when the entirety is fully covered by all the detail I present. Leaving out even one thought standing apart from all the rest of the other thoughts, be it as fundamental as such a thought might be the exclusion of that one thought will nullify the concept as a whole that I present about the fact that the view I bring forward will bring changes to physics and mathematics. When leaving out one argument that is said without the support of the rest of the content in this book will then sound make the rest sound trivial and even mad. In the following there is but a very few examples to try and show my point I wish to make about reducing the concepts entirety.

When I say a line can't start with zero, it sounds as the most trivial statement any one can make, and yet in understanding this concept brings holds the difference between the veracity of a Universe being in place and a Universe being totally absent and being void of ever being established.

The Four Cosmic Pillars; The Result Thereof. 　　　　　　　　　　**In Terms Of Applying Cosmic Physics**

When I declare that pi (Π = 3.1416) is flat then you would think it is the most ridiculous statement I could ever make and when I proceed to prove that pi is flat, you will rubbish the argument without further thought because you would be unable to see the purpose of wasting time on such trivial information. Putting pi down to being flat makes no sense, but after four hundred pages of other arguments I make in support of one another throughout the entire book, you will learn that this statement unlocks our fundamental understanding of the Black Hole and therefore the concept we have of space-time. But when putting the argument on its own in an article to be published in a journal would be wasting paper because if it is published on its own, the merit of doing so evaporates, as there is no purpose of putting forward such an argument.

When I say a graph can't have zero dividing the graph on any axis, by my saying that I clash with thousands of years of mathematical culture and who ever is reading this suggestion, is then disinterested and their repose is to read no more of the rest of my work because such trivial statements can't interest any one ever.., yet mathematics has been incorrectly assessed in this matter for as many years as human memory will allow. My suggestion about this matter alone defies human mathematical understanding of the importance this issue may have and thereby show disregard to mathematical culture completely as it desecrates all mathematical principles formulating physical mathematics. Yet, try to multiply the growth that a graph shows when it breaks through the line of division with zero (zero representing the measured value as the line dividing the graph must have when it is zero as science claims it is) formed by the horizontal axis, then by multiplying or extending the progress in growth on the vertical axis this value of zero should have in further into the graph on the vertical axis, and in doing that it will have the entire graph form a continuing of zero and then all further reading on the graph must be nullified in value.

Multiplying the zero with which any value flowing from what the line of progress must have, must bring more zero to the graph as zero then becomes more because the mathematical principle on multiplying anything with zero insist that the answer must be zero. Anything multiplied by zero becomes zero. No dividing line in a graph can be zero and still this concept has been used for as long as humans used mathematics. The idea of realising this concept sounds trivial but because of this misconception no one ever and that is including Albert Einstein, was ever able to pinpoint singularity, a fact I do on page three or thereabout. However that becomes a prelude of the rest detail as I then from there uncover more about the cosmos than that which is uncovered to this day. Mentioning the zero aspect about the graph not having zero by that fact alone it then changes nothing about the graph by own implication of zero bringing an change in influence of the reading information on the graph, but in the total overview of how I show how everything that changes altogether, then nothing remains unchanged.

When I say the Sun is freezing cold at the temperature of 6500° C, I sound like a loony that escaped from a dangerous asylum and yet, if my theory about gravity is correct, about which I have no doubt that even the smallest detail is correct, then the Sun is freezing cold at 6500° C. If I try to convey the theory about the Sun being cold without the support of the rest of my theory that no line can start at zero, that pi forms the curvature of space-time, that gravity is about condensing heat and mass has nothing to do with gravity applying, that gravity is the movement of space in space and through space and gravity has nothing to do with attraction, then without explaining all these issues, everything I say becomes the mumble of a lunatic, and yet, I challenge everyone to show me where I am mistaken even in one thought, but that can only be clearly understood when the total content supports every suggestion.

If I say that there is no mass found in the Universe and there is not even one shred of evidence to prove that the Universe supports the idea of mass, then everyone concerned with physics rebuffs me as if I have become criminally insane with dangerous tendencies to commit murder. Every reader so far that read this statement I make, when reading my work then in response to what I say threw down the book in disgust at page four without completing the rest of a seven hundred page book...and yet I challenge any person to prove where the Universe does support the factor of mass playing a part in physics! Moreover, I challenge anyone unconditionally to bring any proof that mass does produce gravity and then also show how that is done and indicate the system whereby that is done. Show how mass does bring about gravity as a pulling power! Show me where the proof is that mass is a force of attraction and I will retract every accusation I made towards Newton. I say Newton is wrong and he created mass where mass as a cosmic factor has no role to play in cosmic physics in any shape or form.

The smallest piece of dust in the Asteroid belt spins as fast or as slow as the giant Jupiter does in circling around the Sun and with that being true, then what role does Jupiter's mass play in its circling of the Sun? A giant elephant falls as fast or slow as a mouse does and mass does not play a part in the descending of either. This was what Galileo said and if Galileo is correct about mass not influencing the fall, then Newton is incorrect by putting mass down as the prime factor for being responsible for objects falling! It only shows that as far as cosmology goes, Jupiter and the dust particle has the same mass because they both show the same gravity influences on the attraction the Sun has on both particles and the mouse and the elephant fall down equally evenly. Where is the mass factor prominent in Jupiter's orbit around the Sun when Jupiter spins at the same pace as the tiny dust speck or where the mouse and the elephant falls alike? Gravity does not rely on mass but there are four phenomena working in tandem and in conjunction that forms gravity by measure of the square of Π. The existence or even the presence of the four phenomena are not even entirely recognised by Mainstream Science much less than they are understood by anyone at present holding any position in physics. How can I explain the value of the four phenomena and show how these phenomena interpret singularity without explaining that no line can ever start with zero or how can I implicate the importance of understanding the fact that no line can start at zero and all lines proceed to become a flat Π, if I am unable to extend this understanding to introduce the understanding in supporting of how the four principles form the measured compliment that is behind the movement which is forming gravity. The arguments form a woven blanket and by removing one shred, the blanket fall to ribbons.

I do show where the Universe starts with infinity and where to locate the starting point at infinity as much as I show that the Universe can never end because of eternity holding an absolute relevancy with infinity which then forms time. I show where to find infinity as much as I show precisely where to locate eternity. I show how to find that which can never start and where to locate that which can never end and how to read time forming space between these cosmic limits. I take your finger and place it on the two locations…that are how precise I indicate the positions of each of the limits but that can't be done without accepting that the graph cannot present a line of zero on the horizontal axis as a form of dividing the graph.

This I can only do when I am able to show where the coldest point is in the Universe and that that point is in the Sun as much as it is in all stars. I have to pinpoint singularity at an infinite position before I can explain gravity forming the Black Hole…and this has nothing to do with mass being a factor. In relation to that I then have to show where the hottest part of the Universe is and that I can only show by showing no line can start with zero because if zero is the starting factor of a line, then pi can't start space-time. I challenge any person to show which part of the overall concept that just this small book brings can be left out as that part does not support the rest of my concepts.

I challenge all persons concerned to prove that mass has any function in gravity and then to disprove my proof that gravity is formed by the interaction of the **Titius Bode law**, the **Roche limit**, the **Lagrangian positions** which in conjunction form the **Coanda Effect**, where the **Coanda Effect** is gravity applying. At present in view of Mainstream science policy, these phenomena are hidden so deep in Newtonian misunderstanding that most people never even heard of these phenomena being a part of cosmology and with that in mind it is very likely that most academics reading my work might never even have heard of these phenomena. Since Newtonians can't use Newton to explain the **Titius Bode law**, the **Roche limit**, the **Lagrangian positions** and the **Coanda Effect**, these phenomena are hidden so deep that it is very likely that you as the reader might never even know anything about these phenomena, albeit that they play such a most vital part in translating singularity into space-time.

1) I prove mathematically that everything in the entire Universe we know as a unit starts at a displacement value of 112 (which the displacement value is normally considered as the proton / neutron / electron number). I prove by using mathematics that at 112 the Universe goes dimensional and all above is flat.

2) At a displacement value of 112 it is also the point where singularity being the heat forming outer space (yes, outer space is formed from singularity also known as heat and not nothing as mainstream Science propagates at present) becomes six sided and double three-dimensional and for that there is a very good and well-defined reason.

3) At a displacement value of 112 the atom forms a relevancy unifying the proton, neutron and electron as a sustainable unit, at that point gravity becomes displaced as a liquid heat that we also think of as electricity and the only difference between electricity and gravity is the scale of generating taking place and this happens at 112. Also at 112 the Universe becomes a three-dimensional sphere, this I prove mathematically by not using mass. If you read this you will learn how I mathematically prove all this that I claim.

I prove mathematically that everything in the entire Universe as a unit holds absolute relevancy through singularity to everything else forming even the most insignificant detail in the Universe…but if I wish to achieve this statement and show what I say I show, then also every word in this book holds relevance to everything stated by every other word in this book…which brings me to the question I answer in this book and that is how is that achieved, where everything links by singularity to become one united Universe and to do that I cannot reduce any word forming this book because every word supports the rest of the book by linking the entire concept through every word supporting every other word. It is not conducted in the form you insist on, but just read the content, and when finishing it, then you asses which part were not worth your while and what part I should have dumped to suit your required conditional format. Read the content and then you decide what part could be chucked into the waste paper basket because in your opinion that part is old news to you, and is irrelevant to the rest of the information and is not critical when seen in the light as to how that information is not supporting the rest of the network of information.

The problem I face is getting a publisher willing to publish my work. It seems the only way I could achieve this destiny is to find a means whereby your publishing journal would accommodate my work. In that light I am willing to change the layout (not content) and style (not information or findings) in whatever way is necessary to accommodate your layout requirements. Should this not be achieved, about thirty years of my life's work will go wasted. Therefore, to find your satisfaction I would change to accommodate the required changes whatever style changes you would feel necessary changing, I would oblige.

I just can't show how the ideas flow into each other using twenty-something pages let alone change the essence of science. Normally the only thing I achieve is getting people scared because my ideas are so radical or get all persons resenting my views because I trash what all other persons deem as being holy or being Newton which t science is the same thing applying. My books should be read and that I cant achieve. I am sending you a copy of one of my manuscripts **An Open Letter On The Veracity Of Gravity** so that you can judge the difference there is between an entire book and a note such as this. I have no idea what criteria or format your publishing insist on and would greatly appreciate guidance as to change what ever changes are required in order to accommodate your publishing method.

I am of the opinion that if you do not read the book I present to you through the indicated web sites, and the book being which I wrote in a format representing an open letter, then you would be poorer for not doing so, but moreover, science in the history thereof would altogether be poorer for you not doing so. This book has not been edited by any linguist because I do not have the recourses or the required financial funding in paying for it, but if that is a hurdle in your opinion as it previously was with many of my other books I sent to academics in the past, then if that is how you feel, let it be a hurdle…but it does not take away any of the authenticity of newly introduced information that I put to view on the science table and it does not reduce the veracity of the information I divulge to science which was never yet before presented, linguistically unsupervised or not. Should you think I am boasting about importance of the new information that I bring to your table, then just go on and read the first hundred pages and see whether I am boasting or whether my warning about science missing a great opportunity when you decide not to finish and if I am just vain when warning that you lose out on your personal gain in science knowledge while a great opportunity to further science is going begging. After you have read a hundred pages, then you judge yourself if my saying all this that I am saying is truly boastful or is an honest assessment of reality?

Yours Truly,

Peet (Petrus. Stephanus. Jacobus.) Schutte.

The letter was much longer and showed much more information but since most of the content forms the substance of the book **The Absolute Relevancy Of Singularity in terms of Physics**, therefore I am not going to repeat the thirty or so pages in its full capacity, but I am only supplying the last few pages of my letter to Professor Dr. Ulrich Eckern that should give the trend of my approach I took.

This was the answer I received without Professor Dr. Ulrich Eckern even opening the envelopes that had my work written on to inspect what I wrote which I wished to have published. He did not even read as far as my introduction where I said I could prove and explain gravity by using the four cosmic phenomena no Newtonian ever could understand.

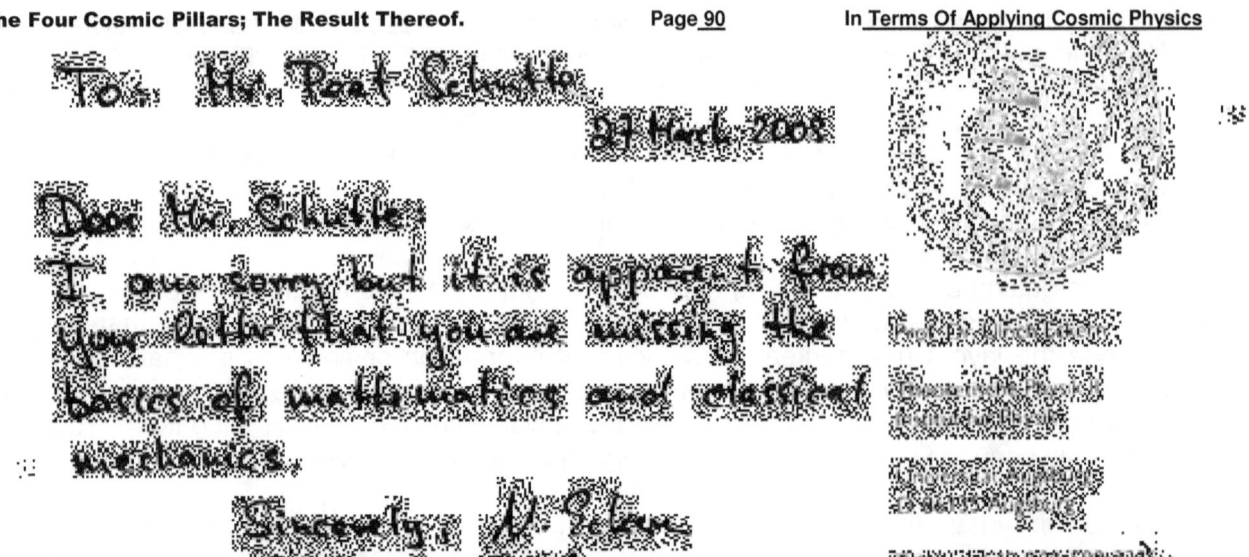

Again no one took the trouble in reading my work. Again my work was rejected without inspecting what I said, which I wished to have printed. Even the fact that I sent an accompanying letter explaining the issue did not initiate the courtesy of Professor Dr. Ulrich Eckern to read the entire letter. My intention was not to draw his attention to Newton's formula, but to read about my remedy I bring to the problem. My intention was to have him read that gravity equates to Π. I challenge all persons concerned to prove that mass has any function in gravity and then to disprove my proof that gravity is formed by the interaction of the **Titius Bode law**, the **Roche limit**, the **Lagrangian positions,** which in conjunction form the **Coanda Effect**, where the **Coanda Effect** is gravity applying. Professor Dr. Ulrich Eckern did not reach this far down when reading my letter. I think Professor Dr. Ulrich Eckern read two or maybe three pages, felt confronted that a mindless weasel such as I am in their eyes has the nerve to tell them they are cheating in science as well as mathematics. I think they detest the fact that someone as small as I could confront mathematical masters such as they are about what is legitimate in mathematical law and when mathematical laws are and to point out when those applying laws are either discarded or ignored. With their loathing of my ill judgement on their mastery they feel insulted and downgraded on a personal level. That will never be my intention but if you whish me to challenge you directly, then please prove that mass forms gravity by using the Newtonian formula $F = G \dfrac{M_1 M_2}{r^2}$ and I will swallow every word I have ever said after eating the kitchen zinc, just before crawling under the bed never to see the light of day again…but without them giving me the proof this will never happen.

In the past I have shone away from getting directly personal in my attack on academics that blew me away. I have complained about their attitude like a baby with a wet nappy, but never did I mention names. Not this time and no longer. Now I am getting personal and direct. I contacted Professor Dr. Ulrich Eckern with the purpose of him reading my work (and not only glance over the first two pages of the letter of introduction I sent with the work meant to be published) and having Professor Dr. Ulrich Eckern evaluate my work and not by merely glancing over the accompanying letter sent as my introduction. At this point and at this time I am going personal in my fight and I am confronting Professor Dr. Ulrich Eckern personally in print. In the past I took these remarks on the chin, absorbed the punch and went crying snot and tears without pointing fingers, but I got a little tired of taking a beating from those in high and mighty positions that are not responsible for any remark they make about lesser humans such as Professor Dr. Ulrich Eckern and others in his class see me to be. Since Professor Dr. Ulrich Eckern is not only connected to a University like all the others I have contacted in the past, but is the head or Editor in Chief of the most prestigious publishing journal on physics in the entire world, I am taking him to account for his remarks. I put a challenge to him in this book to prove what I say in this writing wrong. He did not even have the audacity to read my work, but got annoyed that a little nothing called Peet Schutte had the courage or mental instability (depending from which view point one look at the matter) to confront him on matters science had never been able to explain, much less than prove the facts. The Content of my letter very much condensed would read as follows although I do have the entire letter in detail I named after the man it was addressed too namely ***Annalen der Physics' Professor Dr. Ulrich Eckern***

I confronted Professor Dr. Ulrich Eckern on the issue of delivering proof about what Newton surmised when he claimed that $F = \dfrac{r^2}{M_1 M_2}$ could be equal to and therefore become $F \alpha \dfrac{M_1 M_2}{r^2}$ could be equal to and therefore become $F = \dfrac{M_1 M_2}{r^2}$ could be equal to and therefore become $F = G \dfrac{M_1 M_2}{r^2}$. The idea of proof comes automatically to the door of Newton although Newtonians will deny this fact as if they deny the honour of their Master Newton and that is what they have to do. Please, any Newtonian anywhere, just prove me wrong by replacing the symbols with numerical quantities of equal value and prove there is an equal mathematical outcome!

Think of what planets do... and you think that planets orbit. It is connected to the brain. No one thinks of planets spinning or planets basking in the summer Sun. When hearing about planets the first thing that comes to mind is the rotating of planets while circling around the Sun. No one ever thinks of planets colliding...but when taking Newton's formula $F = G \dfrac{M_1 M_2}{r^2}$ to the letter, the lot had to collide and vanish into the Sun ages ago. They should not be rotating when considering $F = G \dfrac{M_1 M_2}{r^2}$.

However, just using the term orbiting is in total defiance with Newton! Newton said gravity draws or pulls or moves in the direction that is towards... Using this expression would have one understand that the two objects in example the Sun and any of the various planets will be moving directly towards each other. That is what $F = G \dfrac{M_1 M_2}{r^2}$ implies...does it not? The term pulling does not suggest any circling because no one can be pulling towards and do that while circling. When pulling anything it must take place while using the shortest line possible. That serves the term pulling. Then the saying goes that planets orbit indicating they follow a circle. That is not what Newton said. However, wrong that may seem but circling is precisely what planets are doing in spite of what Newton said planets should be doing....

In conversation we speak of we say the planets are orbiting. If Newton was correct we should be speaking of the planets pulling, bumping, colliding, destroying, but talking about pulling without experiencing what should follow the pulling would be blatantly wrong according to the normal spoken word. Never do we refer to the planets pulling the Sun or the Sun pulling the planets, but we speak of seasons coming from orbital positions. Being in orbit has to neutralise the pulling and then cancel the pulling concept that also became culture.

If there was a pulling, and the word orbit cancels such an idea, then there has to be some sort of prevention taking place that disallows the pulling to commit the direction of travel. I know it is said that the orbiting object falls as fast as it circles and by falling while moving to the following side on position it never reaches the Sun, and yes, it makes sense, but there has to be some form of resistance replacing the planet in the next side position and preventing the falling or the pulling from taking place. However when one wants to adopt that line of thought it point to a circle and a circle is Π and I say gravity is Π which then proves me correct and still proves Newton wrong! Using the formula $F = G \dfrac{M_1 M_2}{r^2}$ as Newton provided, disallows any other concept other than moving directly towards. The person Newton got his ideas from and the work he raped completely, that of Johannes Kepler explained this very well, but Johannes Kepler makes no room for any pulling of any sort. In the work of Johannes Kepler he said that the space being the orbiting route a^3 remains at a specific distance k while the orbit T^2 takes place...and in all my other books that addresses more information I take Newton to task on his dismembering of Kepler's formula by corrupting Kepler's work and with what amounts to fraud, Newton takes science on a goose chase that holds no truth. There is no pulling by mass of mass in any way. This is what I wrote to Professor Dr. Ulrich Eckern and he replied only that he is Professor Dr. Ulrich Eckern and I don't understand the basis of mathematics and classical mechanics. Please Professor Dr. Ulrich Eckern or any other Professor Doctor, show me what it is that I don't understand and I will go away very quietly and never bother any one ever again...but until such time as that any one can prove the validity of $F = \dfrac{r^2}{M_1 M_2}$ could be equal to and therefore become $F \propto \dfrac{M_1 M_2}{r^2}$ could be equal to and therefore become $F = \dfrac{M_1 M_2}{r^2}$ could be equal to and therefore become $F = G \dfrac{M_1 M_2}{r^2}$ I will keep on pestering every one I can challenge.

However, this inspired me to write the article to **Annalen der Physics' Professor Dr. Ulrich Eckern,** which I hope Professor Dr. Ulrich Eckern would publish the content of the article.

Then as a result of this letter that I wrote that was in all fairness a much longer letter than this abbreviated letter I publish at this point, to Professor Dr. Ulrich Eckern, it also inspired me to commit myself to the task of writing four short books that are also very much abbreviated in relation to the printed books I hope to publish one day. The four to which I refer I am publishing on www.lulu.com The work I try to introduce has never been thought of where the books hold about more or less information sprout from the content to the followings effect. I hope to get funding from these e-books to help me publish the other books in a Publish-On–Demand from that then may be distributed through Barnes and Noble and Amazon.com.

The Absolute Relevancy of Singularity **called**
Book 1 Absolute Relevancy of Singularity in terms of Applying Physics

The Absolute Relevancy of Singularity in Explaining the Sound Barrier **called**
Book 2 Absolute Relevancy of Singularity in terms of The Sound Barrier

The Absolute Relevancy of Singularity explaining the Four Cosmic Phenomena **called**
Book 3 The Absolute Relevancy of Singularity in terms of The Four Cosmic Phenomena and

The Absolute Relevancy of Singularity used to explain The Cosmic Code **called**
Book 4 Absolute Relevancy of Singularity in terms of The Cosmic Code

Here and now and before the beginning of what this document may be to any potential reader, all parties reading take note that I state it emphatically that all members forming the community of science in physics judges me being not sufficiently educated and certainly not to the level where I am able to form any opinion on matters concerning Sir Isaac Newton or his physics. Any and all of my self-tutoring goes begging in their eyes notwithstanding and regardless of the fact that I did my private and individual studies by which I furthered my insight. In my own time and by my own discipline I studied and questioned what was known as to promote my cognisance and as to quench my thirst for knowledge and improve erudition. This was spawned by my personal interests in researching the issues in hand and through which method I then skipped the manipulating indoctrination of facts and the mind control of academics administered on unsuspecting students that results from professionally tutoring tainted physics where academics purposely place methodical mind control on students.

Their high standards are only furthering their self-promoting values, while my personal studies go unrecognised. They say they require education of unblemished standards being above suspicion. According to them the lack of my education contaminates my judgement. They apparently think I could not gain any insight on matters regarding physics. However, my skipping their methodical and systematic brainwashing by not entering a reputable institution as a student in physics and writing their examinations, and in doing such, circumventing their testing on their mind conditioning, enabled me to see and to express the incorrectness I saw in Newton's teachings.

That allowed me to show with clarity what destructive force Sir Isaac Newton released in order to corrupt the laws of mathematics, contaminating science along the way and mostly raping the work of a great man, Johannes Kepler and what Sir Isaac Newton did to derail the truth and disguise scientific correctness where such violation can only be expressed as being blatant criminal deception. What his deeds amount to, is to corrupt the laws of mathematics, to render the laws of cosmology useless and to rubbish all of science. Should you find this extravagantly unbelievable, then I am glad to announce that this book is written with you in mind and dedicated especially to your thinking and reading it with great care will be more to your advantage than most other persons, so go on and get on with the reading.

By your reading, you will learn what it is that those academics that are guarding science never wanted published and read by the public at large. I challenge anyone that disputes any claim I make to prove me wrong, but then go on by proving me wrong and do not merely make unfounded suggestions in that direction... What I say is don't run and hide from my attack and coward away from my confrontation as so many of the most intellectuals amongst the Physics Paternity did when I confronted their thinking. If you don't read what I wrote, it is the same as you putting your tail between your legs and start running.

On every occasion where I confronted members of the Academic Paternity in the past, those I confronted acted in precisely such a manner, such as cowardly ending all reading by throwing the book down, and then pretending to show the utmost disgust in what I say. Now show your academic worth and your educated dignity and accept the challenge I make to you and to all of your kind: I challenge one and all: **PROVE ME INCORRECT IN ANYTHING I SAY!**

What is it about gravity that I say which no one wants to know? No one wants to listen to my point because I call Newton a cheat. He defrauded science and took all the other suckers running after him like sheep that are / were unable to think by there own ability. Now no one wants to find out how stupid the entire lot was that came after Newton and followed in his misguided footsteps. No one wants to know that Newton went on lying for almost four hundred years. No one in science clearly distinguishes between gravity and mass and everyone in science tries to confuse the two issues by making them one and the same. They are two distinct issues never to be confused.

First you should decide what belongs to the gravity factor and what forms part of mass. Newtonian science is of the opinion that when a body is floating up in outer space the body has micro gravity...that just can't be the case. Newtonian scientists confuse the factors being responsible for mass and for gravity because if not, then please explain which is gravity, the part that tries to move the body to the centre of the Earth, or the preventing thereof? We have to see that mass is created by the pushing of an object onto the Earth and from the pushing (not pulling) comes mass while gravity is what is doing the pushing. While resisting further movement mass comes into the picture and while moving towards the Earth or intending to move towards the centre of the Earth, that movement constitutes as gravity while stopping the movement leaves the object with having mass. Mainstream science loves to confuse the two issues because Mainstream science love to confuse everyone because Mainstream science is completely confused about the science they say they are the Masters of.

Is gravity that factor, which makes all bodies fall to the centre of the Earth, or is gravity that which prevents the further moving of bodies having gravity to fall further down to the centre of the Earth and then by restricting the movement, then forms weight or mass? By restricting movement towards the Earth a mass factor comes about which gives weight! It is presumed that the body has micro gravity because the body is weightless I outer space. This prompts me to ask the question underlying what has never been decided… what is gravity and what is mass. A body floating in outer space has maximum movement because when it moves slower, it starts to fall to the Earth.

At that point the mass (measured as weight / kg) is indefinably small while the movement is maximum in maintaining orbit. However, that is speed measured by distance (meters) travelled in time (seconds). Mass has a value, which is measured in the same currency in which weight is, and then mass is weighed as much as weight is and therefore, undeniably and in contrast to the logic of mainstream science's confusion and frenzy trying to confuse what can't confuse any further, mass and weight is connected as the same thing while gravity is movement notwithstanding mainstream science trying to put mass and weight far apart. If mass was equal to movement as gravity is, then mass must be measured in meters / second. Instead mass has the value which is the same as weight which is measured in grams.

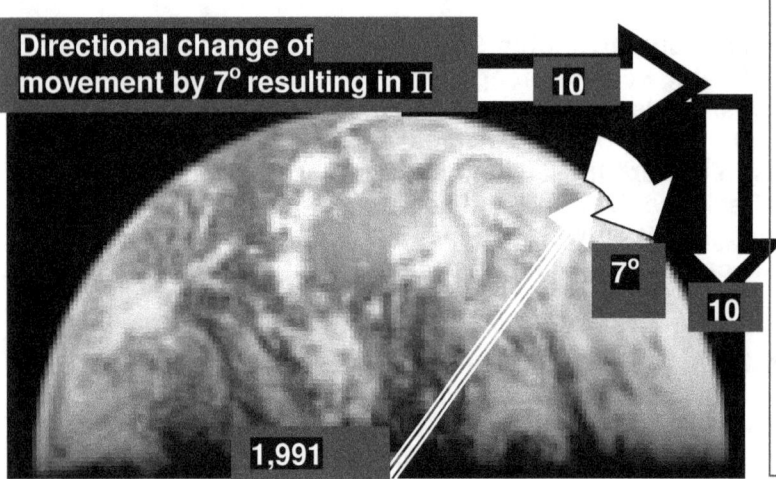

I say gravity is the formulation of Π. Gravity is Π forming singularity in the third dimension. The falling of all orbiting objects results in a directional change that the Earth forms by implementing Π as the gravitational outcome of movement of objects in the Universe. The value of Π being $21.991 \div 7$ has its origins from the redirection of movement that gravity produces. Gravity is the forming of Π.

By orbiting at a specific distance, the distance from the Earth is determined by the rotational speed the object encounters. When the object reduces the orbital rotation (circular velocity), the gravity by slowing down will bring the object to start moving towards the Earth, which is falling and which is what everyone knows is to be gravity.

That puts the falling of the object completely in relation to the speed that the object holds and that places gravity by falling in direct relation to gravity by orbiting.

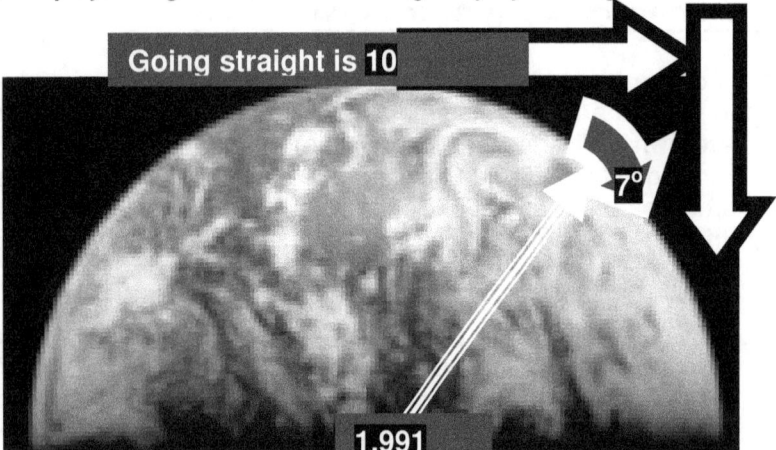

A falling body descends towards the Earth by a directional by 10. Te earth moves forward by 10. The Earth redirects the movement by spinning as it implements 7°, which is the circumference of the bending of the surface of the Earth. Singularity on both sides represents 1.991. In that the gravity in which the body falls is Π

When a body falls there is no mass involved because all objects fall equal and this was accepted long before Newton started fantasizing about his mass involvement in gravity applying. The distance the object orbits measured from the centre of the Earth and the orbit circle holds a direct link to the speed or time in relation to space that the object rotates. If the speed in revolving declines, then the orbit circle declines and this reduces the distance the orbit circle is from the Earth centre or the diameter. The orbit circle is directly associated with the distance the orbit takes place measured from the centre of the Earth in a ratio of time taken versus space travelled through.

The content of my work contain a new view about Cosmology, which I have been working on for the past twenty-seven years and exclusively for the past six years. To give you a little insight into my work, I shall mention the following: I came to realize that lines mathematically couldn't start at zero because there is no evidence of zero as a factor in mathematics. Should you disagree with my statement the question in need of answering is this: What will the length of the shortest hypothetical line imaginable be and moreover, what would the total overall length be in that case?

The shortest possible line (hypothetically) must be so short it must have an initial and ultimate point sharing the same spot. If it used zero as a start, the zero part would not count, because the line will only start at a point past zero where the line then will start forming an infinitely small dot. The dot is in infinity, however small, it is not zero. Zero ultimately means not existing and then that point, as a start does not exist. The smallest line has a beginning and an end at the very same spot located in infinity, and infinity may be beyond human scope, though infinity is still not zero. Infinity may constitute of something we do not yet understand, but we may not define our human misunderstanding as nothing. In this aspect lies the difference there is between arithmetic and mathematical science where arithmetic can have position such as zero since arithmetic excludes the cosmos calculating numbers only.

A man may have that many oxen or so many sheep and even this amount of wives, (in Africa) or not have any therefore having then a total of nothing, but there cannot be nothing between the Sun and its orbiting structures. The having and have-nots are part of arithmetic. Light will indicate a line flowing between the Sun and whatever planet, following dot after dot thereby proving the existing of the possibility of something going about by a straight line, and any straight line in relation to other straight lines will be under the law of Pythagoras. There is no possibility of a straight line not forming in space. Mathematics converts the values of integrating lines according to Pythagoras and arithmetic is about numbers to be added or subtracted.

By mathematically excluding zero from cosmology a new Universe opens to the human mind. For instance the distance between the Sun and Pluto is roughly one hundred times more than the distance between Mercury and the Sun, but both planets mentioned have a vacuum filled with nothing except one atom hear and there occupying the vacuum between them and the Sun. If space supposedly comprises of nothing how can nothing then become plural forming more or be multiplied by a number as to indicate a growth in something not even existing. As the one becomes one hundred the one cannot substitute a value of nothing but then must be part of something.

If the one substituted the nothing, all laws of mathematics will go in disarray because when one multiply any number by zero it becomes zero placing both planets in the Sun. By excluding nothing from the equation space becomes something bringing in a value lying inside the realms of the infinite that must form singularity. As the zero becomes a dot, something else becomes clear about the dot. Looking at the night sky we find darkness overwhelming the space in relation to the stars bringing across light. We can detect the dot because we cannot see darkness since our eyes were only meant to cope with light. With this knowledge, then how can we see the sky as darkness at night? We are only supposed to see the light of the stars and not darkness, yet at night we see a much wider picture than stars alone. One may bring in the argument that the blind see nothing but darkness. We seeing persons do not know what the blind does not see, so we presume it is just about the same darkness, but that is presuming.

When we see a red flower, science knows the flower being all the colours but the red it rejects and this we all know. Therefore the dot we see as darkness also must be light, withholding its light and giving us the darkness we see as light…But the dot must influence the surrounding as well, subtracting the light it claims from the surrounding by casting it as darkness. In the case of stars we see the light the star disassociate itself with, keeping the darkness it has as it pours all the light in excess into the darkness which evidently is then light. From that one may conclude there should be two forms of singularity where one associate with a dark dot being light, and another being matter with flowing light evidently proving to be the dark one. Proving the dot with many such arguments was easy. Naming the dot and its position, value location and proving the influences mathematically was much more complicated and proving the dot has a definite influence on the surroundings was at first seemingly impossible, yet it is done The definitely defined and underlined value of the dot becomes of utmost importance when finding solutions to cosmic factors not yet clearly defined.

My approach might seem unconventional but through the abandoning of the accepted, it enabled me in locating the precise location of a universal singularity forming a connecting basis of the Universe (this I say with some degree of confidence). The smallest figure there can be must be a dot. The only mathematically sensible option about extending a line from the dot will be non-bias progress in all directions equally in order to give a meaningful flow of mathematical equilibrium. The Pythagoras mathematical principle is the proof and that I explain. The obtaining of singularity is in my rejecting of nothing by replacing it with something being the dot.

When observing the connection between the half circle, the straight line and the triangle, be 180^0 and this line of thought has to bring the mind the beginning, the very origins of mathematics. Why and how would a line, a half circle and a triangle be $180°$? There is no form equality bonding the three different shapes! It is these questions that bring answers and not the very spectacular breathtaking, mind-boggling and all impressive mathematic formulas that bring the solutions. When the cosmos started it employed mathematics with which to start both a Universe as well as mathematics. It is there we have to look for the answers.

The Law of Pythagoras is about angles in relation to lines and not one angle that can represent zero because that will reduce all the lines also to zero. The measure of angles between stars at a distance uses parsec as the indicator, but the parsec between the stars indicating an angle has to represent an angle whereby one may measure distance and such a distance cannot be filled with lots and lots of zero because then the parsec will be equal to zero. Again it is multiplying the factor with the measure but if the measure is about a factor of zero, then the factor too becomes zero. That is as basic mathematics as I can present. By relieving zero and replacing zero with singularity the entirety called the Universe becomes clear.

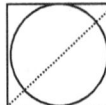 **The value of singularity stems directly from the law of Pythagoras or Pythagoras is the result of the average of singularity. With the shortest line being a dot, all lines must start from a position implicating Π.**

A circle is a square without corners implementing Π and a half circle is therefore a triangle without corners. The corners are the factor that confused every one in the past. When replacing the value we normally attach to circle being r with Π, the law of Pythagoras becomes quite meaningful and mathematical.

By placing a connecting circle on the sides of the triangle half a circle forms. By implicating Π as a relevancy and not the straight-line r, two values of Π applies to each circle and the straight line is no longer r, but is Π^2. This will bring about that each circle holds half the square value implicated to the allocated conditions applying to Π in that specific instance.

By adding the two half squares forming the two half circles and then calculating the square root of the total that then forms the average diameter, an average of Π in the connecting line will come about. As both lines are the straight line forming singularity coming from one line being Π, the connecting line then must be the average of the two lines as Π^2. That is what **the law of Pythagoras says. Gravity is the result of the Earth spinning around its axis as well as around the axis of the Sun and the dimensional change implicates the law of Pythagoras.**

Because every moving line represents one quarter of the sphere in relation to the rest of the sphere and the line also indicate the relevant position between the point indicated and the point in the centre it is a relevancy of singularity in progress. By connecting the line, as Pythagoras will suggest the singularity within the sphere become a specific value indicated representing one half circle.

Gravity comes about as a result of the Earth turning in space and with that it pulls objects from space towards and onto the Earth. This is done by the duplication of the law of Pythagoras.

This has to do with speed or movement and applies to all objects equally holding no specific relation to size or mass. It is a relation between the orbit circle (circumference) and the distance from the Earth centre (circle radius) and if that is the case, then gravity forms by Π having some sort of involvement and that throws any idea of mass playing a part in forming gravity out of the window where I hope it takes all of Newton's ideas of mass-forming-gravity with when going out the window. In forming gravity the centre line (diameter) holds a specific value to the orbit (circle) and with that being the case then we have to search for the part Π plays in the function gravity has and when doing that we can leave mass out of the frame because big or small, all things fall equal. Galileo was the one that proved that.

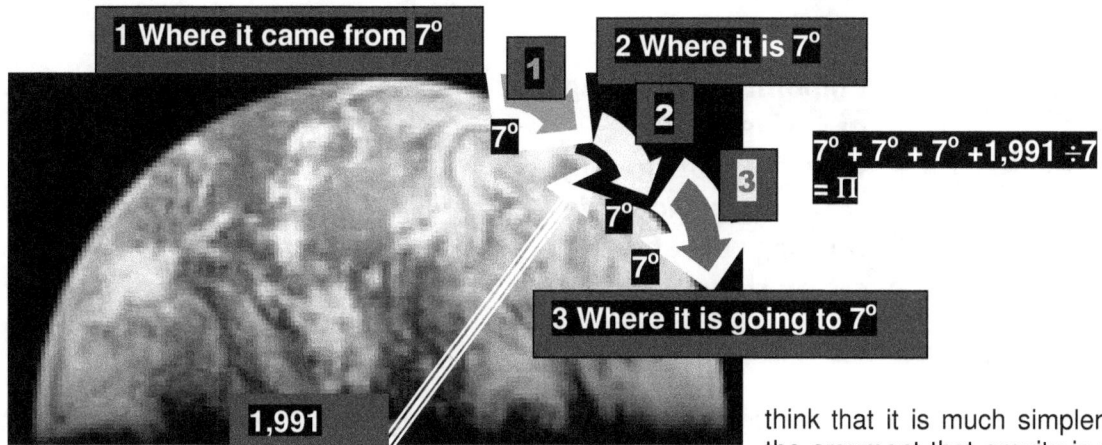

So you think that it is much simpler to maintain the argument that gravity is the force of mass pulling that is pushing the object onto the Earth only when the object moves at the same pace as the Earth rotates…and only then does the object finds mass or weight! It is mass giving gravity giving pulling forces and it all is that simple to understand.

So you still think that explaining gravity remains as simple as putting gravity in a connotation with a force fed to measure by having mass attracting whatever is attracted and this then allows the simplicity of the Newtonian concept to deal with the confusing part of the entire issue!

In the circle we may locate a straight line by reducing r that is symbolising the radius of any circle where such reducing will be indefinitely to the tune of halving r each time, then r would become infinitely small, even beyond human calculating means and become not a line, but a dot. However as mentioned in the case of the smallest dot holding one spot, r would become insignificant beyond human comprehension even, but never reaching zero and still Π would remain intact and dictating form. I believe one can begin too see where my suspicions are heading because the flaw comes about in the manner mathematics are practised for thousands of years. The radius represents the initial line the first that ever was. The line will eventually become r^0 at a point where the line began as a spot that grew into one dot and the dots added eventually to become a line. Finding this line made up of dots is most important when trying to decipher Kepler's formula.

Let us find the smallest possible line first. We already have reached the conclusion that by reducing the line, the reduced line will eventually leave all sides on the same spot. Such a spot must be round in form since it still holds Π as a factor next to r^0. We now are entering the domain of singularity where the visible is no longer traceable and only intellect can bring understanding of the scenario. With the line being the **Singularity** Π^0 smallest line, such a line will start off as a dot that moved away from a spot. With all possible sides being in precisely the same spot we have all possible sides onto one spot. I chose to differentiate the dot and the spot by giving the spot a value of Π^0 while the dot holds Π next to r^0.

Mathematically the spot is placing even form being Π in the single dimension Π^0 where the space is one (1) and holding exponentially zero (1^0). There the space moved over to form the spot Π^0 to the dot Πr^0 forming a circle as a dot. We now are reaching into areas only the human mind can venture by understanding and seeing nothing more than with the eye of intelligence. The understanding of this concept demands our reaching the point where the mind of the animal cannot reach. If it starts with a line where that line only represents two sides being one and as such that is rather a flat Universe. At the dot Π we have roundness because we have Πr^0 while at the spot is not yet round Π^0 because being round are requiring a shape or form and this lies beyond or before space at a time when any form of shape came into the cosmos scenario.

This was in place at a time in a period where shape and form was a part of the distant future hidden in and beyond the developing eternity. The spot is located at a point that entering the spot is also at the same time crossing the spot and landing on the other side where the radius becomes the diameter. It serves us well to realise that the entire Universe was that small at a point where everything started forming because the spot that developed into the dot is still with every spinning circle. With the spot becoming a dot, there must have been a time when everything in the entire Universe was that big and growing in relevance.

Pinpoint positioning of singularity Π^0 with Π positioning space to either side forming the border set by singularity

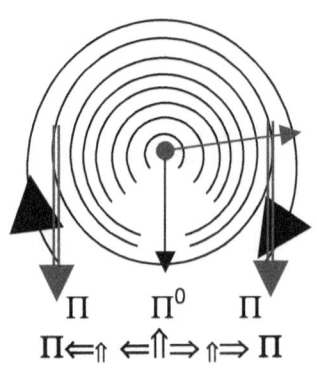

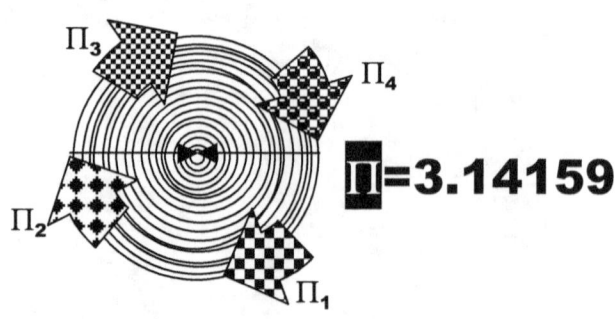

The reducing of the line is one dimension in six and although such reducing is representative of two indicators all the other indicators must still be accounted for two. Therefore the ring or circle is the only way to include all six sides in one aspect. In mathematics there is the formula used in calculating the volumetric inside of the sphere is $a^3 = 4/3\Pi r^3$ which holds two major components that will establish final value where as the rest is indicating ratios. In mathematics there is a line being one quantity and the circle indicator Π being the next circle indicator.

Reducing the line will erode the value of Π by ratio. That will eventually lead to having a circle ratio of Πr^2 and eventually lead to Πr^0 but that is not the point where the circle ends. That is where the ratio applying factor ends but it cannot exclude the circle. The circle as a concept can still reduce when it abolishes form to the single dimension. It is not the radius that is responsible for the circle but the figure value of pi and by abandoning π only then does all the aspects fall back into the single dimension.

$a^3 = (4\Pi\ r^3)/\ 3$

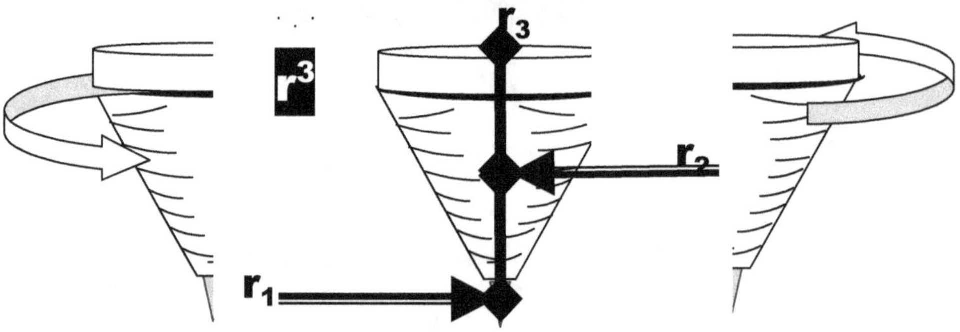

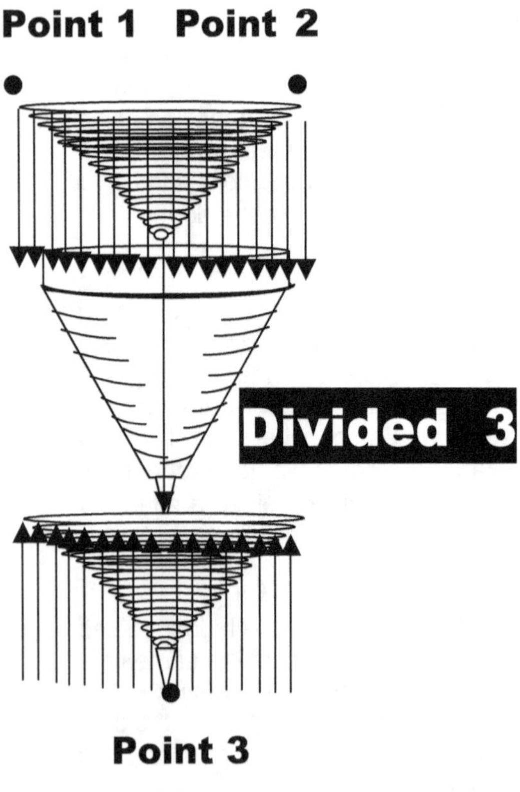

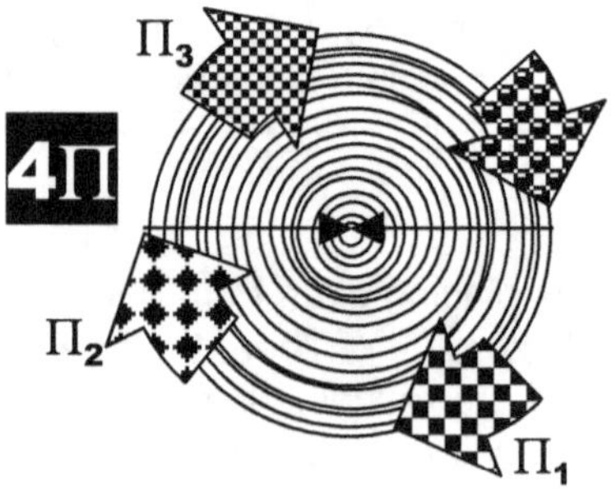

The Four Cosmic Pillars; The Result Thereof.   Page 99   In Terms Of Applying Cosmic Physics

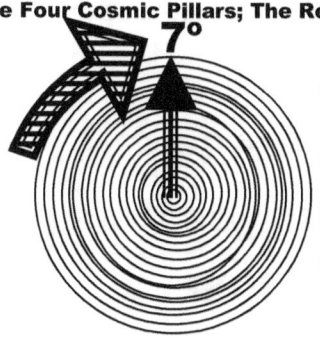

The curving of a circle is 7° and that gives the circle curvature.

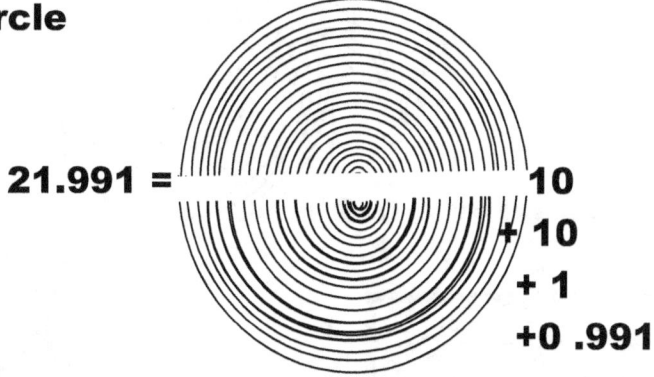

$$21.991 = \begin{array}{r} 10 \\ +10 \\ +1 \\ +0.991 \end{array}$$

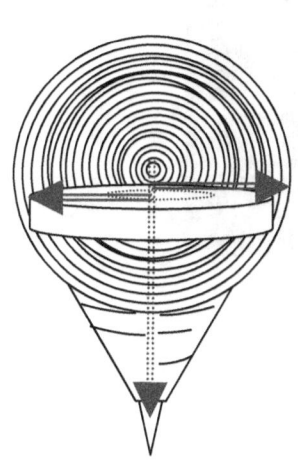

The circle can reduce one step more when the circle eliminated r completely by returning r to a point of singularity r^0, but the elimination of r as the factor reduced the major factor to the single dimension in Π^0. That will not reduce the cosmos to zero, but it will only eliminate all potential lines r^0 to potential circles $\Pi^0 r^0$ and from there the circle Πr^0 will come about by manifesting as a line but that manifesting can firstly only establish a circle Πr^2. The only value that singularity can have although the single dimension may host the entire universe is Π^0. Pick a number and elevate it to the power of zero and in the process one may have established another point holding all points in singularity because that is the value of singularity. Only Π^0 or any other value holding zero as an exponential value can ever be the accurate value to singularity while singularity will then host the rest of all the possibilities in the Universe.

The first value there ever was came in the form of Π. Where mathematics was still an idea in development the universe granted values of the triangle being 3 circles as Π^3, which was 180° and Π^2 which was half a circle also with the value of 180° and finally the straight line also being 180°. Mathematics was not yet established, but the most basic came about and this information is deductible from the most basic form mathematics hold. By going back as far as the Big Bang, Science is not taking the cosmos back as far as possible, they are taking the calculation of mathematics back as far as they can go but mathematics does not go all the way. Mathematics presented as numbers and symbols only became valid (as did all other aspects) later on in development. But the most basic of mathematics was in place when the spot moved on to form the dot by going from Π^0 to Π.

Going across the circle to the very centre the circle would come about from a straight line r growing influencing the appreciation of Π, but to influence Π would lead to a breakdown in r as Π and r are different entities. Looking at the affect of gravity it shows the precise quality of no distinctive point, as gravity never seems to end at a point but flows all over affecting all that holds a position in its sphere of influence. The gravity coming from China meets the gravity coming from America at no particular spot but intermingles without distinction.

The curving of a circle is 7° and that gives the circle curvature.

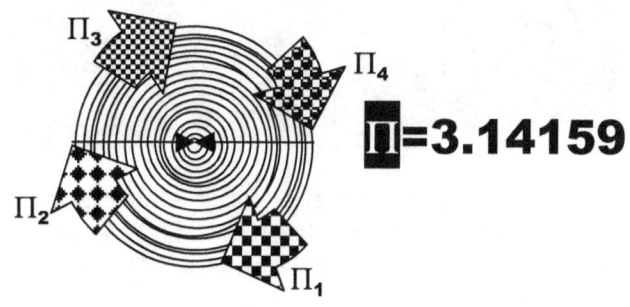

$\Pi=3.14159$

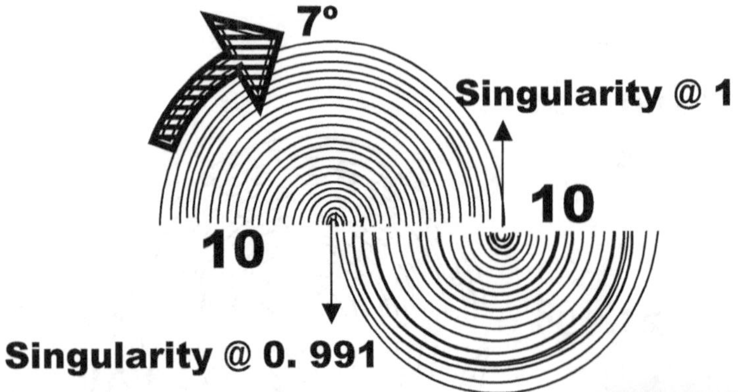

The Four Cosmic Pillars; The Result Thereof.

In Terms Of Applying Cosmic Physics

7°
Singularity @ 1
10
10
Singularity @ 0.991

$21.991 \div 7 =$
$\Pi = 3.14159$
$=$ Gravity or time progressing

5 5
1° + 0.991
7
5 5

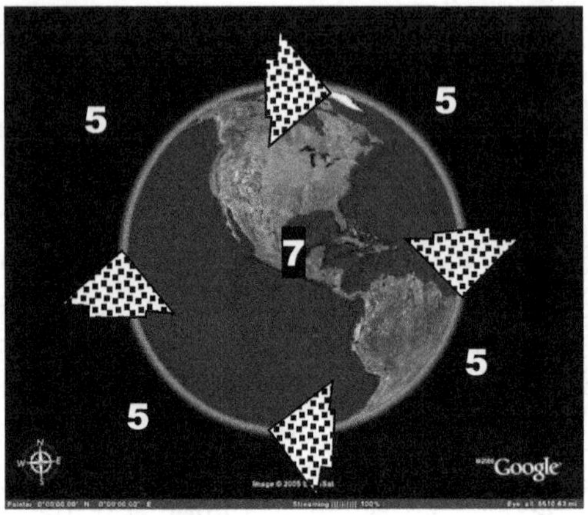

5 5
7
5 5

$\Pi = 3.14159 \times 7$
$= 21.991$

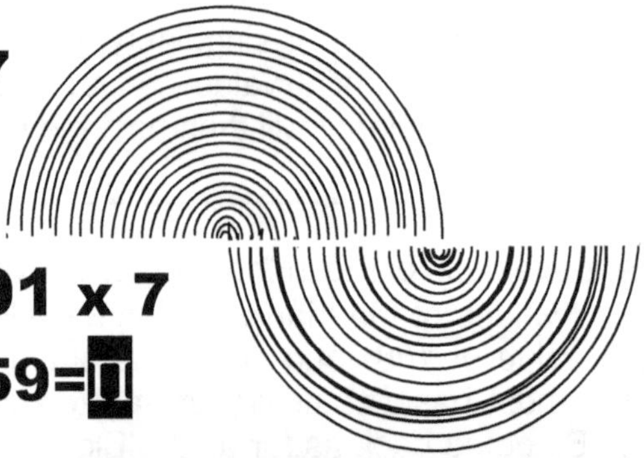

$= 21.991 \times 7$
$= 3.14159 = \Pi$

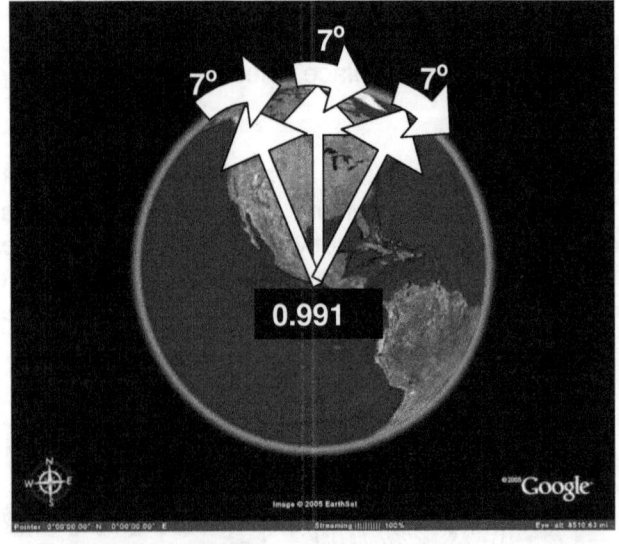

7° 7° 7°
0.991

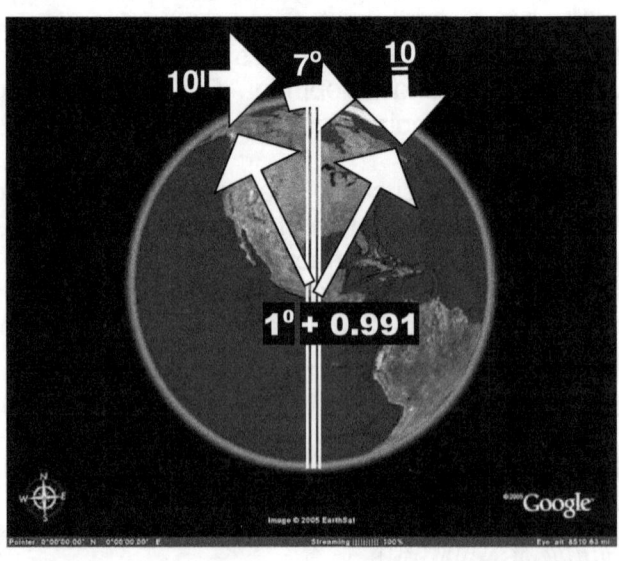

10¹ 7° 10
1° + 0.991

The circle movement gravity solicit is always formed by the double value of Π and that makes gravity Π^2 and when that is put in relation to the relevancy Π coming about in the 3 dimensional sector it forms Π^3. On the one side Π forms in relation to Π^0 ($\Pi \times \Pi = \Pi^2$) and on the other side the relevancy forms in relation to $\Pi = \Pi^3 \div \Pi^2$

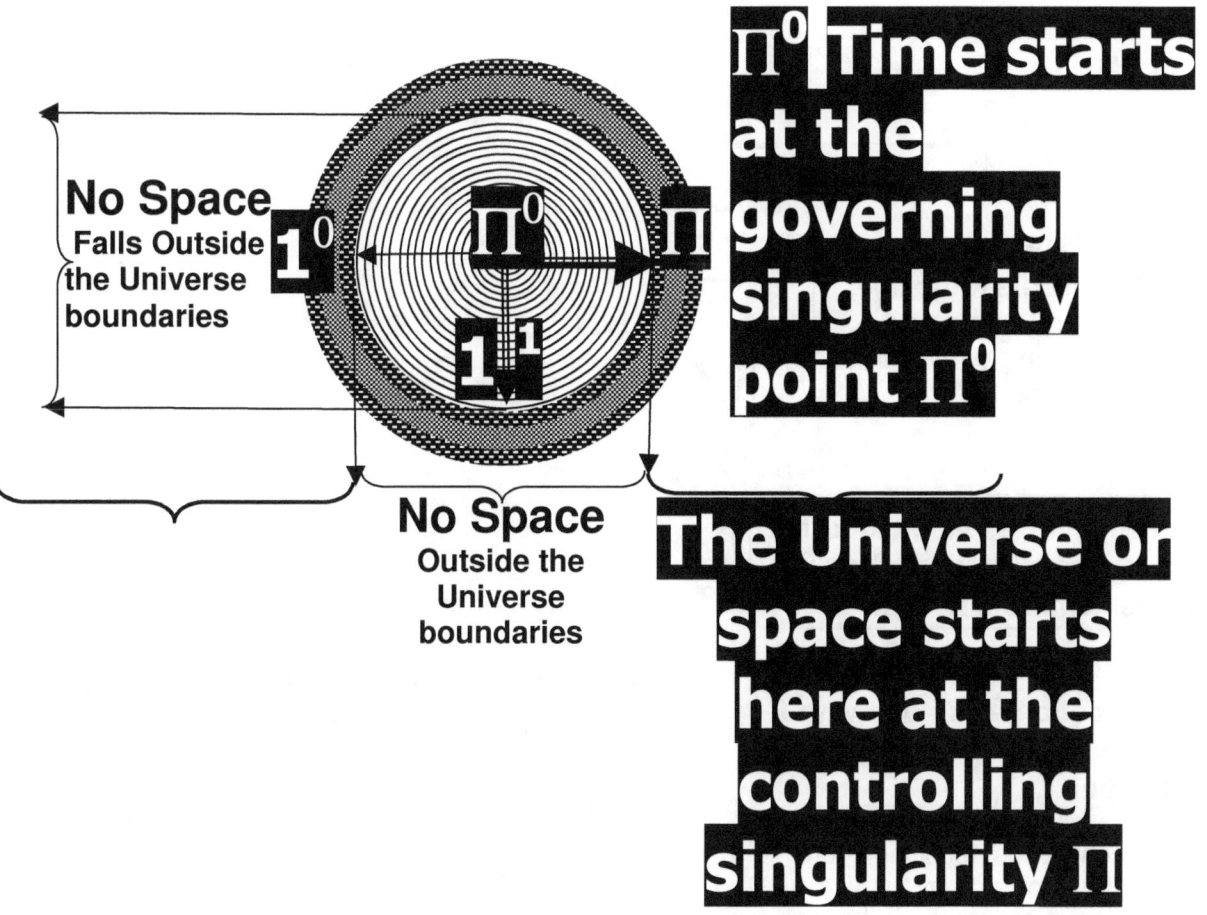

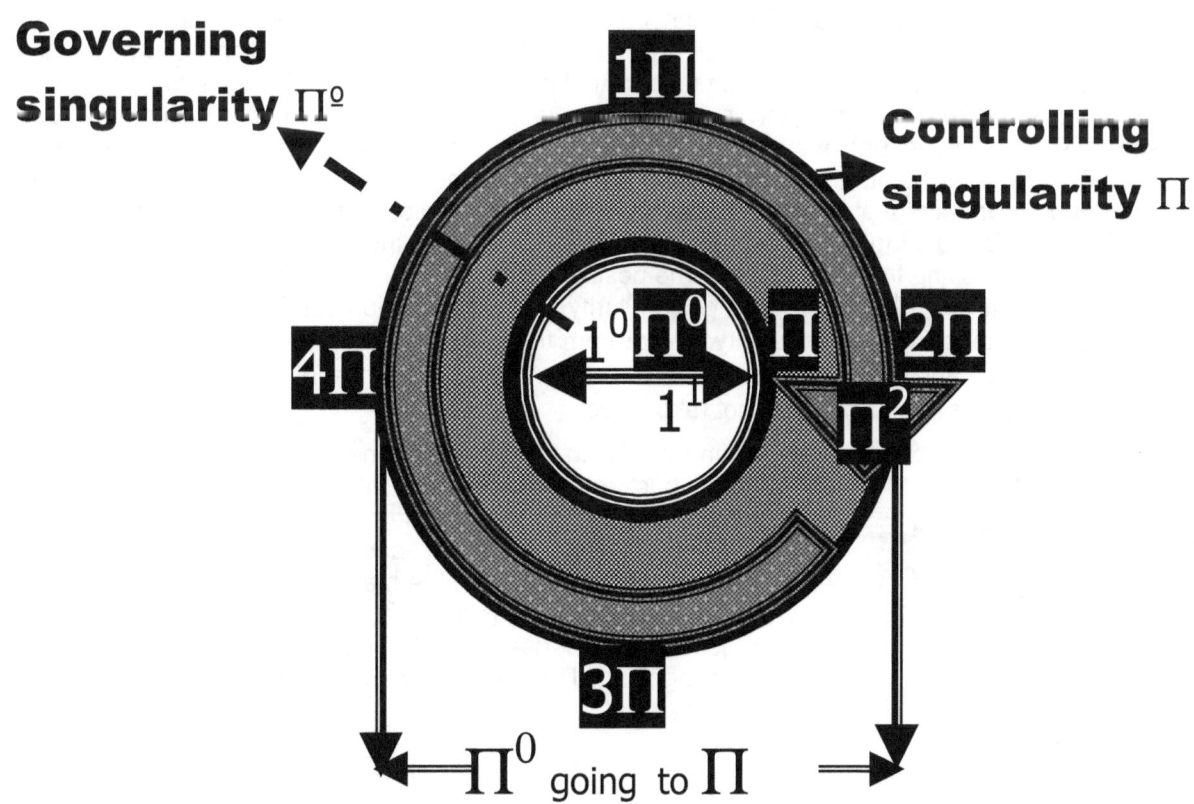

Let's go back once more and reduce the line by half every time. Then repeat the process until it can repeat no more. The reducing of the line by half every time will get to a point where all the ends land on the same position without any possibility if halving the two ends further. The points share one position and moving the points in any direction will lead too an increase of the line once more. This is one of the parts that is most important when understanding the theory about the absolute relevancy of singularity.

The Four Cosmic Pillars; The Result Thereof. Page 102   In Terms Of Applying Cosmic Physics

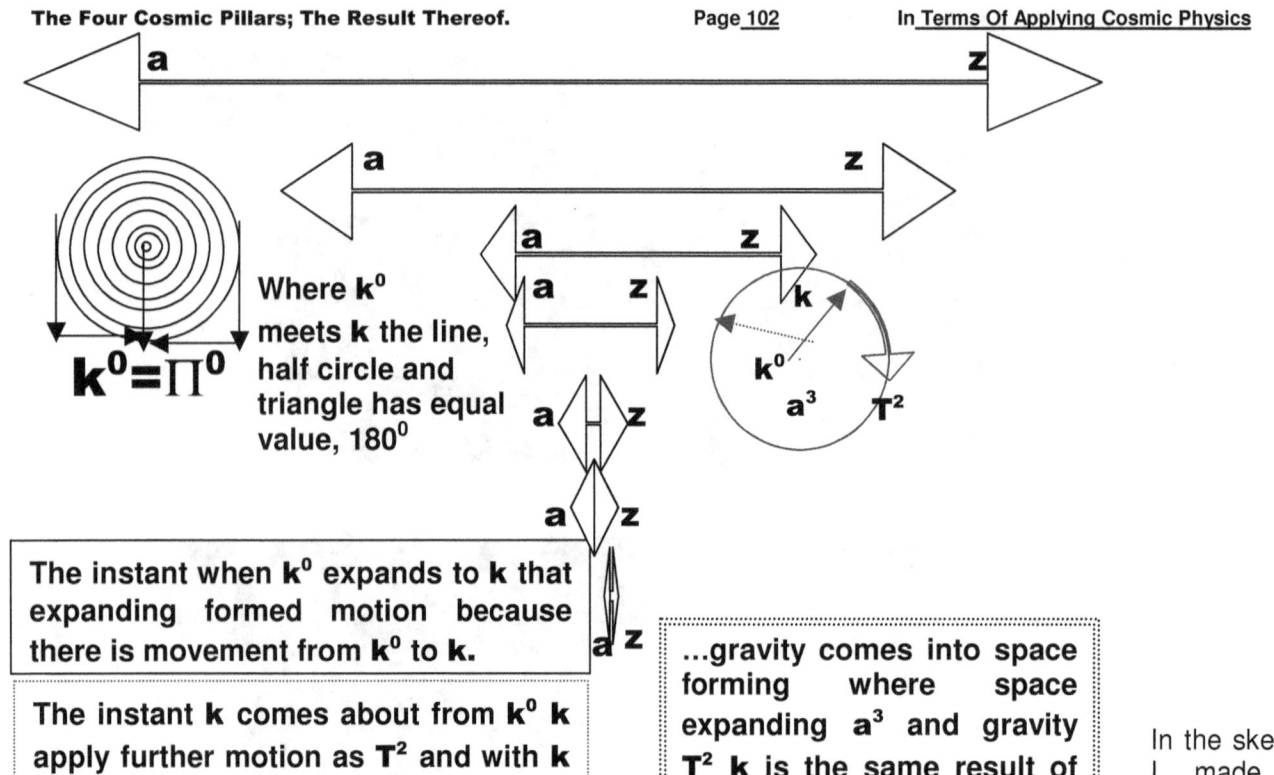

Where k^0 meets k the line, half circle and triangle has equal value, 180^0

$k^0 = \Pi^0$

The instant when k^0 expands to k that expanding formed motion because there is movement from k^0 to k.

The instant k comes about from k^0 k apply further motion as T^2 and with k producing motion by expansion and T^2 by contraction...

...gravity comes into space forming where space expanding a^3 and gravity T^2 k is the same result of singularity k^0 setting motion

In the sketch I made it aims to show below each of the lines space left symbolising two ends will

that with the continuous reducing there is a open between the two ends of the line that is the end of the line in reducing. In the end the share one location even by having one single point holding each one. There is no chance that I can present any sketch reducing the line to a point where the points are sharing one location literally in the single dimension. The points are there and with the points being present they may not be dismissed as nothing. From there no reducing in a natural manner can lead to nothing without changing the rules of mathematics in such reducing. But the two ends has reached a position where any further effort of reducing must bring about the start of extending because every point possible share space with every other possible point at the point of singularity where all points share one common space. By moving any of the points such moving must then bring about an increase of space once more. This also applies to the circle because the circle uses a line to indicate size running from a centre to an edge. By reducing the line and by reducing the circle the reducing will end up having the ends in the same position in the very centre of the circle. It is this fact of the moving of any point from that spot holding singularity that such motion will introduce space as the space exceeds the previous limits of singularity. What I am trying to say is by moving from the spot Π^0 to the dot Πr^0, such movement evoke Π^0 and without the movement of the spot Πr^0 to the dot Πr^0 the allocating or positioning of singularity Π^0 will not take place.

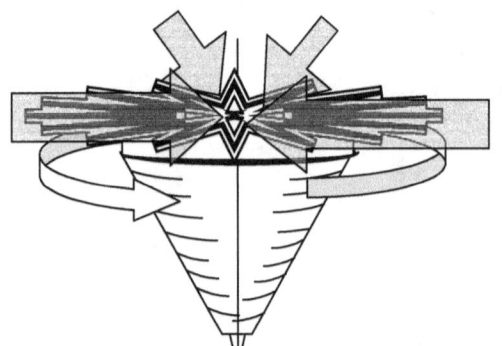

To find validity in my argument one must draw this statement of motion back to the point where singularity is getting sides. When there is singularity there can be no sides. The one forming singularity by measure fills a space. The space that even the dot fills does not really exist in the manner we humans see space to exist. It is a spot that is there without being there. It does not visually exist because it is not filling any substance and it cannot be recognised. The spot and the dot have no dimensional worth of any measure.

It is the point within the Universe I have named as **Infinity** where nothing can go smaller and anything within that point can never reduce. That point is where the entirety called the Universe begins and where everything holding substance begins.

Once one accepts the fact of singularity being present in that location, that accepting of singularity then is contradicting all the things we know and we can measure and we recognise that point being present by merit of the fact that the point referred too is not being formed by any of the things we can recognise. It is made up of everything we don't know and constitutes of everything we are unable to recognise.

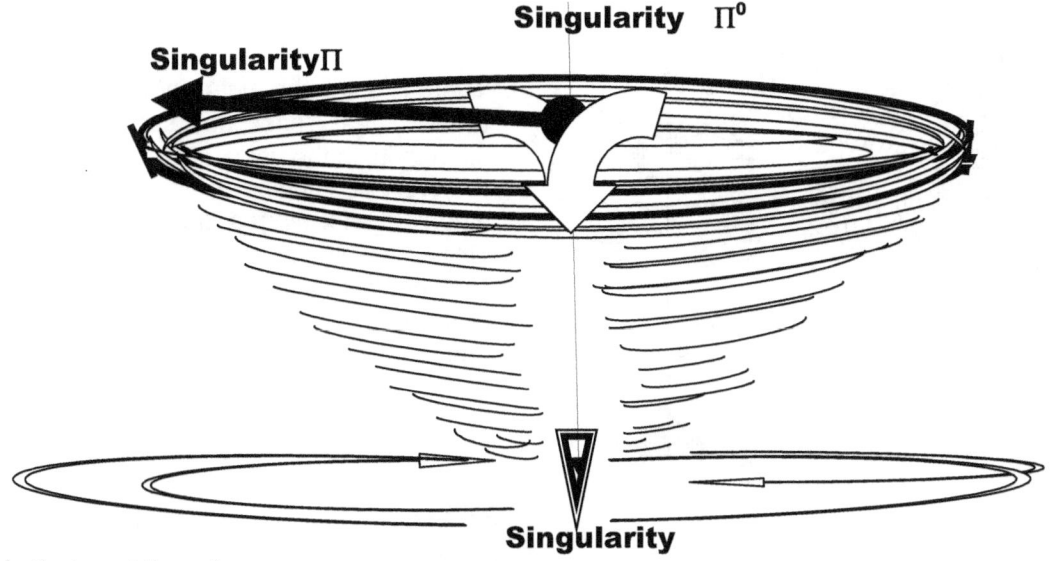

In that spot there is no space.

In that time when only form was in place and when the triangle, the half circle as well as the straight line was equal 180°, the line must have been so small it had reached a point not yet mathematically dividable in any way. The dot that formed was so small during that time that if any further dividing took place such dividing would have brought growth because there then would form space between the sides going in the opposite direction. That is the manner in which the spot Π^0 grows into the dot Πr^0. However it is important to realise that anywhere we might locate Π^0 we also locate 1^0 because Π^0 is 1^0. The dividing brings all there is having all sides moving literally on the precise same spot, and I have located singularity in just such a spot.

I came to the conclusion that the spot I found had to be singularity purely on the grounds that that spot holds only one side to serve as a start to the starting point of all directions possible. In that side is only one spot where there is only one side applicable and one dimension present. With all the factors given one can only come to one conclusion and that is that at that spot in the centre of all spinning circles there can be only singularity. In such a case more dividing of the radius by the continuing halving of the radius as dividing by two applied, such continuing of dividing will land further positions on the other side of the spot being divided. In that spot space ended. That point is serving as a position for all possible points and cannot allow further dividing as it is in the smallest line or spot there may ever be. In the very centre of any and all circles spinning we find this point holding no space and therefore forming Π^0, which is 1, which is singularity.

This spot is the result of a most basic process of reduction as the Hubble constant is a most basic process of expanding during a matter of time. By reducing the line constantly the only value that will eventually remain without dispute from any party arguing about the facts is one followed by an exponential zero. By only having exponential zero instead of a numerical zero and a radius as one in the square (the radius effectively becomes one holding any and all sides on one point) such a point might become any value of any significant measure implicating anything but zero as the radius. By expanding the line, it will be an evenly spaced structure growing into the most perfect round dot ever possible anywhere at the point when it starts to grow.

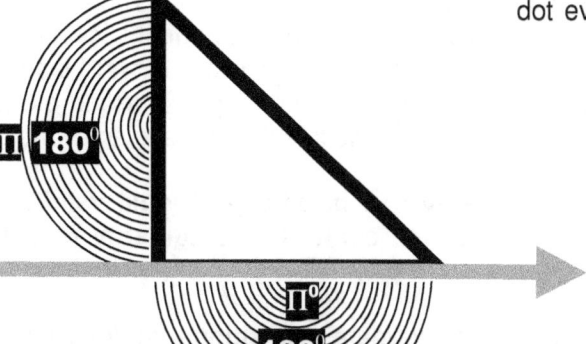

In the sketch below the circle to the right would come about from a straight line r growing influencing the appreciation of Π, but to influence Π would lead to a breakdown in r as Π and r are different entities. The circles to the left shows a continuous growth by extending Π every time and since Π is the same part as the previous Π, only extending that billionth of a millimetre each time, the circle will be truly continuous without any signs of a break

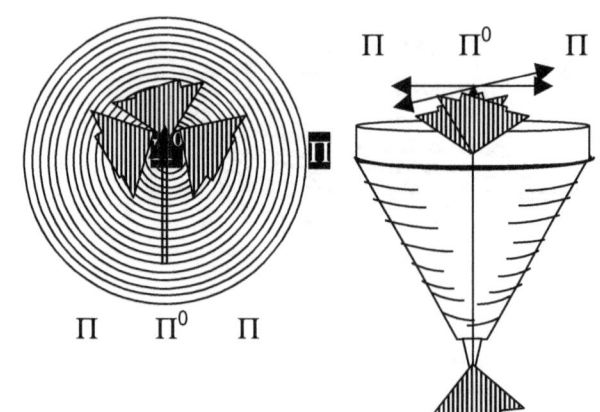

There is a line that puts singularity formed at $Π^o$ that has no space forming singularity to the value of Π that is forming the very beginning of space. That is singularity forming a governing position $Π^o$ that always forms a controlling position Π. Singularity always will be $Π^oΠ$ where Π than will be $Π = Π^3Π^2$ and that forms space-time or the Universe

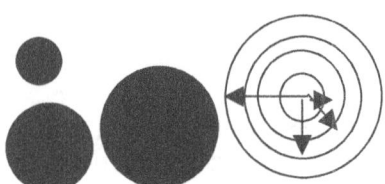   

The triangle, the half circle and the straight –line has two things in common, they share 180^0 as a mutual value and they are part of singularity.

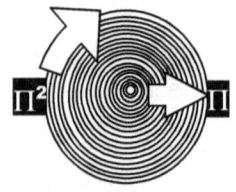

Using the concept that gravity applies Π as the circle factor Π as well as $Π^2$ replacing r^2 the replacing by Π brings two values as Π and $Π^2$. That I found is the case with gravity and will be apparent when explaining the sound barrier as well as the Four Cosmic Pillars. In order to create a distinction I remained using r as the indicator of the cube or non-circle that has vacant space and by vacant space I refer to non-solid structures. In the solid structure I use Π as a value for reasons that will become apparent in due time.

If we put this in terms of singularity ($Π^0$) we find the Earth ($Π^3$) is in relation as viewed from Alfa Centauri (Π) four point six years ($Π^2$). That secures the three dimensional status the Earth has ($Π^3$) in terms of a present ($Π^0$) that depends on a location (Π) secured by a future ($Π^2$) that will come by movement where the future ($Π = Π^3 ÷ Π^2$) also doubles as a past ($Π^1 = Π^2 ÷ Π^3$). That is space formed three dimensionally by keeping time in infinity apart from time in eternity. The relevance (Π) that forms in relation to the present ($Π^0$) will relate to movement ($Π^2$) and the movement is circular which ensures that the relevancy forming is circular (Π) by securing that the movement is circular ($Π^2$) in terms of one specific point ($Π^0$) in infinity which then secures a roundness ($Π^3$) that forms an everlasting eternity ($ΠΠ^2$) which validates an never ending circle. In this time in infinity ($Π^0$) secures that there is an everlasting eternity ($ΠΠ^2$) in space ($Π^3$).

The **governing singularity** ($Π^0$) holds a **positional validity** ($Π^3$) of three dimensions in terms of any **relevance** (Π) formed by the **controlling singularity** ($Π^2$) thus mathematically it equates to $Π^0 = Π^3 ÷ (ΠΠ^2)$.

If a **relevance** (Π) did not validate a **positional validity** ($Π^3$) securing a **governing singularity** ($Π^0$) in terms of movement formed by **the gravity** ($Π^2$) that produces the **controlling singularity** ($Π^2$), a three dimensional status, then space ($Π^3$) would not be obtained and thereby the Universe would not be secured.

Time is the movement of space in relation to any one centralised point not spinning securing such movement. Everything in the Universe moves in relation to any one single point that forms in any location that then has to stand still to form the centre of the Universe wherefrom that point must be motionlessness to allow everything else movement. In that manner the Universe is constructed and there is no valid solid Universe because the Universe is constructed from singularity (Π^0) that holds no valid space (Π^3) other than being in position (Π) while having gravity (Π^2) that forms the time (Π^2), which is also the movement (Π^2) of space (Π^3).

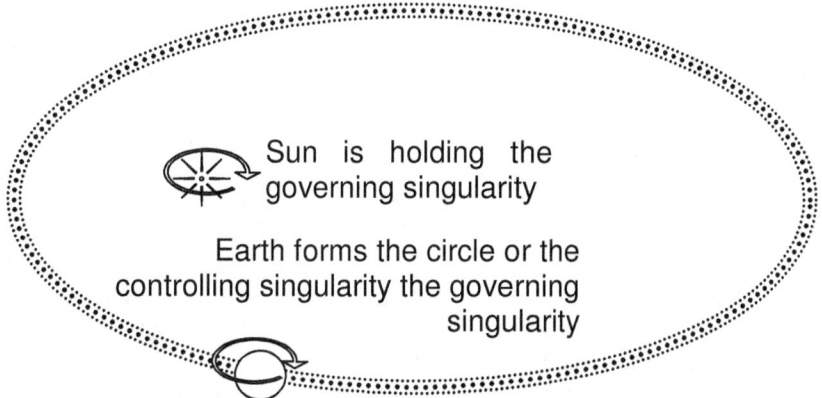

In that space there can be no motion because there can be no space to have the motion within. It is formed as a line that is so small that our human reality by perception declare that point as not being there and the only reason why we know it is there is because of the results it left as an imprint of its not being there. We cannot detect it but notwithstanding our failure to note it we can recognise the dot on the merits of its absence and while in our Universe it is always absent, reality disallows the dot ever to be absent, because it is never absent. It cannot be absent. It cannot go absent but it can never be there where it should be in a place from where the third dimension forms and it is always present if I wish to locate it. It is infinity that can never go away.

If it was absent then it was zero or nothing but since it is there it is not there and that makes it present. The centre spot we cannot see and that we cannot detect has no sides to any side and has no place it fills because it fills all the places that are present while we are unable to witness for we cannot detect. The only way such a spot can fill space is by doubling the space it fills to become more than one place to fill. But the very instant that happens it halves the space it fill because it then cuts the space it has into two parts. From this derives motion and nothing in he Universe is without motion because everything moves in terms of all else filling the Universe. That point instigates the Universe to form by the movement of the Universe in terms of that point's inability to move.

Any motion from such a point in singularity forms the entire Universe by putting everything forming the Universe in sides we do recognise. Anything within the Universe is in one the other side of the Universe because from that point we have dimensions forming by movement. That brings about that the point of not being I call the spot Π^0 is doubling its not being Π^0 and by doubling the not being into being the dot Πr^0 it also cut the not being that became present into half. We have to find this spot as we find religion. It is something that we can only know is there because we cannot disprove it is there but we can never prove it to be there. It is something seen through intellect and not through the eyes in the form of light being visible.

From the smallest ever possible dot will grow a line in every imaginable direction relating to a prospect of Π because only Π will not favour one specific direction and that puts all directions at equilibrium meaning that any form of what ever might develop from such a spot will have the end and the start being in the same position, which will also have to be a sphere as the flow outward will be equal in all directions. This is why we humans show the incentive to acknowledge this fact as we recognise that the sphere is the only shape that can possibly represent the form of the entire Universe hold. From the smallest spot in singularity comes a sphere.

Please think clearly, is that not precisely the commitment we find in gravity, where gravity is flowing from singularity outwards but never favouring any side? We could never explain where the gravity in China meets the Gravity in America. The nature of gravity is to never end and never begin always without favouring and where it seems to favour, there is a valid explaining concerning singularity. This reasoning prompted me to look for singularity in such a spot because if the prime spot from which all came was a spot holding all, then the spot must hold the shortest line but more prominent it will hold the smallest form including the smallest circle or for that matter the smallest sphere.

That leaves the door wide open for the advancing of any radius in all possible directions. With gravity always being in the centre of a sphere where the space is least available in the entire structure (there is not even space left to fill) one finds a flow of gravity from that centre spot outwards in all possible direction even-handedly. The fact that the original gravity will begin as a circle or will be a circle is the direction it will take when being the first spot created. All progress will be evenly in all direction because no direction will stand out or be in favour above any other direction at first. Moreover, what this information brings home is that through motion and only through movement does space develop in terms of a relevancy dividing singularity. I am about to introduce the second form of singularity.

Kepler said that the space a^3 is equal to the motion T^2 of the space a^3 distant from a specific centre k. That then is $a^3 = T^2 k$.

Within the circle $k^0 = a^3 / (T^2 k)$ the centre holding singularity also holds gravity which is centred in the precise middle of the circle. By using mathematics in the way Kepler used it, those rules and laws used correctly in the investigating of the formula that Kepler introduced must form the basis of cosmology. Also such intense investigation then must be without Newton interfering and telling Kepler what he (Kepler) should have found and subsequently Newton's incorrectly correcting Kepler whereas instead Newton should have been looking at what he (Kepler) found because only then he (Newton) could have seen what gravity is. He (Kepler) said that the cosmos said that gravity is $a^3 = T^2 k$. The space is held in check by motion from a centre and that is the way gravity develops. It becomes more than clear that space a^3 is time by dimension T^2 and time is space a^3 without dimension k Gravity is a^3 / k but k is an addition of motion T^2.

This is how I prove mathematically how gravity works. There is no pulling of mass or by mass or even that having mass plays a part in forming gravity. On the contrary, it is the forming of gravity that establishes mass when the space can no longer reduce and the reduced space locks whatever then has mass onto the solid surface of the Earth.

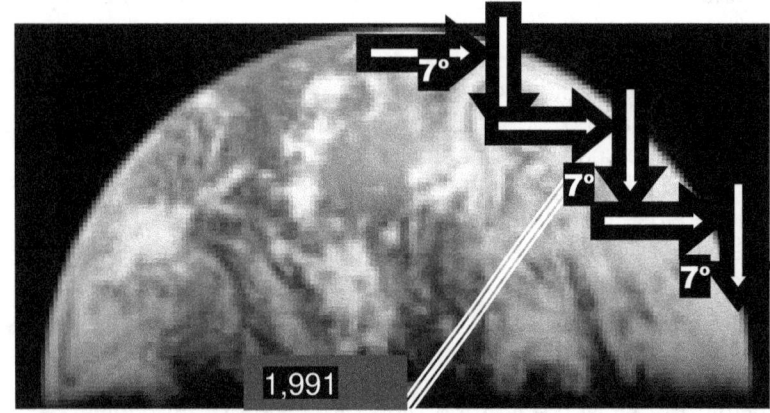

From what is showed this far it should be apparent that spinning is movement in two directions simultaneously. It is moving by implementing singularity 1^0 as well as shifting singularity 1^1 into a new position. It is therefor moving straight by curving and it is turning straight by curving, which puts 7^2 in relation to Pythagoras twice. This is because there are always opposing movement concerning spin and that comes from Pythagoras double actir in gravity.

When the object moves while being in space or in contact (in relevance) with the spinning Earth, the object wishes to continue moving straight ahead while the Earth also moves straight ahead by turning 7°. Therefore, the Earth by spinning is falling away. That clears space or compresses space by the margin of 7° declining (compressing) of air / space.

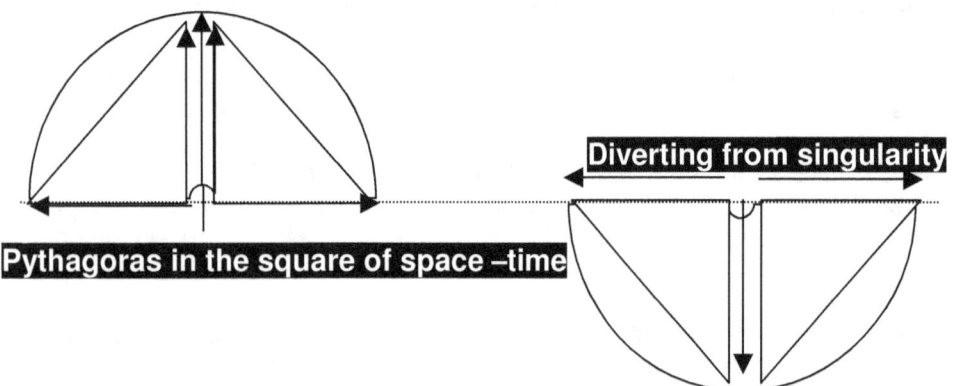

With singularity placed in infinity within the centre of every rotating object every atom and its relation to its surroundings including other atoms form space-time diverting from the point holding singularity as far as rotation goes because every object holds three relative positions in as far as where it was, where it is and where it will be in relation to singularity providing time. I elaborate on this else where.

The Four Cosmic Pillars; The Result Thereof.   In Terms Of Applying Cosmic Physics

Any point will be opposing itself within the **rotating of 180°** where it **then change every aspect** of its **previous flowing** characteristics it had or **will once again have** in 360° from there. While in rotation from the view point of a bystander it all may seem static and never changing but to the object in spin every next instant in time will be diverting from every aspect it had every second passing, and the direction it held in relation to the direction it held the previous mille, mille second will totally be incompatible with the direction it holds the very next mille, mille second of rotation. This is why we can use degrees measuring the circle by (6^2) (forming the square relating to matter through singularity) X 10 (square if space) = 360° however it is always in motion. That proves no point can be static or constant, though it may seem that way to outsiders. Although matter is matter, matter can also be anti-matter and moreover form its own anti-matter at the same time. This degeneration of structure is very likely to occur with overheating.

Revaluing Π to Π^2 will bring about a new contact point where Π meets **r** forming another relation in Π^2 **Time is** the **changes in relation** where Π **contacts a different r** not withstanding the many r points there may form because **every r constitutes a different value** to the Universe through other ratios and relevancies brought about **by heat and light. Time is the duration it takes Π to rotate between any two given points of r** and therefore must always amount to **a square (T^2)** moving from point to point through the **cube of space (a^3)** in that **duration of time (k)**. With that it proves **Kepler's a^3 (space) =T^2 k (time in the instant of motion)** but motion must continue through a specific value in space where the space-time is maintaining relevant equilibriums throughout singularity connecting. Every quarter is directly opposing the next as well as the previous quarters thereby starting a new set of principles it has to adhere too, but breaks by moving through time anyhow.

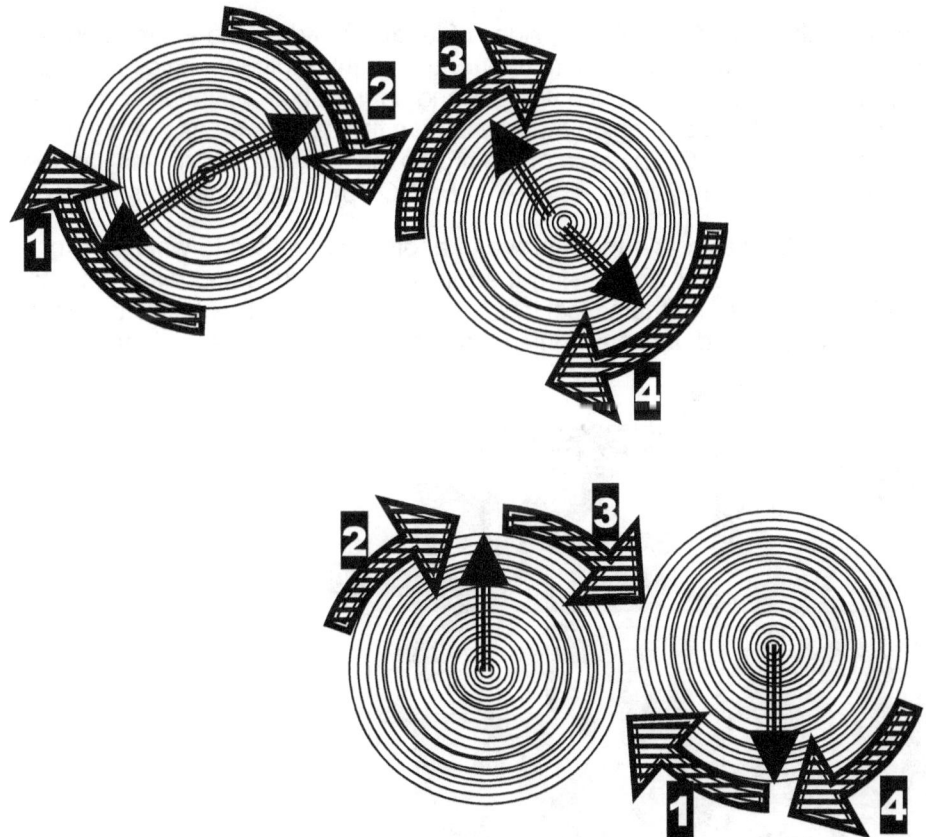

With this opposing rotation in mind, we have to translate this knowledge to gravity also. Gravity composes of movement and the movement will be at two points in space or three points in time where a double square seven in relation to singularity will transform into forming space at a value of 10.

In that movement comes about destruction of the self-preserving because any change of what ever small proportions will lead to destruction coming about as if with a snow ball effect. The top can be its personal matter and anti matter just by changing the speed of rotation where the points does not precisely meet the previous points and deformation stars by overheating bringing about an altogether change in relevancies to itself as well as other matter in the same orbiting time. With all matter having the same start from the same singularity, all matter should therefore be synchronised in growth and in rotation, where the matter is in support of all surrounding matter spinning around the common and original singularity that produced similar growth and rotation speed since time began to the present day.

The accepted methods and strategies used to construct a convincing mathematical argument have evolved since ancient times and continue to change. Consider Pythagoras' theorem, named after the 6th century BC Greek mathematician and philosopher Pythagoras, which states that in a right-angled triangle, the square of the hypotenuse is equal to the sum of the squares of the other two sides. Many early civilizations considered this theorem true because it agreed with their observations in practical situations. But the early Greeks, among others, realized that observation and commonly held opinion do not guarantee mathematical truth. For example, before the 5th century BC it was widely believed that all lengths could be expressed as the ratio of two whole numbers. But an unknown Greek mathematician proved that this was not so for the length of the diagonal of a square with an area of 1.

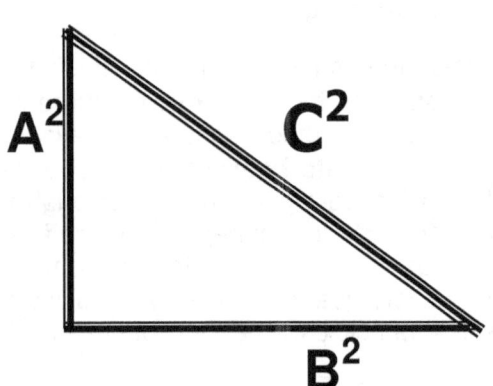

The Pythagoras' theorem. Figure 1 and figure 2 demonstrate that the relationship $A^2 + B^2 = C^2$ holds in a right-angled triangle with sides A and B and hypotenuse C. Figure 1 shows that a square of side $A + B$ can be divided into four of the right-angled triangles, a square of side A, and a square of side B. Since the two squares of side $A + B$ have the same area, they must still have the same area once the four triangles are removed from each of them. The total area of the squares that remain on the left side is $A^2 + B^2$, and the area of the square remaining on the right side is C^2. Thus $A^2 + B^2 = C^2$.

Pythagoras' theorem states that if a right-angled triangle has sides of length A and B, and a hypotenuse of length C, then $A^2 + B^2 = C^2$. Figure 1 and figure 2 each contain four equal right-angled triangles with sides of length A and B, and a hypotenuse of length C. Since figure 1 and figure 2 both have the same area, removing the four triangles from figure 1 leaves a region that must have the same area as the region left when the four triangles are removed from figure 2. The area of the region left in figure 1 is $A^2 + B^2$, and the area of the region left in figure 2 is C^2. Thus $A^2 + B^2 = C^2$, proving Pythagoras' theorem.

Pythagoras personifies where mathematics begin and that I prove in another book I wrote where I explain how the Universe began. I am able to do just that when using the four Cosmic Pillars and dissecting how the four Cosmic Pillars in conjunction with other mathematical laws including the law of Pythagoras had the Universe started. What I do promise is that the idea of Newtonian mass does not feature anywhere.

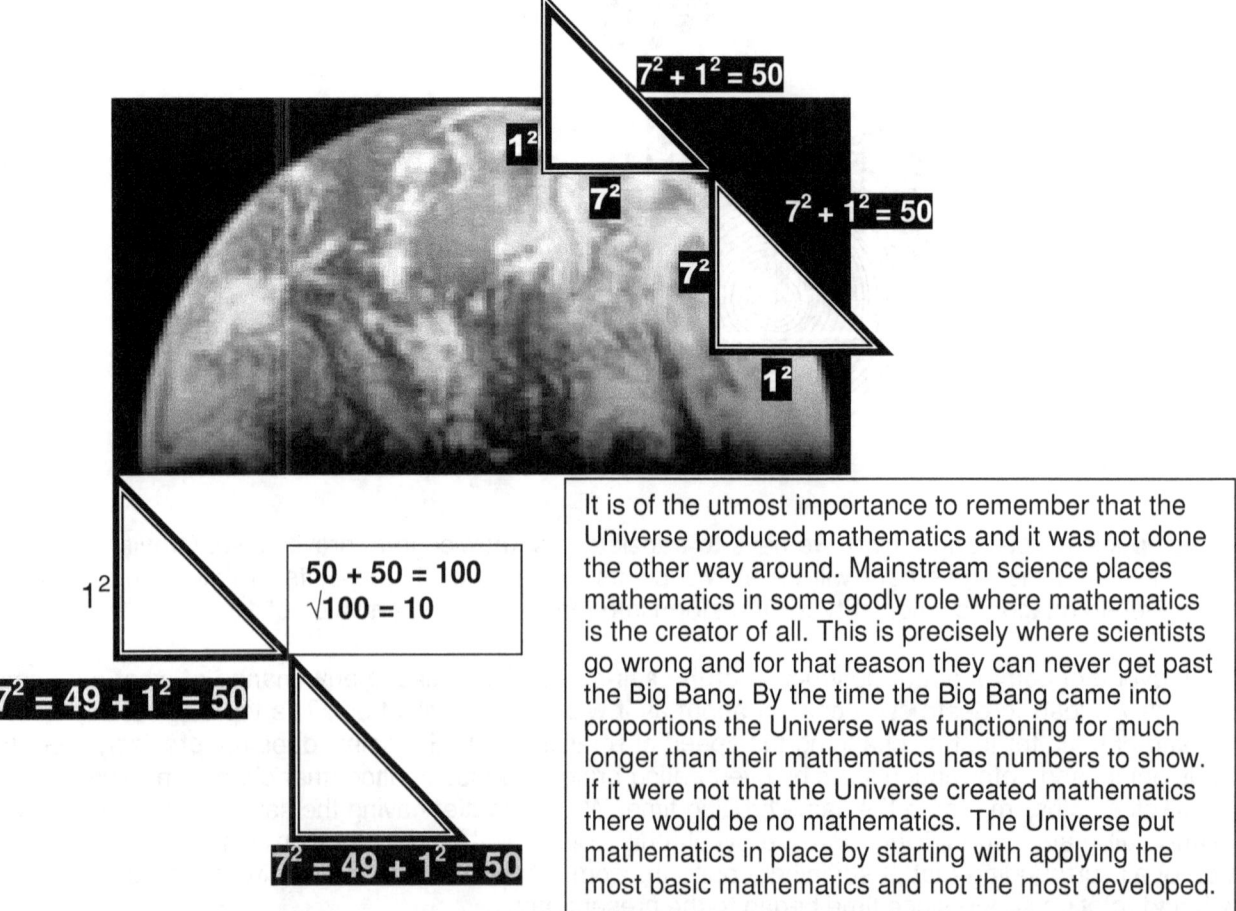

It is of the utmost importance to remember that the Universe produced mathematics and it was not done the other way around. Mainstream science places mathematics in some godly role where mathematics is the creator of all. This is precisely where scientists go wrong and for that reason they can never get past the Big Bang. By the time the Big Bang came into proportions the Universe was functioning for much longer than their mathematics has numbers to show. If it were not that the Universe created mathematics there would be no mathematics. The Universe put mathematics in place by starting with applying the most basic mathematics and not the most developed.

The Four Cosmic Pillars; The Result Thereof.   **In Terms Of Applying Cosmic Physics**

The Earth is moving, constantly spinning and in this is contracting space by compression (we call this contracting of space in air the atmosphere) and while the air is getting more compact, it takes whatever is filling with space towards the Earth constantly at a rate of 7°. By the Earth rotating, it is compressing space and with space compressing it is moving objects in the direction of the Earth. That is why objects that is falling, has no mass and only the stupidity of the simple Newtonian mind will force scholars to accept that it is mass that is pulling gravity. There is nothing in the Universe that ever could remain still because everything cosmic that is filling the Universe is spinning while it is also at the same time moving in a straight line. The Earth is only moving straight ahead because the Sun is spinning and while the Sun is spinning, it is compressing space, which allows the Earth and all other rotating objects to spin around the Sun in a perfect synchronised fashion. This process is going on throughout the entire Universe. However, explaining why this process is going on throughout the entire Universe requires a lot of volumes of pages and that are why the book **an Open Letter On Gravity** has **two Parts** with each part having **two volumes**.

That is what gravity is. Gravity is space moving or changing position in time and when an object can retreat no further towards the Earth centre, it only then forms a solid that aligns with the spinning solid material and with that then receives mass... Gravity is the movement of space in regard to any one specific point...and that is also precisely what time is. Nothing is standing still in the entire Universe. There is not one fragment of a sub-atomic particle standing still in relation to any other particle through out the entire Universe that is standing still. Having mass is when one object is standing still in relation to the Earth forming a part of the Earth while the Earth does all the moving on behalf of the particle having mass as well as the Earth and only happens when through having mass the object becomes part of the rotating Earth.

What does this all of this controversy mean...it means the way **the Brilliant-Master-mind-Newtonian** say the Sun and all the planets formed is total rubbish. The way **the Brilliant-Master-mind-Newtonian** say the Universe came about is hogwash. The age **the Brilliant-Master-mind-Newtonian** gives the Universe is proof of their total incompetence and total ignorance. The Universe is something **the Brilliant-Master-mind-Newtonian** can't dream to fathom...or begin to understand and then **the Brilliant-Master-mind-Newtonian** wish us to consider their positions they hold in society as the wise experts that can explain it all, while all along they can't even explain gravity. ...And for me not applauding their corrupted incompetence and academic on physics there is not one that is prepared to read my work

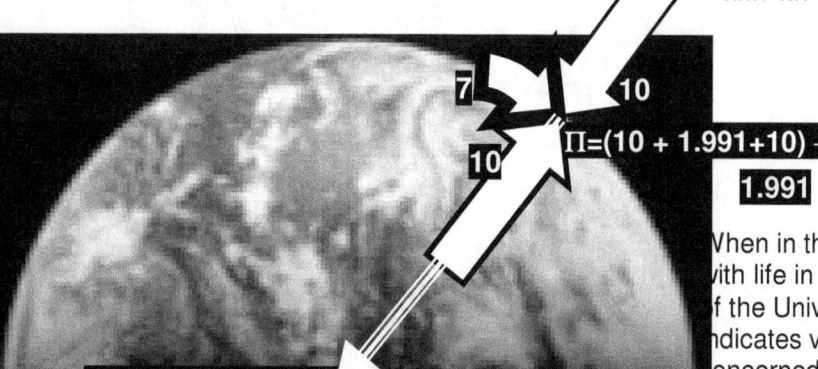

This explaining flush down the toilet Newton's idea that gravity is being formed by mass that through some form of magical intervention is pulling on other mass and this is forming gravitational contraction, which is madness. This idea is going down the toilet and seeing it flow down the drain into the sewerage where it belongs. If one takes the formula $a^3 = kT^2$ Kepler introduced, which Kepler received from no less than the cosmos at large, one find the **space a^3** is equal to the movement of the defined space in a **straight line k** as well as a **circle T^2**. In the cosmos no line

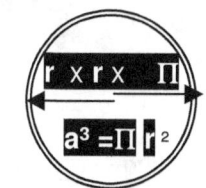

When in the three-dimensional state anything with life in the Universe can only be on one side of the Universe. The mathematical formula Πr^2 indicates volume but where singularity is concerned volume is in singularity.
This shows the value of pi in forming a double 10 in relation to 7

In this way the controlling singularity Π or 21.991 / 7 forms where the 10 going double is space that time forms and the 7 is time becoming space in relation to singularity growing from 0.991 to 1.

This brings clarity about the concept named as the Coanda effect where Π forming the controlling singularity connects to the governing singularity Π^0 growing from 0.991. However this includes the Titius Bode law, the Roche limit as well as the Lagrangian effect. In other words for the first time in your life, notwithstanding you ac academic qualifications or the lack thereof as it is in my case you have been physically introduced to gravity and therefore you have been introduced to physic s, as I promise you would be in my web site: **www.singularityrelevancy.com**

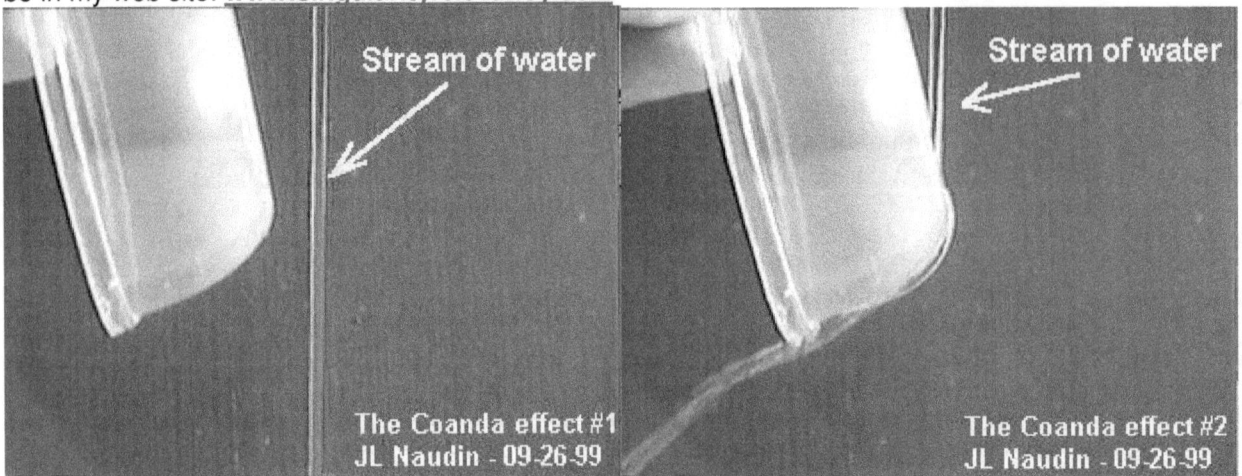

The condition for the presence of this centralised singularity $k^0 = a^3 / (T^2 k)$ **is movement** $T^2 = a^3 / k$ **of space** $a^3 = k T^2$ **in relevancy** $k = a^3 / T^2$ going both ways $k^{-1} = T^2 / a^3$ (Newton's 3rd law.)

The movement of space $k^{-1} = T^2 / a^3$ flowing towards the centre is the same as the trust of the solid $k = a^3 / T^2$ moving outwards from the centre. That is gravity where the spin of the solid is replenished by the condensing of the liquid and the liquefaction process is solidifying the liquid at the same time. That is what gravity is and mass is only a small by-product of the process we call the Coanda effect. I elaborate much more about the process in the books:
The Absolute Relevancy of Singularity in terms of The Sound Barrier
The Absolute Relevancy of Singularity in terms of The Four Cosmic Phenomena and
The Absolute Relevancy of Singularity in terms of The Cosmic Code

This explains the Coanda effect and the Coanda effect is gravity and gravity "glues" the water to the glass!

In considering the spinning motion in the fraction of time in the detailed instant every aspect of rotation will turn in every instant of

change in time by putting every spot there is in another location in accordance with the centre point that is unable to spin. While spinning the points will change direction every 90° of spinning and will oppose what it was every 180°. Although the points had the same characteristics only one instant before, they oppose the characteristics it had just before and just after the very instant in which they are and to which they relate by similar points also in rotation. The fact of the graph proves my point in quarterly opposing dimensions and values. As every point relocates, therefore every point completely changes its attitude from what it was to what it is in terms of what it will be when it is going there.

Going down to the centre, as the rotating direction moves inwards, the rings will become smaller and smaller. In dimensional terms, which I explain later on the value of **2k** relates to T^2. That relation extends to the next value where T^2 relates to k, which relates to T^2. The first space in the circle will then be T^2k. From the centre being in infinity one can realise by applying mental power the single dimension factor not seen but present all the same. Extending that into the 3D comes six k and any one of the six will further extend to form a seventh point as T^2 All this is a multiplying of $k^0 = a^3 / (T^2 k) = 7$.

This is how Newtonians explain the curve ball effect.

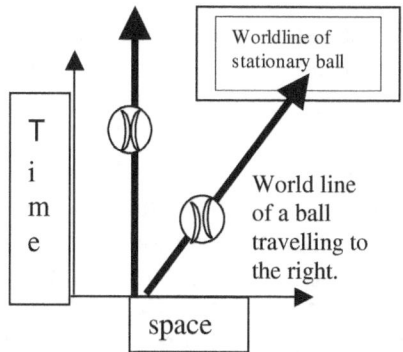

Time is always in control of space by the motion that time exerts on space when forming space in a new location at a new allocation. Time positions the location of space in the relation to time allocating new centre where time places all that

It is common knowledge that soccer players, golfers, softball players and tennis players sharpen their skill to kick, hit, pitch and throw what is commonly known as a curve ball. A curve ball is when a ball does not follow the straight directory line as the ball should follow but go off line and follow a curving directory. But as usual Newtonians go ballistic in being hypothetical when they start to philosophy about the magic they think science is and invent what is needed in order to remain and sound to be clever. The picture to the left is how the Newtonians see the curve ball action come about and the suggestion they have on the matter is to say the least very weakly thought through. The Newtonian got very clever by introducing a world line and a time line in relation to a space line and only God knows where the Newtonian gets all that crap. The ball it seems has to become somehow divided between a world line the Newtonian invented that is somehow half way between the time line and the space line going 90° directional opposing to the time line.

holds space by movement in relation of all space that was formed and that then forms which is proportionally to time related to the rest of the Universal space. If science wishes to put space in a motionless stance science should produce evidence where space is motionless or bring proof where time can find the ability to stand still...otherwise I am correct and Newton is wrong.

Here is the correct way to explain the curve ball according to the Coanda effect principle.

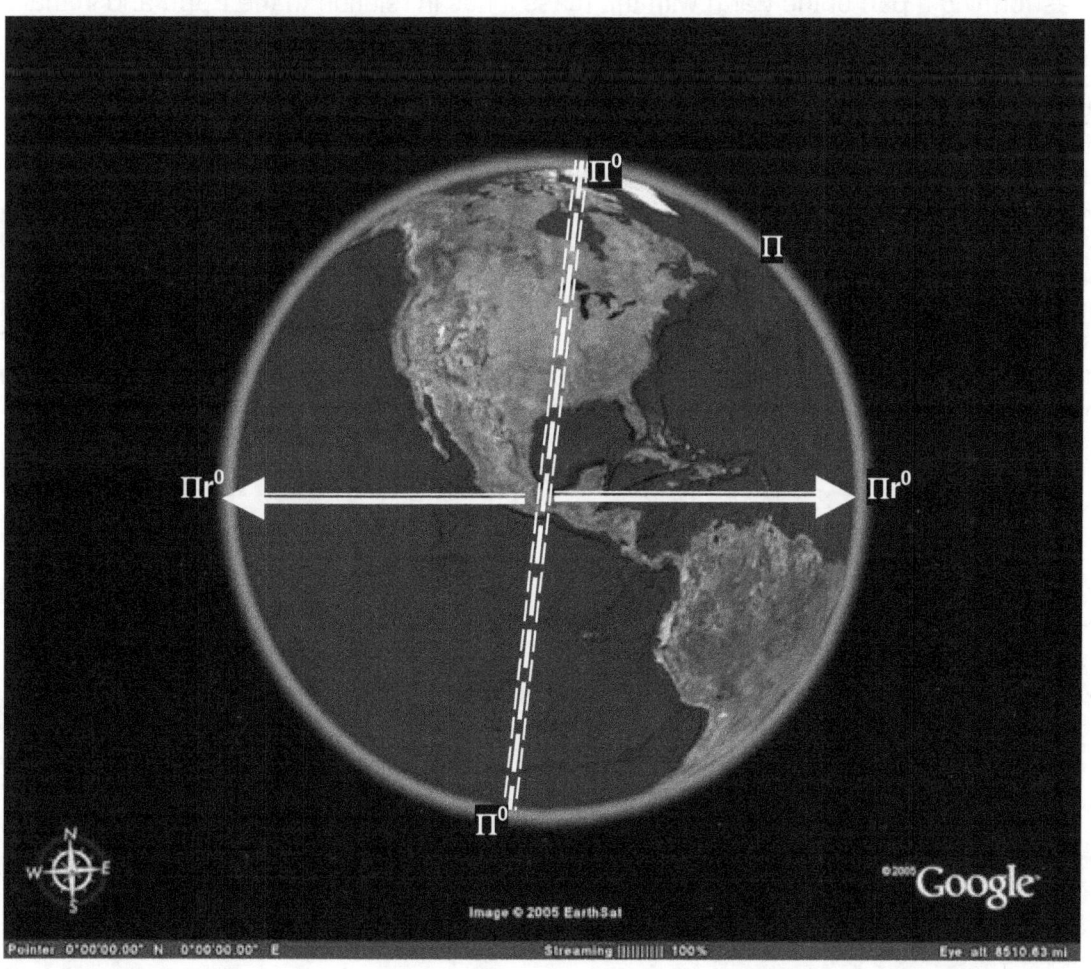

As the Earth spins, the Earth forms a line that is representing singularity at Π^0. It is what we call the Earth's axis and it is what the entire Earth turns around. From the axis to the end crust or the surface there is a continuous line of atoms where all of the atoms each also have an individual axis to the value of 1^0.

That makes the singularity within every atom centre having an equal value as the centre of the Earth has being 1^0. That means there is a line forming 1^0 that will in the end of the line form Π because that is the formula by which one uses in calculating a circle. A circle is Πr^2, but in this case $r = 1^0$ and the square of that is 1^0 by which then the formula is $\Pi(r^0)^2$ and that becomes Πr^0. By being in the air a falling body will not have Πr^0 because there are no atoms directly linking the falling body to the surface of the Earth. However when the body touches the surface of the Earth, the body will come to a halt and receive mass because the body then has Πr^0. Having Π^0 renders that point a governing singularity and if there is a governing singularity that point must then have a controlling singularity Πr^0 or better written as Π.

No body can ever fall straight towards the Earth but will always fall at an angle because $\Pi^3 = \Pi^2 \Pi$. That makes all falling bodies fall by the value of $\Pi \Pi^2$ giving the body independence of the Earth's structure just as Galileo predicted. That is why all things will fall equal irrespective of mass, which makes Newton's claims that gravity is formed by mass attracting a lot of hogwash. When the object holds mass, it has to be on the Earth surface that forms $\Pi^0 \Pi$, which puts the object on the ridge of the Earth in terms of the Earth's controlling singularity or Π putting in place the mass formed by the Earth by ratio of the Earth.

When a body moves independent of the Earth that body no longer holds a governing singularity attachment to the Earth but it holds the Earth in place as a primary singularity while the object moving in rotation will have independence by measure of $\Pi^3 = \Pi^2 \Pi$. This is connected to movement. Also what must be accepted is that where there is $\Pi^3 = \Pi^2 \Pi$, there also has to be $\Pi^0 \Pi$ to put $\Pi^3 = \Pi^2 \Pi$ in place.

That means when an object is standing with mass only rotating in relation to the Earth, then that object with mass holds the governing singularity of the Earth Π^0 as a reference and the object then forms the reference of the controlling singularity being in Π. The object forms the relevance Π but it is the relevance of the Earth Π and it is related to the Earth's singularity in as far as the governing singularity goes. The object therefore is the Earth as far as forming a part of the Earth with the mass it has in relation to the Earth and in that it is holding a position in the controlling singularity spinning around the governing singularity.

If there is Π^0 forming the governing singularity, then there is Π forming the controlling singularity and if there is Π then there is $\Pi = \Pi^3 \div \Pi^2$ and then relevancy places space-time in relation to time where time is movement of space. So everything is not as simple as Newton's idea that it is mass pulling on mass that then becomes gravity!

There are innumerable lines criss-crossing the Earth running around the Earth and running towards the Earth. To get a missile to go straight or to get a bullet to fire accurately or to get a satellite to orbit truly there has to be a spin that lines up with the lines awarded by singularity and it is these lines that are responsible for the sound barrier amongst many other things. It is the creating of Π^0 to make contact with Π that forms the Coanda effect.

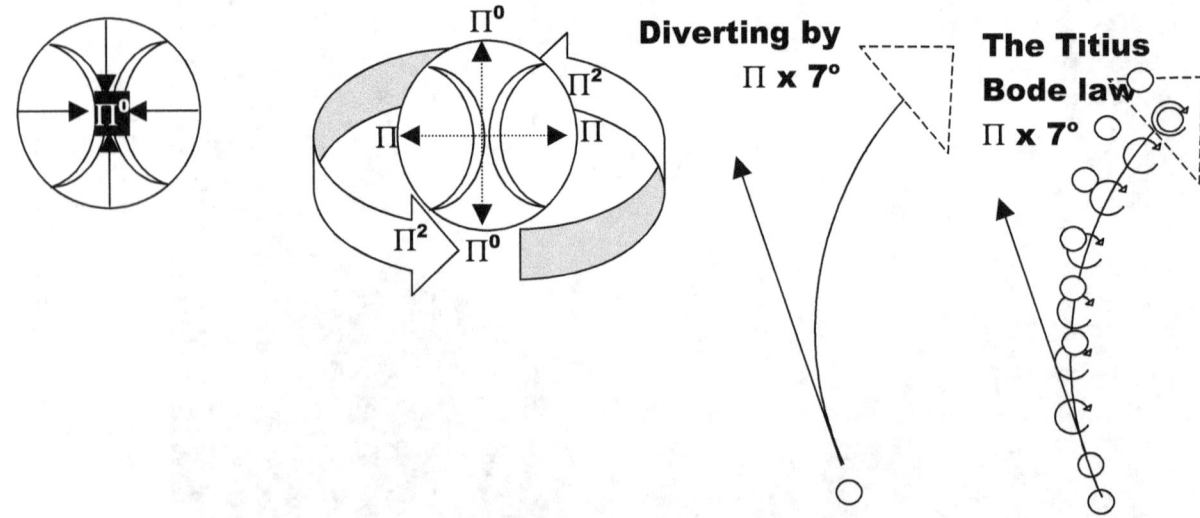

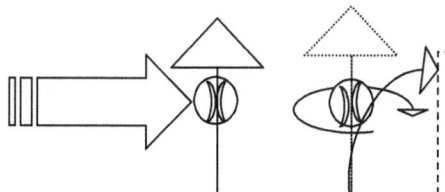

At the same time the ball is spinning with the Earth around the axis of the Earth. At no time can the ball ever be stationary in relation to the rest of the Universe and that relation in motion is time.

When a top or a ball rotates the ball establishes $\Pi^3=\Pi^2\Pi$ because $\Pi =\Pi^3\div\Pi^2$ and when relevancy establishes time in space then $\Pi^0\Pi$ by producing a governing singularity around which a controlling singularity will form $\Pi^0\Pi$.

The principle applying is the very same principle guiding the top to stand erect while spinning.

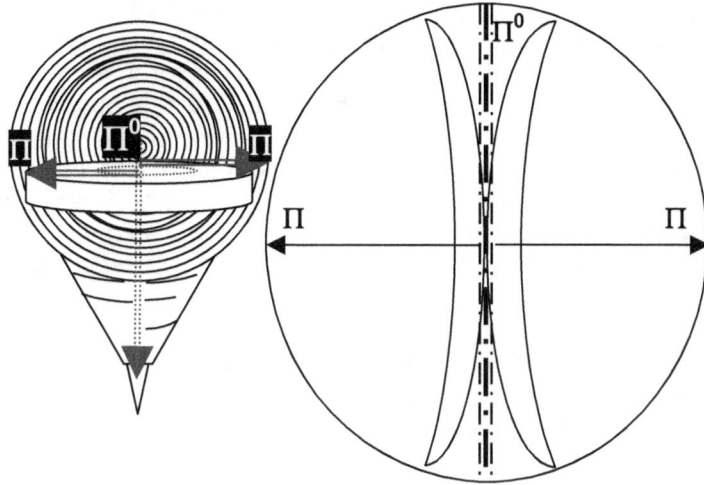

That explains the curve ball since that explains the Coanda effect since that explains gravity since that explains Π.

The ball establishes $\Pi^0\Pi$ but since all Π^0 is equal to 1^0 the difference applying is vested in Π. The ball establishes a governing singularity by measure of 1^0 that latches onto the earth's governing singularity measured at 1^1 and because $1^0 = 1^1$, the two acts on each other. The ball finds that although it has a governing singularity at Π^0 the balls governing singularity is still attached to the Earth's governing singularity in terms of the Roche limit. Therefore since the Roche limit is in place, the ball's governing singularity $\Pi^0\Pi$ presumes as Π being the Earth's controlling singularity. Therefore by the influence of the Earth's $\Pi^0\Pi$ singularity it takes the directional charge of the ball's moving direction and it curves the ball the same way as gravity curves light by redirecting the flow of light.

Looking at the ball it seems that the body structure of the ball is solid and the air surrounding the ball is liquid. The ball as a structure composes of solid particles that light cannot penetrate and that material cannot pass through. In that sense it seems to fit all the conditions we set for solidness. The ball spins and it spins through the air that allows the ball to spin seeing that the ball has much more density than the air has.

What can move is liquid and stands related to what cannot move being singularity. Since everything is singularity everything is immovable but also since everything is moving in time with time.

Every thing outside the ball is liquid with the ball forming a solid or so it seems to us. Well yes in a way and not that much either. The ball is a pump that pumps heat from the outside inwards just like a turbine engine. Every atom that is rotating inside the structure of the ball is keeping the centre erect. The centre is totally motionless because all the atoms in the ball are moving and the moving of the ball circle is extending the singularity of the ball to the edge where the ball meets eternity. The extending of singularity is holding the air as a liquid and being the liquid the flow of the liquid keeps the ball floating in air while spinning. The spin produces a cold in relation to the hot that the liquid is. What is moving is liquid and what is not moving is solid. Everything has a reference in relation to another point. That which is capable of relocating is forming a liquid in relation to that which is securing the position of rotation. Everything in the cosmos can move and yet not one particle in the cosmos can move. The cosmos stands divided between the eternal moving of eternity and the immovability of infinity

Everything around the ball is liquid with the centre being a solid. However the solidness and liquid has cosmic standards and just as it is in the case of hot and cold, big and small, fast and slow, our standards and cosmic standards do not share any measurements. So too does cosmic notions about liquid and solids have a totally different meaning in cosmic terms.

There is a pumping interaction of space-time flowing towards singularity through every point that confirms singularity. Every thing in the ball that forms the material is also liquid. By providing motion the matter in the ball serves as the liquid factor that extends the space that singularity provide. The structure is composed of atoms. In the atom there are a governing generated singularity around which all material rotate. In the case of the atom all the rotating material forms the heat while the generated centre, which is incapable of rotating, forms the solid factor. Every aspect that is without motion stands in a relation of 1^0 and that which is relatively moving or changing location or find a new position holds 1^1. Everything that is standing still is 1^0 and everything that is moving is 1^1.

Gravity or motion is a constant relation that solids have with heat where heat forms the liquid and solids form space. There is the rotation but part of the rotation is the lateral progressing by rotation to confirm the generated centre. The generation is in the rotation but the flow towards is the lateral and just as electricity produce a flow of time in relation to space collapsing, space-time by measure of gravity is using the same system to do the very same.

There is no substance difference between 1^0 and 1^1 and it is a relation where one moves as the liquid partner and the other is the solid factor. Both are not as much equal as they are precisely the same. Infinity cannot move and eternity cannot stop moving. By parting infinity had to move and eternity had to introduce as part of the cycle a point where it stops moving in relation to the other side that cannot move but does start moving. The factor that shows motion forms the liquid while at that moment the factor that does not show motion forms the solid. The measure of 1^0 is transformed to 1^0 and which ever are 1^1 is passing the extending of space on to 1^0. Time spin because everything spins in order to secure the centre singularity. But also time moves and in that there is the linear that always are part of cosmic motion. The centre is referred to by heat but heat also secure the centre by reconfirming the centre in the lateral. But in both cases singularity is reinstating singularity by confirming as it is referring one another. In the manner that 1^0 confirms a position in singularity 1^0 is supporting 1^0 by generating 1^0. By generating 1^0 it is repositioning and reallocating a position by confirming 1^1.

In the books **The Dissertation On Gravity** or the more informing **The Veracity Of Gravity** I explain the process in much detail. Gravity is the Coanda effect as the above picture indicates. Should you wish to find more information on **The Dissertation On Gravity** or **The Veracity Of Gravity** please visit the web site called www.questioneblescience.net

At this point so far after all my numerous attempts in trying to establish some contact with academics world wide I wrote seven books in a combination I titled **"Matters Time In Space: The Thesis"** covering the entire issue of my work plus the mentioned books below books wherein I combine all the various letters I wrote to academics through out the six years of ardent trying to establish some line of communication. The last letter I addressed to academics I include as part of the content of my web page called www.sirnewtonsfraud.com for your insight and which forms part of this and other of my books where I join and elaborate on the letters that I combine to form a unit as a book.

The Books holding the letters are entitled
1) **Newton's Mythology**
2) **Newton's Fraud**
3) **Sir Isaac Newton: A Conspiracy to Defraud Science**
4) **An open letter On Gravity Part 1 Volume 1 + 2**
5) **An open letter On Gravity Part 2 Volume 1 + 2**
6) **An open letter Announcing Gravity's Recipe**
7) **An open letter Addressing Gravity's Formula**
8) **An open letter About Gravity's Prescription**
9) **An open letter Explaining Gravity's Rules**
10) **An open letter To Selected Academics**
11) **A Cosmic Birth Dismissing Nothing**
12) **The Veracity Of Gravity**
13) **An Open Letter About Investigating Kepler.**
14) **The Dissertation On Gravity.**

In those letters mentioned above I call Creation by name and prove with science that we are in Creation. I employ science to prove that that which resembles the Biblical view of how Creation started. I prove that what controls Creation, is also which is not in the Universe yet is noticeably because it is not in the Universe. In the light of all proof and when facing evidence I bring I dare an atheist to prove me wrong about Creation. In mentioning this word Creation by name in a science book I break a ground rule enforced by the atheistic dominated world of science. I overstep all boundaries because I prove mathematically that Creation (the entire Universe) came about in the manner exactly and precisely as the Bible states…to the letter).

But in my work I do some things no one should do. I break rules never broken by man in three hundred and fifty years or more. I cross a line that is forbidden to cross by any man not dead or insane. I go into the darkness of the foreboding chambers of insane madness and mental instability. I disagree with Newton on the subject of physics and not only that... I dare to call Newton a fraudster and prove it mathematically...and that is one thing that is never tolerated...blemishing the religiosity of Newton.

To all those that feel disgusted by me accusing the greatest name in science that ever lived being **Sir Isaac Newton** of fraud, please go on and prove me wrong! $F = \dfrac{r^2}{M_1 M_2}$ This is the formula Newton used with which Newton proved gravity. Now prove gravity by using this formula. Do the following to prove me wrong. To find the force of gravity one has to multiply the mass of the Earth (M_1) with your personal mass (M_2) and then divide the distance there is between you and the Earth (r^2). Using these factors by multiplying (M_1) and (M_2) and dividing with (r^2) should present gravity coming from mass. But science uses a fixed value to calculate gravity. Now, convince your mind about my correctness. Do the simple calculations.

Take the mass of the Earth (M_1). Multiply the Earth mass by your personal mass that any scale should indicate (M_2). After multiplying the two mass factors, then proceed to the following step by dividing the multiplied mass factors with the square of the radius there is between your feet and the Earth (r^2), which should not amount to more than a few billionth of a millimetre. If the answer in front of you is not 9.81 Nm/s^2 then there is something very wrong. The incorrectness has to be either one of two possibilities presented: The measured value of gravity is not 9.81 Nm/s^2 as science uses it, or

Sir Isaac Newton's formula suggested as $F = \dfrac{r^2}{M_1 M_2}$ is complete fraud...Now which is it...you can decide...the force of gravity that the world of physics uses to do measurements is 9.81 Nm/s^2. If the answer you have in calculating your force of gravity is not 9.81 Nm/s^2, then it is either this measuring value of gravity that is wrong or it is Newton's formula that is wrong because by the calculation you did, the calculated answer you got could not possibly have delivered a measured value of 9.81 Nm/s^2. After all, science maintains it is the pulling of the combined mass in relation to the boosting that the radius would present to the force created that delivers the force of gravity! If by using the factors of mass and the radius does not accumulate to 9.81 Nm/s^2, then how can mass deliver gravity? Multiplying the mass of the Earth with the mass of a person and then bringing this answer in relation to the radius by dividing must be 9.81 Nm/s^2. If not, something is wrong with either the prescribed value science puts in place or Newton's suggestions.

To teach students that $F = \dfrac{r^2}{M_1 M_2}$ are the measuring formula in determining gravity, while knowing very well it is not totalling gravity at 9.81 Nm/s^2, then doing that to students while enforcing a thinking pattern in the minds of a student is committing brainwashing because by forcing examinations on students, expecting them to confirm the falsified statements used that the tutors present as correct, is brainwashing, a way of enforcing mind control and it is manipulating the thinking process of students.

If you can't prove that my manner of thinking is incorrect and you keep surmising that science is correct then recalculate the formula or start reading the rest of their fraud.

Gravity is a constant of 9.81 Nm/s^2. This is used in all cases of scientific calculations. Mass is an individual factor that is different on anything on which it is applied as a measuring factor. How could something as different as mass that is never constant even on Earth form a constant such as the force of gravity and still be the same in all cases?

In these following books forming a series of four and parts of this series about gravity there are the books entitled:

An open letter Announcing Gravity's Recipe
ISBN 978-0-9802725-6-7

In this letter I call Creation by name and prove with science that we are in Creation. I employ science to prove that that controls Creation, which is not in the Universe but is noticeably because it is not in the Universe. In the light of all proof and when facing evidence I bring I dare an atheist to prove me wrong about Creation.

In mentioning this word in a science book I break a ground rule enforced by the atheistic dominated world of science. I challenge any person to bring proof about any part where any of my theory might be incorrect and furthermore I challenge any Academic in physics to prove that Newton's mass pulling mass is anything other than fraud. I charge any one to bring proof that the cosmos is contracting by the force of mass and that mass produce gravity as Newton advocated when he committed the biggest fraud of all times.

2) An open letter Addressing Gravity's Formula
ISBN 978-0-9802725-7-4

Where the author explains how the Universe came about at the first instant the Universe came about in using evidence on the four cosmic pillars and it matches the Biblical explanation in explicit detail

3) An open letter About Gravity's Prescription
ISBN 978-0-9802725-8-1

Where the author explains how the Universe came about at the Solar system, as we know it took place. It explains why there are four solid planets, four gas planets and one cold structure. It also explains mathematically why all the debris is encircling the planets and where they come from. This is one part of another book entitled the Seven Days of Creation

4) An open letter explaining Gravity's Rules
ISBN 978-0-9802725-9-8

As the Author goes into detail about a new cosmos theory where the four cosmic pillars produce a cosmos everyone ca understand. It puts time in relation to space and discovers what space is in relation to time. Never yet before was either time or space understood because everyone drooled on the misconception about mass and incorrectly interoperating that mass produces gravity.

This series forms as a unit with four individual titles forms a prologue to a Thesis that introduces a whole new concept about Creation.

Matter's Time In Space: The Thesis in seven parts
ISBN 0984410-8-1

Questioning Questionable Science as we Question the formula $F = G \frac{M_1 M_2}{r^2}$ and in that it totally

Disputes the correctness of the formula $F = \frac{r^2}{M_1 M_2}$

Using the formula above as Newton did does not imply a suggestion or carry an idea across as a thought but must be seen to be acting as confirmation about a fact because one cannot suggest anything mathematically, one can only confirm a fact mathematically. There is no mere suggesting of any possible movement in a specific direction of any suspected behaviour by an object moving from and to a point as suggested, but this formula says the gravity of the Earth measured in mass at it's totality is colliding with the falling body's measured mass as the two factor's diminish the radius from both ends. This mathematical formula as it stands is no mere suggestion, but in its use it must back up or prove a fact!

This Is a Book That Is Not Afraid To Show How The Paternity of Newtonian Science in Physics Openly Cheats To Cover Their Oversight In an All Out Effort to Hide Newton's Misjudgement

Newtonians say the force F of gravity is $9.81 Nm/s^2$
Newtonians also say the force F of gravity is

$$F = \frac{r^2}{M_1 M_2}$$

Then in terns of mathematical principles

Newtonians say $F = \frac{r^2}{M_1 M_2}$ is $9.81 Nm/s^2$ where both are equal to the force F of gravity

$F = \frac{r^2}{M_1 M_2} = 9.81 Nm/s^2$. That is the way that proving with mathematics is done and what does it prove?

SIR ISAAC NEWTON'S
FRAUD

The force which Newton saw GRAVITY DOES NOT EXIST, IT MERELY IS A MEDIEVAL HOAX

Let's take the implication of $F = \frac{r^2}{M_1 M_2}$ under observation. Newton started off with the mathematical idea that $F = \frac{r^2}{M_1 M_2}$ where mass is. Newton saw the apple fall and came to a conclusion that since the apple had mass that could be measured, then the Earth also had to have mass that could be measured. He then gave mass a very prominent place in physics and concluded that whatever there was n physics, had its reason for being there, totally being the responsibility of the mass factor. I am getting to the mass factor in a short while.

The Four Cosmic Pillars; The Result Thereof.   **In Terms Of Applying Cosmic Physics**

We have $F = \dfrac{r^2}{M_1 M_2}$ that then inexplicably it could, just like a butterfly, turn into coming from a cocoon become $F \alpha \dfrac{M_1 M_2}{r^2}$ and then as if coming from a caterpillar it could unfold to become $F = G \dfrac{M_1 M_2}{r^2}$. If anything could be classified as magic then part of physics that Newton introduced, that is magic. Forget soothsaying and witchcraft because if you are in for magic, try Newton's magic and come into contact with true magic!

IN THE BOOK: "MATTER'S SPACE IN TIME". Each value stands in relation to space-time. $a^3 / T^2 = 1$ which is (a^3 = space) (T^2 = time) depends on the relevant of **k**. With this the facts about cosmology is changed completely. The only vision the author and science at present share common ground about cosmology is the definition to what the Universe is. Other than that, ALL universal aspect portrays in A NEW LIGHT AS NEVER YET SEEN BEFORE. IDEAS AT PRESENT WICH IS SACRELIGIOUS, ARE PROVEN VASTLY INCORRECT

Newton said that $F = G \dfrac{M_1 M_2}{r^2}$ which puts mass at the pinnacle of the entire Universe

How does M_1 connect to m_2 forming $\dfrac{G}{r^2}$ when the connecting medium constitutes of nothing that then has the value of zero as the Cosmologists wish to value outer space? The cosmologist insists there is nothing holding outer space in place, because Newton said outer space or the space outside material is nothing. This is noted in Newton's proof that the factor of relevancy Kepler named **k** by declaring that the spin illuminates the reference put in place by $k \dfrac{dJ}{dt} = 0$. Newton was not altogether wrong and also Newton was miles from the truth, but I shall explain that as I go along.

A comet follows a designated route around the Sun, without colliding with the Sun

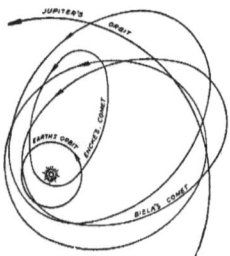

The comet travels around the Sun, but it steers clear of the Sun, at a time when the force of gravity should be at its strongest. Then quite the opposite happens when the comet is at its outward point therefore gravity should be at the weakest position of influence.

This is not all.

Because the radius between the Sun and its planets change during one orbit, the systems should draw closer. Alternatively, move father apart. However, they do not.

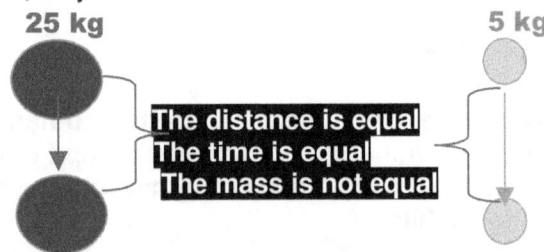

The distance is equal
The time is equal
The mass is not equal

GALILEO argued, undisputedly that two objects of different weight will reach the Earth simultaneously, if dropped from the same distance.

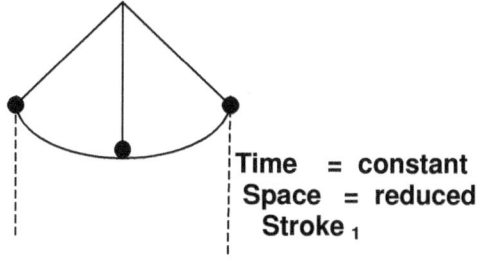
Time = constant
Space = reduced
Stroke $_1$

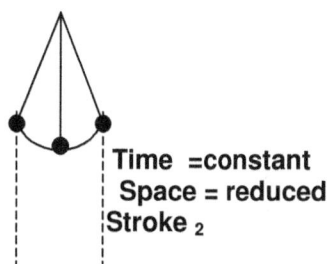
Time = constant
Space = reduced
Stroke $_2$

Time remains the same. The swing distance tarnishes. Period$_1$ = Period$_2$
Swing distance $_1$ ≠ Swing distance$_2$

In the pendulum principle that brought Galileo his everlasting fame, the pendulum swings at an even interval. As the pendulum swings, the space tarnishes while the time (period) remains the same. This is the principle on which all clocks work.

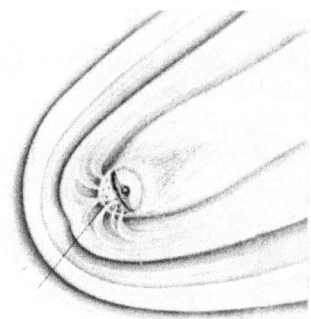

This means there is **NO FORCE between objects in the cosmos** and therefore there is **NO GRAVITY that is enticed by mass**. There is only space (stroke) time (period) = space-time.

Gravity is based on **Newton's presumptions**, which are **all incorrect**. **My work is based on** the findings of **Galileo and Kepler,** and **their findings** are totally **inconceivable** with the **findings of Newton**.

I **DO NOT EXPECT ANY PERSON TO BELIEVE** THIS STATEMENT OUTRIGHT, SO I INVITE YOU TO SPEND **THE NEXT FEW MINUTES** OF YOUR LIFE TO **READ THE FOLLOWING** ARGUMENTS, THEN SEE IF YOU STILL ARE **AS CONVINCED ABOUT GRAVITY AS BEFORE.**

We may start by determining the influence of gravity on planets as we find them in the solar system. **First of all let us concern ourselves with a comet**.

It is common knowledge how the comet stands related to the Sun's gravity. **Firstly picture the comet at its farthest point, away from the Sun.**

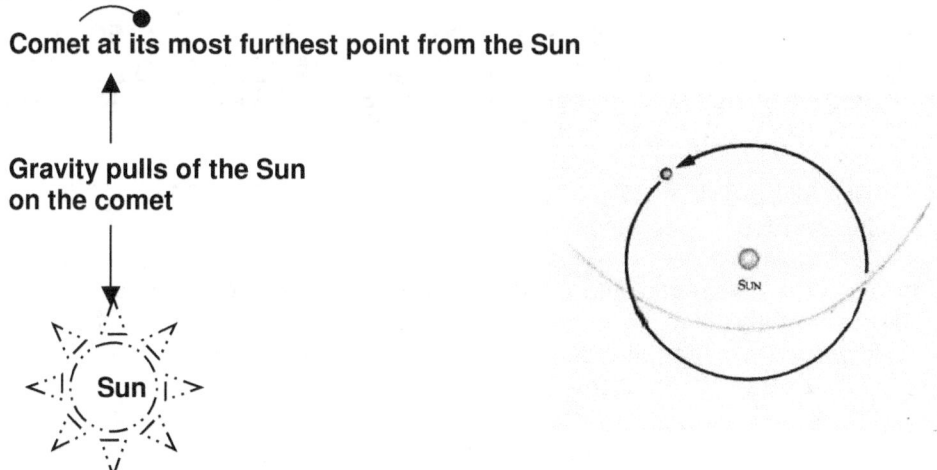

Comet at its most furthest point from the Sun

Gravity pulls of the Sun on the comet

Sun

The **gravity** of the **Sun pulls** the **comet straight towards the Sun**, this we all know. Gravity always pulls an **object directly towards** the **centre of a cosmic body**: that too is common knowledge. Therefore, the comet is drawn directly towards the centre of the Sun and throughout its journey the comet is picking up momentum directly related to the gravity that is centred in the middle of the Sun, (**gravity is always centred in the middle of a cosmic body**). As the comet is increasing its speed, the comet comes closer to the Sun and therefore the Sun's gravity pull is simultaneously increasing as the distance between the two cosmic bodies is reducing. Each instance the comet is drawn towards the Sun, the gravity that the Sun applies to the comet becomes larger progressively. When the comet is at its <u>**closest point to the Sun, something odd happens which cannot be explained by Newton's gravity at all! Remember gravity should now be at its strongest point because of the proximity of the two objects.**</u>

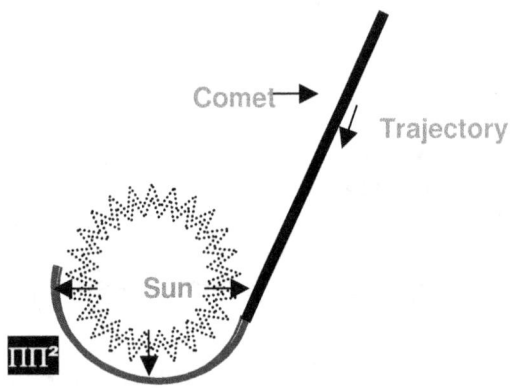

1. The comet remains at an even distance encircling the Sun.
2. No longer does the gravity of the Sun pull the comet towards the centre of the Sun.

3. At this very point the gravity that the Sun applies on the comet does not pull the comet towards the centre of the Sun any longer; in fact, it seems as if the effect of the gravity has been neutralized.

4. The comet stays at an even space from the Sun as it goes around completing a half circle's orbit around the Sun. It only completes a part of its rotation around the Sun.

5. After this, an even more peculiar event takes place. **The Sun, at the point where gravity should be at its most dominant, suddenly loses its complete grip on the comet.**

6. The comet brakes free from the Sun's pull of gravity and speeds off towards its destiny into the vastness of the cosmic space, undeterred by the gravity of the Sun.

Then after a pre-determinate and pre-calculated time the Sun starts applying its gravity on the comet once more. At a point where the comet is at its farthest point, the gravity of the Sun becomes strong enough to bring about a complete turn around to the comet's direction of travel. **However, the gravity between the Sun and the comet is at this point, at its weakest point of influence.**

So, when the Sun's gravity is at its strongest, the comet manages to brake loose and neutralize the Sun's gravity pull in order to avoid its fatal collision with the Sun and when the Sun's gravity is at its weakest, the comet cannot escape the pull of gravity. There is definitely something very wrong, either with the comet's behaviour or the laws made up by Newton.

NEWTON SEEMED TO BE ABLE TO CONVINCE EVERYBODY ABOUT HIS COSMIC LAWS, EXCEPT THE COSMIC BODIED THEMSELVES, WHICH PREFER TO BEHAVE AS IF NEWTON'S LAWS NEVER EXISTED. ARE THE COSMIC BODIES REBELLING AGAINST THE LAWS OF NEWTON OR ARE THEY SIMPLY IGNORING THEM AS IF THEY NEVER EXISTED.

MY GUESS IS IT IS BECAUSE THEY DO NOT EXIST, EXCEPT IN THE IMAGINATION OF ISAAC NEWTON AND HIS FOLLOWERS.

However, **this is not all**. When we regard the planets as they stand related to the Sun, the effect is the same, but not as obvious. All the planets follow an oval orbit around the Sun and therefore the same factors concerning gravity applies to the letter as it does in the case of the comet. Let us now investigate the one planet we relate the best to, which of course is the Earth.

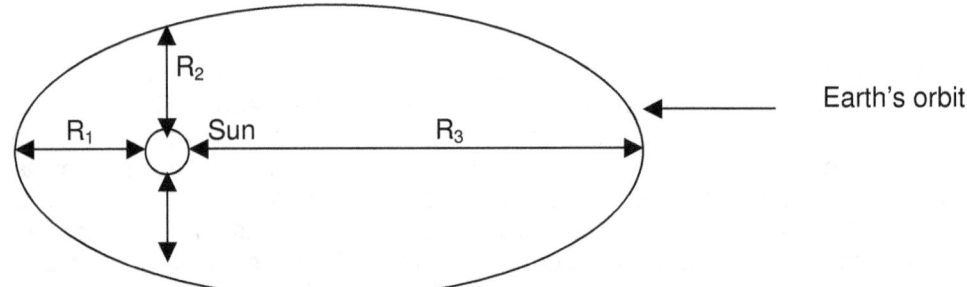

This illustration does exaggerate the radius of the Earth's orbit around the Sun, but since it has taken place 4 500 000 000 times, it has no real effect on the validity of the next statement.

The Four Cosmic Pillars; The Result Thereof.   Page 121   In Terms Of Applying Cosmic Physics

At one point (R_1) the distance between the Sun and the Earth **is less than** at another point we call R_3. Let us put a value of $R_1 = 1$ and $R_3 = 3$. This means that each year, for the past 4 500 000 000 years the effect of the common gravity between the Earth and the Sun has a greater effect than at another point six months later. **At one point the Earth should be drawn or pulled closer to the Sun** and **after another six months** interval **the Earth should stand less effected by the Sun's gravity**, therefore it should move away from the Sun. Each cycle of twelve months would have one point where the gravity pulls the Earth closer and exactly the opposite must apply six months later when the gravity is at its least. So, for the past 4 500 000 000 years the Earth has been re-establishing its seasonal swing towards the Sun and away from the Sun, which means by now the Earth has to collide with the Sun in midsummer or escape from the Sun in midwinter, as it may then drift away into the unknown.

For the more mathematical minded person the argument is as follows. May I remind you THAT NEWTON'S OWN LAWS ARE APLIED and again the planets disobey these laws completely!!

THIS IS THE SUGGESTED FORMULA THAT PLANETS APPLY WICH ENABLE THEM TO MAINTAIN THE ORBITS AROUND THE SUN WICH THEY DO.

We know that $F_1 \neq F_2 \neq F_3 \neq F_4 \neq F_1$ because that is what seasons is all about.

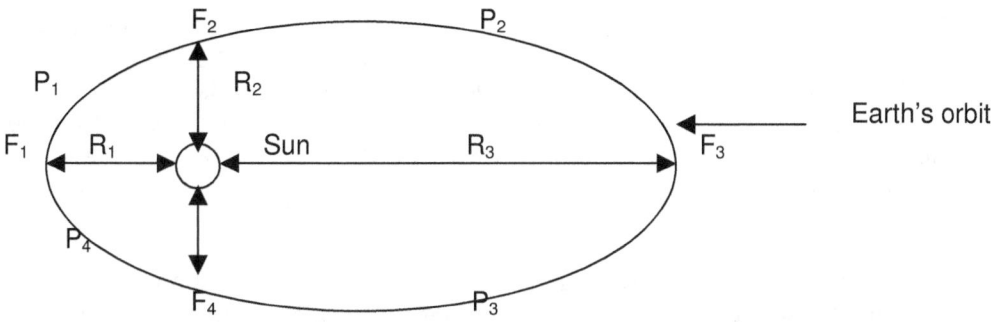

Even if $F_1 \neq F_2 \neq F_3 \neq F_4$, we know that $P_1 = P_2 = P_3 = P_4$.

How can I dare to even try to prove this? The answer is child's play. **When two balls**, one made of cast iron and the other made of wood, **is dropped from a building**, we all know that **Galileo** stated that both balls **will hit the ground at precisely the same instant:** a fact that has been proven even as far off as on the surface of the moon.

IN THE CASE WHERE TWO BALLS ARE DROPPED VERTICALLY, GRAVITY, AS A FORCE DOES NOT APPLY AND THEREFORE GRAVITY DOES NOT COME INTO EFFECT BECAUSE THERE IS NO DIFFERENCE IN SPEED OR DURATION.

The two objects should have their own value of gravity and gravitons and in comparison with the gravitons of the Earth; their value is insignificant. However, these two balls are in their own individual deuce to see who reaches the Earth first, and the iron ball's graviton should give it a superior advantage. This comes about because the two objects are put in a position where they are compared in relation to one another and share a common second factor, which is the Earth. In relation to the Earth, the gravitons of the two balls do not come into consideration, but this do not play a part since the Earth is a common factor. The balls, however, is put in a situation where they stand in relation to each other. When compared to one another, the gravitons should give the heavier ball a sizable advantage.

HOWEVER, IT DOES NOT. SCIENCE HAS IGNORED THIS MATTER EVER SINCE GRAVITY WAS "DISCOVERED". WHY DO THE MIGHTIEST OF ALL KNOWN THINKING POWER AND THE MOST BRILLIANT INTELLIGENCIA ON EARTH AVOID THIS FACT AS IT HAS RABIES?

The reason why the **mass of the smaller object do not apply**, is because **it is not the object** that is **drawn** to the Earth, but **the space-time in which the mass of the object finds itself in,** that is being **drawn towards the Earth**.

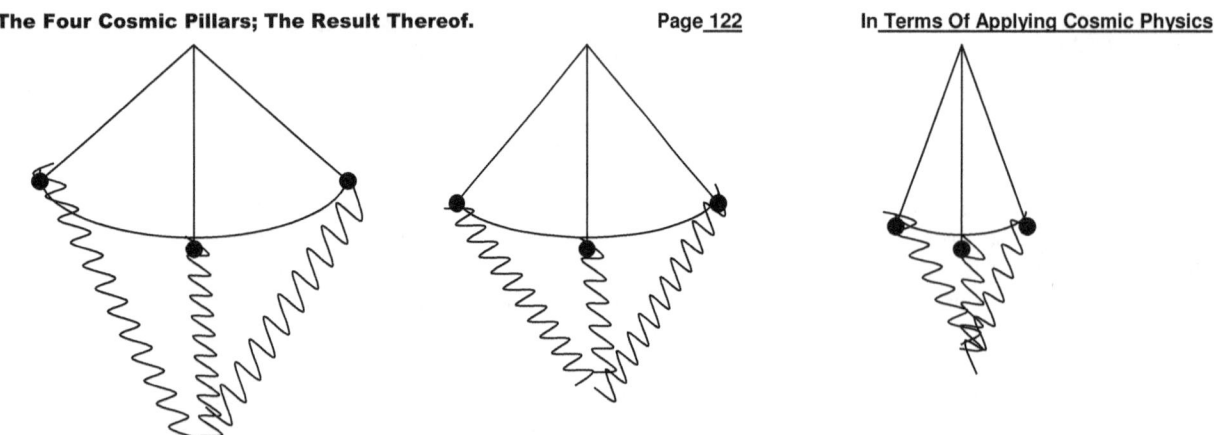

When mounting a spring of 9,81 Nm, to a pendulum, which is an equal force to that of gravity, both the time and the swing distance would be equally affected, but to a lesser degree as the swing, distance declines.

This means there is **NO FORCE** and therefore there is **NO GRAVITY**. There is only space (stroke) time (period) = space-time.

ALL SCIENTISTS GO INTO FRENZY BECAUSE GALILEO WAS PUT IN HOUSE ARREST FOR TEN YEARS!! **HOWEVER, THESE VERY SAME SCIENTISTS ARE STILL HAVING GALILEO'S WORK KEPT IN HOUSE ARREST AFTER ALMOST 500 YEARS. HOW DO THEY EXPLAIN THAT?**
If gravity was a force, one should get the same result by applying a pull spring to the pendulum. With the spring connected to the bottom of the pendulum, both time as well as space would be compromised. However, we know that the result proves the opposite, which proves that there is no force that is applied of the pendulum.

In the pendulum principle that brought Galileo his everlasting fame, the pendulum swings at an even interval. As the **pendulum swings**, the **space tarnishes** while the **time (period) remains** the same. This is the principle on which all clocks work.

You may ask yourself: **"What is the person trying to say?"** This is what I am saying:

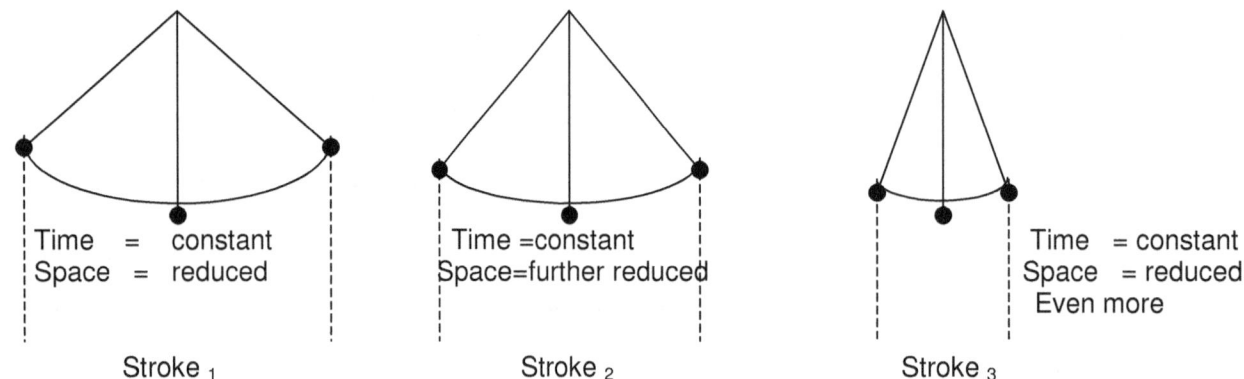

Time remains the same. The swing distance tarnishes. Period $_1$ = Period $_2$ = Period $_3$
Swing distance $_1$ ≠ Swing distance $_2$ ≠ Swing distance $_3$

If gravity, which is a force, did apply, then it would be as if a spring was fixed to the bottom of the pendulum and to an unmovable object below the pendulum.

You might think: **Very well, but what has all this has to do with the cosmos?**

Here is but a very few changes that comes about.

What is the Universe?

The Universe consists of two factors, which is space and time. These two factors are inseparable, undividable and one single unit. The gain to one factor is the loss to the other factor. There is occupied space-time and unoccupied space-time, but there is NEVER NOTHING. What is presumed to be empty space is unoccupied space-time! That is the value of the Universe, which I named geodesic space-time, and is the current value of time evenly distributed through space. The fact of "Nothing" can only exist in a person's understanding and perspective, but not in the Universe. Three factors rule the Universe:

1. Densified space-time (matter). My word I created in order to describe material.
2. Occupied space-time (atom of elements) is the position material holds in space.
3. Unoccupied space-time (The value of time in space) is space not filled with material and vacant.

There are only two energy forms in the Universe. The first is heat and the second is life. No other form of energy exists in the four dimensions that the Universe exists of. No force is to be found in the Universe, only balancing values.

There can be no such a thing as empty space. The Universe is time contained in space, which makes it space-time. Space has only one value, and this is to contain time and time provides space with a definite value.

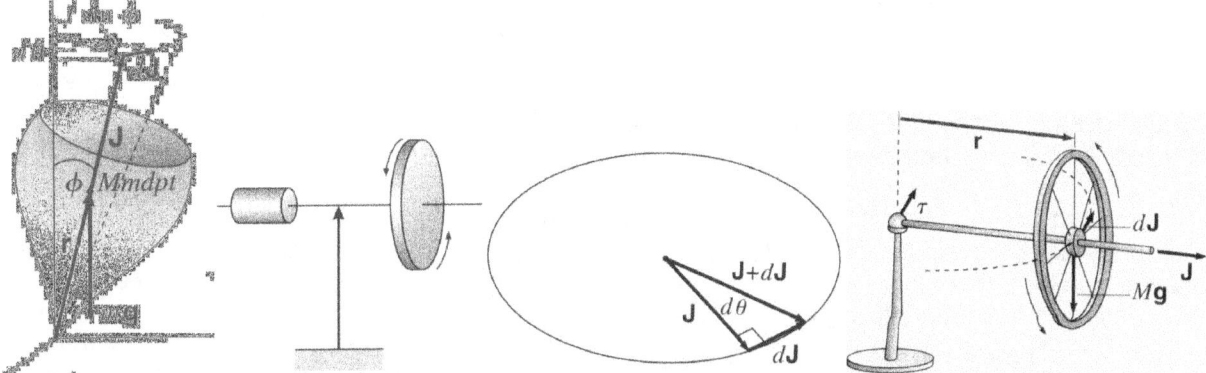

Sir Isaac Newton's says that $a^3 = T^2$. I have to believe **Sir Isaac Newton** when it is said that three dimensions are equal to two dimensions or in mathematical terms that $a^3 = T^2$ on no more grounds than that **Sir Isaac Newton** said so and without having any other proof to back the statement. Remember, Kepler never said $a^3 = T^2$, that is the part coming from the fantasy of **Sir Isaac Newton**. Kepler said $a^3 = kT^2$ which places three dimensions on one side holding three dimensions equal on the other side of the equation. There is a^3 on the one side of = and then there is kT^2, which is $k^1 \times T^2$ which is $k \times T^{2(1+2=3)}$ and that makes $a^3 = kT^2$ having three dimensions on the one side being equal to three dimensions on the other side. There is no way in heaven or hell that one can have the third power being equal to the second power or have a cube that is equal to a square, even if you are **Sir Isaac Newton**. There is no one on Earth that will tell me that $10^3 = 10^2$. There is a case that $10^3 = 10^2 \times 10$ or that $2^3 = 2^2 \times 2$ but never can it be that $2^3 = 2^2$. Not even when **Sir Isaac Newton** is doing the saying so. If one says that in the event where $a^3 = kT^2$ one may assume that $a^3 = a \times a^2$ or $k^3 = k \times k^2$ or even that using $T^3 = T \times T^2$ will also bring equality but never can $a^3 = T^2$...and then there are academics who try to convince me that $a^3 = T^2$ because **Sir Isaac Newton** was of the opinion that $a^3 = T^2$ and furthermore they expect me to also believe that it is true that **Sir Isaac Newton** has never been wrong on any suggestion and because no one could ever find **Sir Isaac Newton** to be wrong, I have to accept that $a^3 = T^2$ and take it as the absolute truth without questioning this abnormality!

  The one image is a cube with three sides. The other totally different image is a square having two sides. **Sir Isaac Newton** said the two are equal while they can never be equal since they are one dimension apart. **Sir Isaac Newton** convinced so many generations of idiots considered as being the wise amongst the wise and fooled those to the point where these stooges are willing to believe they are wise enough to believe that a cube is equal to a square and only on the ground that **Sir Isaac Newton** said so.

Sir Isaac Newton proposed and moreover convinced the world of science, and this includes every one and all members that should be the most intellectual bunch living on Earth in human form, that they and the entire world should accept that the inexplicable $a^3 = T^2$ is correct and that the biggest trick in fraud can be played on a bunch of fools all willing to be stupid enough to pretend they are clever enough to see that $a^3 = T^2$ and they are so stupid they pretend to be so clever that they will accept that $a^3 = T^2$ which when translated in words means that two dimensions are equal to three dimensions. This is the same as stating that a person's reflection coming back from the mirror is the same as the person filling reality while standing and looking at his image in the mirror. In this group hosting the most advanced minds man can produce there are a big enough bunch of zombies pretending to be mentally superior while being big enough idiots that are foolish enough not to think and not to ask questions but be small minded to the point that they will accept that a cube is equal to a square $a^3 = T^2$ just simply going on the say so of **Sir Isaac Newton's** Let's bring what **Sir Isaac Newton** interpreted as to what Kepler said in relation to Kepler's formula and see how much **Sir Isaac Newton** defrauded the work of Johannes Kepler and what in fact Johannes Kepler did say.

If it is true that $a^3 = T^2 k$ and we dismiss Newton's obscenity while going back to basic mathematical principles we find that: $a^3 = T^2 k$.

That means the space will move in a straight line while it circles. That can only indicate a controlled expanding because $a^3 = T^2 k$ (also is); that the line between the centre k^0 and a^3 / T^2 is increasing at a rate of $k = a^3 / T^2$ which indicates that expanding is happening at this very moment. The line k is not zero as Newtonian madness would suggest or the line is not 1 ($a^3 = T^2$) but the line is gaining by one dimension every rotation that is completed.

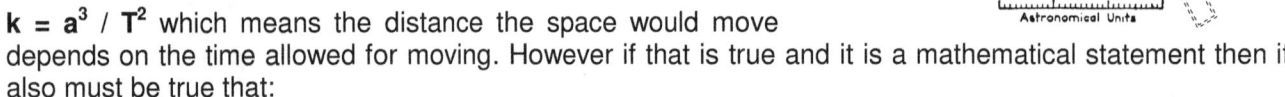

$T^2 = a^3 / k$ which means the time it takes to move is in ratio with the distance the space will move.

$k = a^3 / T^2$ which means the distance the space would move depends on the time allowed for moving. However if that is true and it is a mathematical statement then it also must be true that:

$k^0 = a^3 / T^2 k$ there is a appropriated centre

Taking the argument back to Kepler's law,

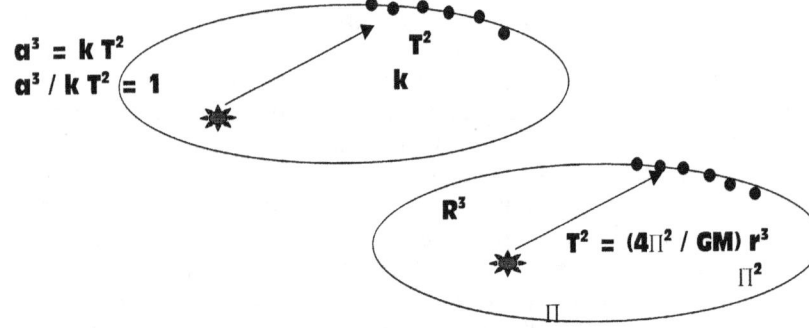

$a^3 = k T^2$
$a^3 / k T^2 = 1$

$T^{-2} = k / a^3$ Time can increase by manipulation as well as time that can reduce by manipulation. One can travel by increasing or decreasing the valid time.

$k^{-1} = T^2 / a^3$ The distance is moving between the objects. If the objects are not coming closer it must be the substance holding the objects that is coming closer. When we have a table indicating a shift towards the centre $k^{-1} = T^2 / a^3$ that does not involve any of the planets coming towards the Sun it would then show the space holding the planets are moving inwards because it shows definitely that something about the radius is decreasing and that means something is shifting towards the centre.

The space a^3 is rotating T^2 and the only other counter action could be when the space holding the rotating space is reducing $T^2 / a^3 = 299$ by the margin space is expanding $a^3 = T^2 k$ and $k = a^3 / T^2$ that proves the Big Bang is in progress. Kepler told so much about so many by using so few syllabifications that a mathematician such as Newton was unable to comprehend the full implication

PLANET	SEMIMAJOR AXIS $A(10^{10}m)$	PERIOD T (y)	T^2/a^3 $(10^{-34} y^2/m^3)$
Mercury	5.79	0.241	2.99
Venus	10.8	0.615	3.00
Earth	15.0	1.00	2.96
Mars	22.8	1.88	2.98
Jupiter	77.8	11.9	3.01
Saturn	143	29.5	2.98
Uranus	287	84.0	2.98
Neptune	450	165	2.99
Pluto	590	248	2.99

What I could never understand is how do Newtonians justify their believing that Newton is correct when Newton says $T^2 = a^3$ while there is a long line on the table giving T^2/a^3 values ranging between **2.99** and **3.01**. The values are there, it is not as if one could whish the values away or believe that the magic of mass will remove the table from the sky. The values ranging between **2.99** and **3.01** cannot indicate a factor such as mass because clearly it does not.

From Kepler's space-time $a^3 = T^2k$ formula we find that the relevancy of all planets $k = T^2/a^3$ in relation to the Sun is alike. This is only possible when the planets are floating in buoyancy because when in buoyancy all objects are equal in relation to the water holding them. There is no big or small but just those having specific density in relation. If there was no buoyancy mass or size would form some sort of resistance that would allow more and less restriction in some or other form to be present. In the sea and to the sea in that case there is no big fish or small fish but there is only fish. To us humans we think in perception of distance but in the cosmos space in the form of distance is the measure of time developed. The same goes for temperature and mass. Mass is good and mass is a product of the Human mind to put perception when it is needed but mass is dysfunctional in relation to the cosmos. All the planets are more or less 299 in ratio from the Sun, which makes the time affecting all the planets $T^2 = a^3 / k$ which is then $T^2 = a^3 / 299$ and in relation to space $a^3 = 2\ 99 \times T^2$. What this does is it puts all the space in ratio at an equal distance to the centre of the Sun and it puts the time in motion rotating at an even period around the Sun.

The Sun is at a level with all the planets parading past the Sun in the given ratio of 299. All the planets are precisely the same "distance" **299** $= T^2/a^3$ from the Sun and has precisely the same "mass" $a^3 = 2\ 99 \times T^2$ in relation to the Sun and the rest also floating about the Sun while they all travel at the same velocity $T^2 = a^3 /$ **299** around the Sun. The lot is in a bowl of liquid and it is the liquid that keeps a regard to the Sun where the Sun forms the solid as the regard to the liquid.

Johannes Kepler said $a^3 = T^2k$ which is the space a^3 in orbit T^2 is equal to the distance T^2 travelled in terms of a centre k that is in place.

Kepler said that the space a^3 is equal to the motion T^2 of the space a^3 distant from a specific centre k. That then is $a^3 = T^2 k$.

Reading this mathematically encrypted coded formula of the cosmos given to Kepler and keeping it removed from Newton it reads as the space a^3 is equal to = the motion T^2 of the space a^3 in ratio to a centre k.

What this proves is that gravity is the motion of space provided by time being the liquid.

Please allow me to explain. In the formula $a^3 = T^2 k$ the space forms as the space is in motion. Newton suggested that $\dfrac{dJ}{dt} = 0$ where he stopped time to have the motion of the circle demolish the work that the circle does.

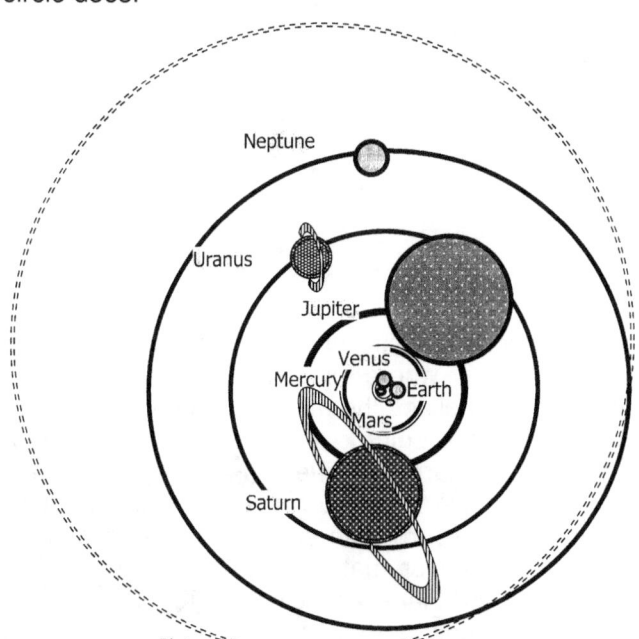

That means he got time standing still or being T^1 and the motion $T = 0$. Let us ponder on that thought for a while, while remaining with the formula Kepler suggested it will seem that according to Newton $a^3 = T^2k$ and in that T^2 then becomes **1**.

Newton changed the symbol of k by using the mathematical equated symbols G ($m + m_p$). This is just a longer and probably a more detailed manner of indicating k and better defining of k but it symbolises precisely to the point what k stands for nonetheless. I wish to draw your attention to the matter of Johannes Kepler's findings that Mainstream science considers as resolved and closed for many a century while it is not. My investigating Kepler helped me to resolve other unresolved matters but it was only possible by using Kepler's work. This changed the aspect of gravity in cosmology fundamentally and as I am about to show most and totally incorrectly.

Let us for one minute leave Newton's surmising about Kepler's failure out of the picture and concern us with what Kepler found long before Newton thought about what Kepler found.

Kepler said that the space a^3 is equal to the motion T^2 of the space a^3 distant k from a specific centre k^0. That then is $a^3 = T^2 k$.

Reading this mathematically encrypted coded formula of the cosmos given to Kepler and keeping it removed from Newton it reads as the space a^3 is equal to = the motion T^2 of the space a^3 in ratio to a centre k.

What this proves is that gravity is the motion of space provided by time being the liquid.

Please allow me to explain. In the formula $a^3 = T^2 k$ the space forms as the space is in motion.

Newton suggested that $\frac{dJ}{dt} = 0$ where he stopped time to have the motion of the circle demolish the work that the circle does.

That means he got time standing still or being T^1 and the motion $T = 0$. Let us ponder on that thought for a while, while remaining with the formula Kepler suggested it will seem that according to Newton $a^3 = T^2 k$ and in that T^2 then becomes **1**.

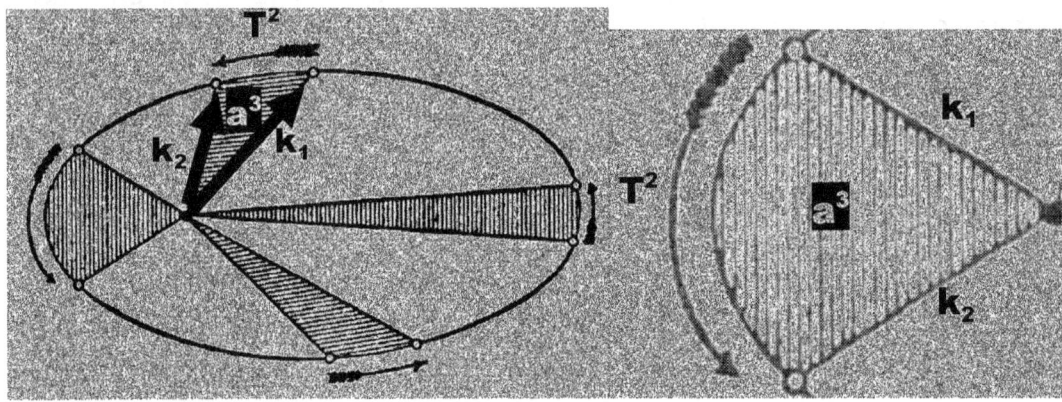

When Kepler said $a^3 = T^2 k$ it mathematically reads that the space a^3 is equal = to the movement T^2 of the space in relation to a radius k connecting to a specific centre.

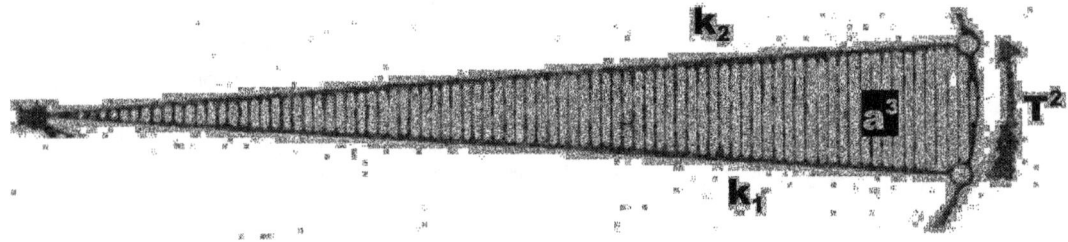

This indicates the value of space-time putting space a^3 equal = to the time $T^2 k$ moving.

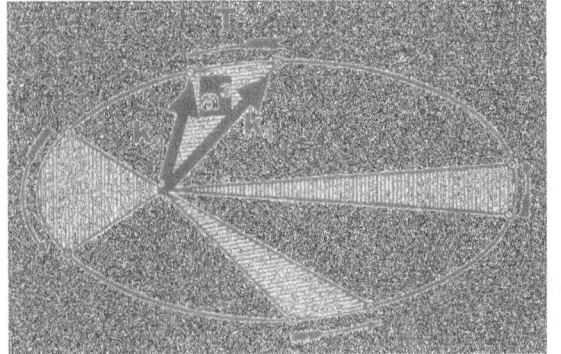

Two lines connect the value of T coming from one point and ending at another where this using of T is forming the line T^2 and that symbolises the movement between a point that k_1 indicates and a point k_2 indicates and that gives a value T^2. Nobody on Earth, not even Sir Isaac Newton has the authority to remove that factor k on any grounds there can ever be thought of from an equation Kepler produced as forming space-time or mathematically said $a^3 = T^2 k$. When any person and that even includes Sir Isaac Newton goes about and takes onto him the authority to make changes to a mathematical discipline in terms of the authority such a person claims to have and as that person wishes, then the result is catastrophic. Certain things can't be changed by man and one of the things is the law on mathematics.

PLANET	PERIOD (Years) (T)	MOVEMENT (T^2)	DISTANCE k	SPACE (a^3)	RATIO
Mercury	0.241	0.058	0.39	0.059	0.983
Venus	0.615	0.378	0.728	0.381	0.992
Earth	1.000	1.000	1.000	1.000	1.000
Mars	1.881	3.54	1.524	3.54	1.000
Jupiter	11.86	140.66	5.20	140.6	1.000
Saturn	29.46	867.9	9.54	868.25	0.999
Uranus	84.008	7069	19.19	7067	1.000
Neptune	164.8	27159	30.07	27189	0.999
Pluto	248.4	61703	39.46	61443	1.004

In the above table that Kepler configured as $a^3 = T^2k$ we have three distinct factors combining to form a specific value that indicates space-time $a^3 = T^2k$ and moreover shows that the Universe structurally is composed of in terms of space –time $a^3 = T^2k$ and every factor as much as a^3 and T^2 as well as **k** has a part and a role in forming the eventual value of space - time $a^3 = T^2k$. What did **Sir Isaac Newton** say happened to all the values under the column reserved for distance or then the symbol **k**? How did **Sir Isaac Newton** explain the values just disappearing? Why did no one ever think of questioning this unless every one participated in covering up fraud? The values are there and no person may ever discard the values in terms of then only applying the others unchanged.

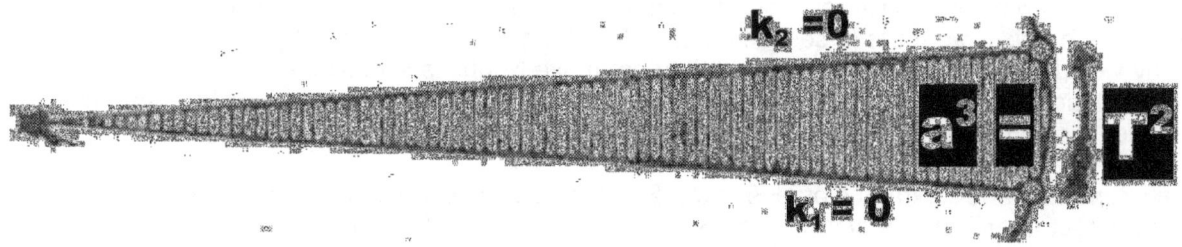

Those that wish to resort to using the excuse that **Sir Isaac Newton** implicated a ratio was in place, well that would not stand up either as a reason because **Sir Isaac Newton** used $a^3 = T^2$ which says the space a^3 is equal = not in ratio with but equal = to T^2. If one puts an equation it will also involve a ratio but the ratio then can show the symbol used to indicate equality = as being a ratio. **Sir Isaac Newton** put the space a^3 in terms of equality = to the distance travelled T^2 by disregarding the distance from a centre such as the illustration above will show and the illustration shows complete madness on the part of Newton because the distances **k** is in place and holds a value other than zero.

This rubbish excuse Newtonians contemplate to cover the fraud their Master committed of only indicating a ratio is nonsense because **Sir Isaac Newton** clearly says $a^3 = T^2$ and that puts every value Kepler had in the column devoted to a^3 as being precisely the same (equal to =) the value dedicated to T^2 being in the column devoted to T^2. He never meant it to be seen as a ratio as Newtonians wish to prove. There is a ratio, yes, but that ratio hinges on **k** and removing **k** removes all the validity such a ratio presents. If $a^3 = T^2$ the cosmos would have corrected this statement by removing **k** altogether and then the cosmos would not have had to leave the removing of **k** up to Newton to do it. Surprising as it may be to science, but Newton is not God.

When **Sir Isaac Newton** says $a^3 = T^2$ that does not prove that $a^3 = T^2$. It only proves **Sir Isaac Newton** was the worlds biggest and best silver tongue devil and cheated an entire Earth load of scientists for almost four hundred years. He fooled the supposedly wisest humans we all think there can be to pretend to be wise so that they can hide their stupidity while they only focus on their stupidity by not questioning the validity of $a^3 = T^2$. Can you bring me one other con artist and fraudster that can manage such fraud as to fool four centuries of scientists in believing horse dung is fig jam?

Those pretending to be wise showed how big fools they are because those fools tried to pretend they think $a^3 = T^2$ is correct just because **Sir Isaac Newton** said $a^3 = T^2$ and only those as equally wise as **Sir Isaac Newton** can be able to see that $a^3 = T^2$. It takes some doing to fool so many people for so long and leave all those fooled feeling good about themselves in that they are fooled while believing they are the wisest there could ever be born from a woman. **Sir Isaac Newton** was the biggest con artist ever to live and never again will the world experience an equal to **Sir Isaac Newton**. It is no small wonder that science is infested with atheism because science upholds disdainful lies based on mediocre understanding about truth applying as a reality and crooked science! The cosmos is dependent on motion and if motion disappears then the entire cosmos collapses. It is funny that Newton was the one that based all his theories on motion and he is the one that missed the most important part about gravity, which is motion.

We with life take motion as for granted. We never consider why objects in orbit would have motion and what would be the result when such an object is deprived of motion. Having life means moving and we as humans are on the move from birth. We even call an infant that has died at birth being still born meaning it did not move in the manner we accept life would move. The last thing life would have before capitulating to death is to have the blood circulation terminated. In that way to our reason motion is a fundamental or a basic which is true but that makes it even less common.

Why would all things move? The quickness of the Newtonian mind would remind me about the galactica inner core that shows no motion. That is a time concern and not a result of inadequate motion. There is lots of motion but the motion is more delayed and therefore we are unable to witness it.

Before I allow myself to get tangled into that explaining (I do explain this issue on another occasion in **The Cosmic Code,**) I wish to return to the basic concept of motion. Kepler quite correctly stated in his formula that the cosmos holding gravity is motion. Motion of space in time as well as through time stated as space $a^3 = T^2k$ moving in time T^2 and through time k. In that we find the birth of the cosmos. Newtonians have this grand dilution of a Birth of a Cosmos that could be instituted by using one formula that the entire birth's information is held in this one formula, which can crack the jaws of a hurricane.

Kepler's formula shows how the Universe started. When space a^3 started it moved in time and through time $= T^2k$ where time holds two parts in securing space. Realising this birth we must get to terms why stars move. Why would a star move and why would they not move? I guess it would be fair to put any and all activity concerning motion of the star in relation to the activity within the star. The motion of the star must be related to the time component the space uses to move within.

As humans tend to do we look at the brooder picture and try to assess from that what applies. That makes us miss the target by miles. The star moves with everything in it but it is everything in it that moves and therefore the star moves not with everything in it but because of everything in it. The star does move everything in the star but that within the star moves more and therefore by creating more motion the star establishes an accumulative all including motion of which all individual motion takes part in and therefore is part of such an accumulative motion.

The atom is vacating the allocated place it occupies and moves in time and through time to claim and a new position in which it holds space for one point in infinity. It vacates to fill and while vacating to fill it claims a spot where it is. Let's put it in the manner Kepler introduced it. Material a^3 vacates T^2k an allocated position to fill $k = a^3 / T^2$ a predestined position and while vacating $k^{-1} = T^2/a^3$ to fill another spot it claims $T^{-2} = a^3 / k$ a spot where in it is $a^3 = T^2k$ at the present $T^2 = a^3 / k$. Well normally we would leave it at that but then that would be that and we would be left with no results. What makes the atom move?

In the first factor of singularity a line indicates direction and by forming a relation the control becomes a volatile by-product. But singularity also provides space a^3 and duplicates by motion k to destroy space by rotating T^2. In dismissing the space the proton grows by accumulating space that the other side lost. Following the process and seeing the influence of singularity should bring about a pattern that may lead one to a pattern of how the required heat formed and how the intended heat transformed to space.

Density depends more on proton number arrangement producing specific form in relevancy as to merely and only having mass as factor that contributes to the forming and development of stars in the cosmos. The evidence is so clear that mass has nothing to do with gravity but density has everything to do with gravity. Density is the volume of space in numbers used to fill material in ratio with numbers of space per volume not filled with space. It is matter versus space in every sense there are. This came about before the Big Bang took place and before space was formerly space and time was formally motion. It was a time when singularity set relevancies moving from Π^o to Π

In that manner we know that that was the way particles formed combinations just after the arriving of moment-Alfa. Singularity brought the Universe but also singularity brought the divisions between the many Universes or atom combinations that followed and the immeasurable many Universes that came after the flooding of atoms forming societies and grouping became structural Universes. The term "moment-Alfa" is the way I refer to the moment when singularity changed, where infinity parted from eternity and a differentiation came about. This is still not when space formed or time began or space exploded but is when even before anything including mathematics became definitive.

At that point I named "moment-Alfa" is the point that mathematics is rendered useless. There was no space or time to calculate because relevancies came in place. Form took shape but space there still was not because Π^O moved to Π and that does not require space. Every slightest point in space became an opportunity of establishing a Universe with most different functions and ingredients there might form. This is apparent from the fact that it still takes place at the present moment by motion attaching new singularity through duplication and through duplication releases previously attached singularity from serving the purpose of duplicating by motion.

When the cosmos came about by motion, motion was not yet defined. When the cosmos brought about motion, the first motion was relevancies. Cold parted from hot. Eternity parted from infinity. Motion parted from motion absence. Infinity broke the laboriousness of eternity for the duration of infinity. The spot became and grew into the dot. From what the spot was to what the dot now is might be just a mathematical implication of going from 1^0 to 1^1 but in reality that first motion was the creating of and establishing of an entire Universe with all possibilities now in it. Never again can that much growth become a reality, although to us the growth is beyond what we ever can notice. But it is because the growth is so massive and we are so small that we are unable to notice such almighty growth.

When the spot Π^0 became functional and established all relevancies possible, heat parted from cold as eternity parted from infinity. The expansion was not clear motion but more a parting of relevancies where a centre formed a relevancy because the centre could not provide motion. Without being capable of motion, the centre established four points, which also served singularity. From the inverse square law we know that the centre doubled by producing the four points holding singularity.

By exciting the centre spot, the centre spot came to be because of the heat that formed in relevancy as heat parted from the cold bringing about the division that followed and that was the motion that formed. Heat grows space and cold dissolves space. Therefore the heat had to move but being singularity it could not get singularity to move. In an attempt to establish growth, singularity activated six spots of which four was having motion drawn into relevance four spots that was providing what was to be motion and three that was to be securing the position the centre holds. There were four forming a ring around singularity with two forming in locations we will refer to as above and as below or north and south. We still use these factors when calculating the measure of the sphere.

The three in line was in singularity not being able to move but the four was also in singularity and just as incapable of moving. All the points came as relevancies applying the forming of more of what was to come but only the four committed to time were expected to move. The four points that came as a result of discrepancies (4Π) that became time that produced form and that established the relation with the one but had to perform the motion by expanding. This it did, as it was as much incapable of motion as the centre was that charged the four with motion in the first place. As the four points were incapable of motion, it still required a tendency to apply motion that did separate Π^O from Π. This not only involved form but it involved all relevancies that did come or may in the future come about as a result of the attempt to commit motion. If mass was a factor contributing to gravity the cosmos would have frozen back to singularity without ever releasing singularity to relevancy.

Mass does not establish gravity. There is no magical graviton. In the beginning there was no mass but boy was there gravity! The only means that the cosmos could find a way to break from the grip of eternal eternity was to expand into relevancies. Such a feat can only go to task by forming opposing hot and cold. In that we can see the heat that applied in the instant of the Big Bang. Becoming hot produces more of what is heating. That implies motion or a moving away from where it was by generating more of what is available. Only where hot released from cold could whatever was repeated once again and duplicate what was before into what then is more. Secured by motion T^2 in relation to a specific centre k from where singularity holds the Universe true to form. The k was an intention to place apart and by today's standards will not even qualify any noticing. Going from Π^0 to Π constitutes too moving. If there is Π^0 there also has to be Π and that expanding is growth, which is movement.

All that are is in singularity. From singularity comes the motion and the space we call space-time. Singularity is dimensionless, time less and space less and because of all this features, it carries the value of Π^0. By expanding, singularity applies a relation coming about that reforms singularity from Π^0 to Π. Only when extending Π^0 to Π, the extending creates motion and the motion creates space that then doubles through motion applying which cuts the space in motion in half by matching the space as a duplicate. Motion creates another dimension or another level reforming singularity from Π^0 to Π or from Π to Π^2 or from Π^2 to Π^3

As said before we now know Π came about since Π is achieving form and not space. Only r can establish space as size will accumulate and as it had with everything else singularity had r covered by one as in being $r^0 = 1$. By reducing the circle radius r by half continuously his process will lead to an infinite small circle and an infinite number holding r would place r^0 to the power of one as a factor. Then as a factor r^0 would not contest any change when change is introduced into any future equation but Π will remain because the circle as a form remains even being infinitely small. By reducing r indefinitely to the tune of half each time, r would become infinitely small, beyond human calculating means, however as mentioned in the case of the smallest dot holding one spot, r would become insignificant beyond human comprehension even, but never reaching zero and still Π would remain intact and dictating form. To amplify by dimension a value has to be set to r but if r^0 remained covered by singularity all alterations that could possibly come about was in the form, which was Π.

This expanding can be a problem one can wrestle with for one lifetime and never reach any conclusion. How can something grow without getting more that what was before? Then it hit me like a ton of bricks. The answer is in heat but not heat, as we know heat. It is heat in getting relevancies between outer limits. Only heat could break the monotony of singularity. Heat in the form we now know heat as heat is now. Since the Big Bang heat is material transforming from one state to another state.

The change that took place involved singularity but singularity was 1^0 and being $Π^0$ forming 1 it could not grow. However growth came about. Heat rose from singularity, but if heat rose from singularity then singularity as a factor changed from 1^0 to 1^1, which means a relevancy came in place that no one could detect. It is true that 1^1 are still one, but one could then escape from singularity by producing factors other than 1. Heat came about but only as a relevancy to utter cold. If there is heat, there is cold or if there is no heat there can be no cold. Space came into forming a relevancy that brought form. Since it is a relevancy and not a generation by accumulation, the form produced was Π.

The spot formed a dot by heat and cold establishing relevancies and from that singularity was broken to allow all other forms of relevancies to come about. The cosmos did not start because of gravity. The cosmos started with heat and cold coming into a relevancy and in the cosmos there is no hot as much as there is no cold. The cosmos broke, put from the confinement of singularity by establishing a singularity in a relation of heat and cold. The heat that came about was beyond measure because the cold that held the heat was also beyond measure. The immeasurable heat was on the outside of the dot that formed and the cold was on the inside of the dot that formed. The cold contracted because in nature cold contracts.

The heat expanded into a dimension of form and heat by expansion is in nature about motion. Motion is duplicating that which is and heat is what is duplicating by motion. But only heat by expansion was possible because in affect singularity cannot move. This is because $1 \times 1 = 1$ and therefore 1 cannot duplicate. The motion became contraction, as the motion was the result of heat expanding which was forming four points in the rim of the dot. The expanding of the points created motion in relevance of a centre that formed because of the motion, which established an immovable centre as the Coanda effect, placed more dots in relation to more dots that formed.

Every dot was Π and every dot formed $Π^3$ because of the expanding heat, which produced $Π^2$, just because Π came in place as a result of $Π^0$. With that a new relevancy came about forming a centre $Π^0$ in between the four points of expansion that was resulting in time. But since the points were in themselves singularity, which is immovable and space-less, they still heated forming a cold centre with the heat bringing about motion. It became a repetition where infinity broke eternity by producing a centre because of space (or rather form) forming the motion to enable the space to form in relation to the heat applying motion.

This brought about a Cosmos being conceived. The spot forms a full circle, but the line running through the circle is forever present because that is the future radius of the circle that will one day develop the circle, which is equal to the present diameter. The fact of the presence of such a possible line in such a possible circle dividing the possible circle into two parts makes the centre line equal to the half circle. The line forms the half circle but not only that the line presents the half circle as much as the line is the half circle. The line then is 180^0 and the half circle is 180^0 because in singularity the two factors are the same.

The same value is of course $Π^0 = 1$. The issue of concern is to understand that singularity cannot move. Singularity has no space. Singularity is no part of the Universe but singularity is the Universe. By establishing motion singularity has to be charged with the time delay we find space to be. The space is time taking a period or a duration while moving from one singularity point to another singularity point while conducting the heat and the accumulation of heat that built up due to the retarding of the time to conduct the heat that forms the space that is conductor to bring about the motion of the space.

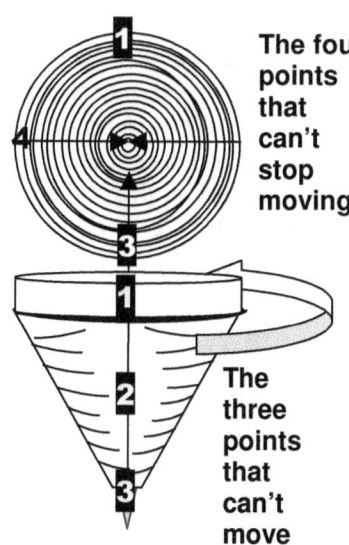

The four points that can't stop moving

The three points that can't move

Three points formed a line covering singularity where the centre singularity recovered heat to grow and two points served as an axis to allow the rotation and to assist the duplication. There is one centre connecting the duplication of three as well as the recovery of one (the fourth one) that is applying the tie aspect. In that we have time formed by three dots always motionless forming a line and two half circles forming the four of space. Therefore, motion consists of three positions in relation to a centre, which forms as space in relevancy to the motion and the space receive a controlling centre.

The duplication comes about as singularity holding three immovable points is exciting a relevance holding singularity in four unstoppable points of rotation, but the three points confirming infinity charged is as space less and as motionless as only the centre singularity is. The heat it requires to carry the exciting between points forming space and the space excites heat to expand and then to cool and contract and the time delay it takes to excite singularity between points forms space-time pulsating.

The development came into eras as the relevancies brought about new relevancies that spawned even newer relevancies that all remained in touch with the original singularity centres. Every one focused a new time delay that eventually brought about space and every distortion of time brought more distortions that later formed space. That concentrated between singularity points that charged the points to form space. When the charging became overheated in some sectors it erupted in forming the Big Bang. By the time the Big Bang erupted there was such a huge backlog in heat and time corrupted and delayed the next result was the employing of space as a commodity in the Universe. The relevancy was C the gravity was C^2 and the space was C^3.

That left what was inside atom still spinning faster than the speed of light applying the relevancy of **k** > C where the electron applied the relevancy of $T^2 = C^2$ and that formed the atom which then became the cube of the speed of light $a^3 = C^3$. That left the atom at the relevant size of what the speed of light permitted at the time but since the Universe from that relevancy expanded as the atom grew in space to the extent it has now. The purpose of the atom is to form cluster-consortiums that we call stars. The purpose of the star is to recapture the space the atom grows into and from there dismiss the space by spinning faster than whet the speed of light will be on the outside of the star.

This form came about when only form was present in the cosmos. It was in a time era where form featured in relevancies that would lead to one day becoming the atom and the Big Bang was not yet a thought. The atom forms a dual purpose of duplicating as well as dismissing and some in science prefers the one concept more to the other. This relevancy came in place when time was just a line and space was yet to form. Time is forever eternity being interrupted by form in infinity to bring about eternity flickering as infinity ticks. Before that singularity took on stages in forming relevancies between duplicating and dismissing space-time, which incidentally was not yet truly space-time in the sense we think of as space-time. At first a dot moved from the spot leaving the spot but taking with the spot as part of the dot to remain in the dot. The two never separated but the one allowed the other to be.

As the dot confirmed a discrepancy between infinity and eternity by defining infinity as an interruption of eternity cold and hot parted a union. The dot that formed was not space but a relaying of time to form a new point of singularity where eternity was interrupted by infinity. Time took form from 1^0 to 1^1 or from Π^0 to Π. It brought form into differentiating between interrupted eternities with infinity doing the interrupting. This is the **Absolute Relevancy of Singularity.** At this point I can introduce my theory on the **Absolute Relevancy of Singularity** At the point in the centre of the circle a line must start. In the beginning when I explained the way I figured how the line starts I said a lot of dots has to continue in order to form a line.

It would be 1 + 1 + 1 etc. because the line must form by holding singularity. After that point does mathematics begin but in the line that forms representing space as other all factors, then time holds 1. The line can only form when all the points forming the line have the value of 1 being 1^0. In that conclusion one realises something must separate singularity from all other factors because singularity hosts all other factors but is by own initiative Π^0. Only when singularity meets the end value can the end value have Π where the final ring of the spinning circle forms Π. That will be the spot of origin forming the relevance in Π. That will hold the eternal spot…the smallest spot ever because all spots that ever can be were secured in a position in the centre of that spot that must continue as a line that forms. Because of the progress singularity follows from the single dimension singularity only allows mathematics a start at Π^0 progressing further onto Π^0 and from there the line is born as $\Pi^0\Pi^0\Pi^0$ and to $\Pi^0\Pi^0\Pi^0\Pi^0$ etc. where Π^0 then may form the concept and value of r. But the line starts at $\Pi^0 = r^0$. This forms because cosmology is singularity based and the value is $\Pi\Pi^0$. This line $\Pi^0\Pi^0\Pi^0$ of singularity can only continue because every spinning atom preserves Π^0 in the very centre and since $\Pi^0 = \Pi^0 = \Pi^0$ the line is the same without finding conclusion except at the end where it forms mass at Π. At the point where Π forms, the movement Π^2 of the circle defines the space Π^3 of the circle and it confirms the centre Π^0 of the circle through the rotation. Let's call this the solid forming or if you wish, let's call it Kepler's singularity. After that singularity forms a line $\Pi^0 = \Pi^0 = \Pi^0$ where this forms another line again as Newton stipulated it by $\frac{dJ}{dt} = 1^0$.

Let's call that the liquid singularity or Newton's singularity and the relevance of singularity having a solid base compared to the singularity holding a liquid base comes about by the movement of gravity.

From these conclusions I prove that gravity is the result of four cosmic phenomena interacting to form the value of Π which by movement becomes the value of gravity Π^2 and gravity is equal to cosmic time applying. In order to understand the development of the cosmos and moreover the start of the cosmos and the progress in the cosmos as the cosmos formed, one has to understand the measure of Π. One has to see that Π is not merely 22 over 7 or that Π is a ratio that no one ever bothered to clarify, but Π is the key that unlocks every lock that hides a secret in the Universe.

One has to microscopically dissect the measure of Π to find the cosmos in measure. One has to understand where 7 fit in Π. The fact that Π is 7 at the bottom and that 7 relates to a double value of 10 is a key issue. Furthermore, it is very important to see why Π is 10 times two by adding 1.991 on the top part of the equation. In this measured value is what holds the building blocks of the entirety we call the Universe. It is behind Π that we will find the four phenomena, which I named the four pillars performing as gravity as they form gravity. It is by the actions of Π that the Universe develops. The Hubble expanding goes by implementing gravity as Π in the square through the four pillars on which gravity and time rests. It is behind Π we discover the meaning of singularity and how singularity forms the absolute and only building block as a form that forms the Universe. It is in Π we find the Cosmic Code unlocking the meaning of the Universe. Time is centralised in Π^0 that forms Π as space's limit that becomes space by gravity being Π^2.

Space is time gone to the past in which time confirms its presence it had in the cosmos by moving from the present time into space and then onto the future leaving space behind as the past. By forming a present, time is in infinity forming singularity that then has to move on and in doing so it leaves a legacy behind being space. Time is the movement of everything forming the Universe where in time the movement of time relocates everything in space by moving from the present onto the past leaving behind space. As time becomes the past by going to the future it forms space as it confirms the past, and in that space is what time forms by going to the past leaving space behind. Space becomes what time was at the point where time formed the particular space in relation to Π.

As time becomes the present coming from the past, time has to move on to the future at the same time and as time moved on it left space that represents that instant in time in relation to other space that was in some position at a specific location at such a point in time wherever that point in relevancy might be. The fact of Π not only refers to form but also validates the Universe by splitting infinity from eternity. By forming space when creating Π, time is using Π^0 in establishing movement Π^2. It is in the process of relocating Π to new positions by establishing Π^2 and connecting this as it forms a network consisting of Π° by forming space Π^3 in relation Π that establishes infinity Π° that always stays motionless.

If not for movement, the Universe would be one line holding time by repeating singularity Π° uninterrupted and it is in the diverting of eternity to a position away from infinity that the Universe comes about. This is what happens in a Black Hole where no movement within the Black Hole places eternity that always moves in a standing position to infinity that never moves. Without movement the entire Universe will fall back into and onto one point and everything we thought is real and solid will disappear into that one point holding infinity onto eternity where infinity and eternity then reunites. The Universe is an unreal concept with nothing being a reality but for the movement whereby Π confirms everything in a location in relevancy to all other things in a specific time slot or space.

When I, as a person forms a part of the Earth by the virtue of having mass that connects me Π to the Earth Π^2, stands on the Earth Π^3, my position in relation to the Earth gives me a specific positional relation to time Π^0 and the Earth. That gives the Moon a future of say one point five seconds being the past in relation to the Earth and that gives the Earth a past in reference to the Moon's future of one point five seconds. Where I am at any specific point in the present, that point I am holding is that which secures my present point in time. The Sun is eight and a half minutes into my past with all the space being in between the Earth and the Sun and by my view of the Sun I have a present time slot, as it also gives me a past of eight and a half minutes in relation to the Sun since the light travelled eight and a half minutes through space to confirm my past during that present instant.

That secures my past by eight and a half minutes at the point of giving me a present location in time. However, that also secures my future I have from the point I now have in the present by the margin of eight and a half minutes because that establishes a flow of light that would last another eight and a half minutes of filling a presence worth eight and a half minutes while travelling through space by moving with time and every spot filled on the way would secure a position that I will have in a future presence for the next eight and a half minutes, which then becomes my future as it fills my past.

Looking at this scenario in a view from Alfa Centauri the allocated position Alfa Centauri holds in space relating to the Earth, gives the Earth a past of say four point six years while this secures the present and having that present secure the Earth to a future of say four point six years by forming time as space between Alfa Centauri and the Earth and this is confirming time to the tune of four point six years. By securing movement it forms time in having a past in relation to the present that by the same margin also secures a future in relation to a definite past. This is how the Universe builds space in establishing time. This applies to all allocated positions of rotating objects throughout the Universe. This means that every point away from Π^o serving as Π, wherever that might be, secures my past I have by giving me a future in terms of the present Π^o.

Take this in relation to Kepler's formula we then find the Earth (a^3), which is in relation as viewed from Alfa Centauri (**k**) four point six years (**T^2**). That secures the three dimensional status the Earth has ($a^3 = T^2 k$) within the space from the Earth to Alfa Centauri (a^3) forming the Universe in terms of a present (k^0) being in the Earth centre which then depends on a location (**k**) secured by a future (**T^2**) that will come by movement where the future also doubles as a past ($k = a^3 \div T^2$ and $k^{-1} = T^2 \div a^3$). That is time and that is how time forms space and that is how space-time forms the Universe and that is the ***Absolute Relevancy of Singularity***.

That then forms time in the centre in infinity in relation to space in eternity in singularity where time that moves forms space by holding time that does not move secured in positions in relevance to where every point was in time gone by. Π **Divides infinity from eternity where infinity can't move and eternity** eternally moves as time. Life is in infinity leaving the body to part infinity from eternity. The body is what parts life being in infinity from where life walks when forming part of eternity. Life is a second form of time occupying material for a short purpose filled period.

When we look at the night sky we see images of stars. I am of the opinion that our vision of stars and our interest we show in stars is just what sets us apart from other species that we share the Earth with in the first place. We have the ability to recognise what we cannot experience and we are able to detect what is not in our immediate sphere. We have what makes us able to look beyond what touch and smell can confirm and acknowledge and to my opinion, that is what other species are unable to see. It might have something to do with the darkness of the night that is dark to us but is a clear light to most other species. We are able see what we never can touch though we can appreciate what we never can have. We interpret what we see without ever making contact to confirm and that gives us external knowledge and insight. Our vision about that, which we see tell us that there is more than the animal's concept of a plain survival on Earth where it is that you can eat or you can be eaten. Fathers show their children the constellations and although we no longer attach religion to our stargazing it never subdued the bliss we find in our astonishment about stars. That ability gives us religion as much as it gives us science and music, art and an ability to somehow reason by accumulative thought that brings about culture. We can build on the past by reason ad not by genes alone. We are what we generate and not what our accumulative genes of our species generate.

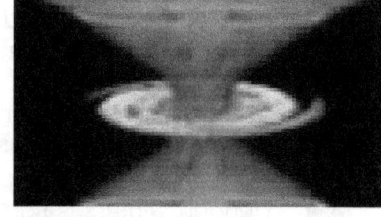

Lets begin where all living things begin. I shall begin at the ego of man because in all men (including women) that is where the Universe starts. I am going to start to show what man is and why man is what man is... or at least as far as the physics of man goes. Every human look at his or her body and refers to "me". Every person sees the body and say: that is I. It is what we see in the mirror that we call that which we think of as I. If the body were I or "Me" then I or "Me" would begin life as a skeleton and not finally end as a skeleton. Then if the body was in control of life the body begin by at first being the skeleton.

The body would start to furnish the skeleton by providing flesh to cover the skeleton and as it furnished the flesh the skeleton would provide the gain in flesh with suitable life. But the process starts where life begins as a cell uniting with another cell already filled with life. Life starts way before the male and female cells unite. However, when life disappears from the body the body is unable to replenish the life that went missing and starts to detach in composition and structure.

The process whereby life leaves the body we call death and the body ends as a skeleton a very long time after life left the body and the body then eventually ended as a skeleton. It is fine to look at DNA but it is life that controls DNA and when life abandons DNA it is not possible for DNA to replace life. That makes DNA the footprints life leaves behind and DNA does not fill the cells with life. DNA is space that is occupying material that is not containing life and definitely not representing life. All DNA is representing is it forms a trace where life walked through space and left DNA as markers or footprints. Life leaves DNA filling space just like time leaves space filled with time gone by. Life can manipulate and reprocess material because life is a second form of time manipulating cosmic time.

Life stars at a point where a cell starts to grow. You can't see the point directly where life is as much as you can't see singularity. You can't see the moment holding time because time in the moment is in singularity. You can see space but that is merely an image time left as a mirage of what time was when time formed the space. You can't see life but you can see when life is no more part of time. If life at any point leaves the cell, then the cell will decompose and lose form and structure.

Therefore the first condition for life to be is that a cell has to hold life and not the cell being present. It is life that provides the cell its structure and not the cell that provides life with anything. The cell is merely a vehicle in which life will ride for a time. When life departs from the cell the cell returns to an atomic structure or an atomic order. It is life gathering the body in the totally unique fashion that bodies are in and the body does not render life with unique qualities. I can't touch any person or any thing other that make contact to a few molecules set on the edge of a small point of the body structure that I am touching. If what I would think of as I am not permanent but a hallucination giving an interpretation of what could be a reality, think what a mirage the Universe must be.

Should I think of my body as me then where am I? The factor called life gathers in a place by using a process we have no idea about, but life gathers a cell to hold life. It is life that gains by duplicating and replacing the cell and not the cell that gains duplicity by giving more cells to accommodate life. If the cell was the factor representing the ability of life then as soon as life left the cell the cell should find the ability to again to fill the cell with new life. That would put the cell holding life in control of time. But that is not the case because as soon as life leaves the cell, the cell starts to decay and rot until it is in a structure composed by atoms.

Again, if it were the body that held life and the body was I, and then as soon as life left the body then the body would replace what was is lost with life again. While it is true that a cell filled with life procreates and assembles more cells as need arrives and as time moves on, that proves that life is in charge of the task in performing the collecting of more cells and not the collection forming more life by forming more cells to hold life. Life is spawning off by collecting new material, which will hold life, and that process is purely based on cosmology and moreover it is using time that we think of in terms of what is named as the Hubble constant.

Some Hoggenheimers have the saying that time is money but they are immensely more stupid than Newtonians are and Newtonians are already immensely stupid to think mass brings about gravity! A Hoggenheimers is someone such as a banker or a money broker that lives for money and collects money, as money becomes the banker's god. A Hoggenheimer will sell off his sister or his wife or daughter for prostitution as long as there is money to be made, such as the mentality is of all bankers and moneylenders. You will find any stock exchange filled with the mentally impaired Hoggenheimers. To the Hoggenheimer everything is turned into money and everything is in terms of money.

Time is equal to life…no time is life. If you steal someone's time you are stealing that person's life. It is time that life uses to collect cells to last life a lifetime. Time is the essence of life because time is life and I prove that time is within singularity and then afterwards time becomes space as the history of time. I am going to show this statement in the following but brief argument.

Money is replaceable any day of the year but time is irreplaceable life. It takes time to live an entire life whereby during that life I as a person should not gain money of which I eventually will have no purpose for, but I have the purpose to gain information so that my psychic who is I can gain from the time I had on Earth. It is my body I leave behind and life I take with me because it is life that is the driving engine of the body I leave. Life is the energy and according to Newtonians energy can never be destroyed but can only go from one form to another form. If the body becomes a cadaver, the energy that drove the body went missing. That energy can't destroy so I have to realise in terms of physics that the energy I call life went from one form I could see to another form that I can't see. What leaves the cadaver when life leaves is movement, intellect, emotion, personality individuality and ability to think which are all thing that forms me much more than the body forms the "me" in me.

No person can touch another person. That is a misconception mesmerising all that can think. I can't see another person nor can I touch another person. I can at best touch a selected few atoms that forms a smaller part of the larger construction that represents that person's body. I can't touch the Earth but at best I make contact with a few atoms that is part of the overall structure my intellect think of in terms of as being the body and life Earth. I can't ever see another person. I can see light reflecting from a position that the body of another person was in when the light was reflected by the body of the other person. Life uses time to gather a body and when life leaves the body, then the body disintegrates to become mere scattered atoms. It is life bonding the body it is not the body containing life. Therefore, placing the emphasis on the body when thinking in terms of life is totally Newtonian which means it is wholly backwards. I can never be the body that holds me because as soon as I holding removes from the body, the time component that gathered the structure we think of as forming the body completely disintegrates. My life leaving the body will desecrate the structure and life leaving the body will allow the body too decompose into eventually becoming scattered atoms that only portraits total fiasco.

The body I use as a vehicle to get me around during the time of my life is in some way connected to that body making the body not really exist. If life did not accumulate the body and drive the body the body would never be and without the life the body can't go on remaining to be what we think it is but it is life compiling the structure allowing the body to be what it can and will never be. The body is not except for life allowing the body to be. When life is present there is a body but without life the body destructs back to atoms. The body I think is I am not I because that body will disappear.

That body is a hallucination life creates to fore fill a function and serve a purpose and then when the purpose is served that body will disappear once more. Once life is removed from connecting to the carbon cell the structure of the cell disintegrates totally. That makes life the reality while it was holding the cell structure together and that makes the cell structure a hallucination without proportions. The human or animal body does not truly exist at all but is only present in the mind of the person holding life. Even when I ate the body of the animal it serves me as a compliment extending my life for a very short duration, then it is no more and no traces of it is left.

I can't prove the other person has a body since I only see the light being reflected and never see the body. I have to use interpretation about some hallucination I think of as being another person to convince me it is the other person I think it is. I can never touch the person to find conformation that it is the other person because I can merely touch a few cells and that would be the hallucination that I will use to form an interpretation that it is the other person that I touched. The same goes for the smelling and the sound. Therefore I have to depend on being foul witted to believe I see what I see, I smell what I smell and I hear I hear b y thought or life and not by reality in the cosmos. Every aspect of what I think is I or what I suppose is another person in the Universe is a fantasy that either I or in the case of another person then the Universe creates on my behalf. At this point I wish to show how an important factor **Sir Isaac Newton** just chucked away because **Sir Isaac Newton** had no idea what the work of Johannes Kepler consisted of. There is no true Universe that we see. We create a Universe in our minds by perception of what we think we see ands not what is there to see.

The Universe is one large hologram with only make-believe filling it.

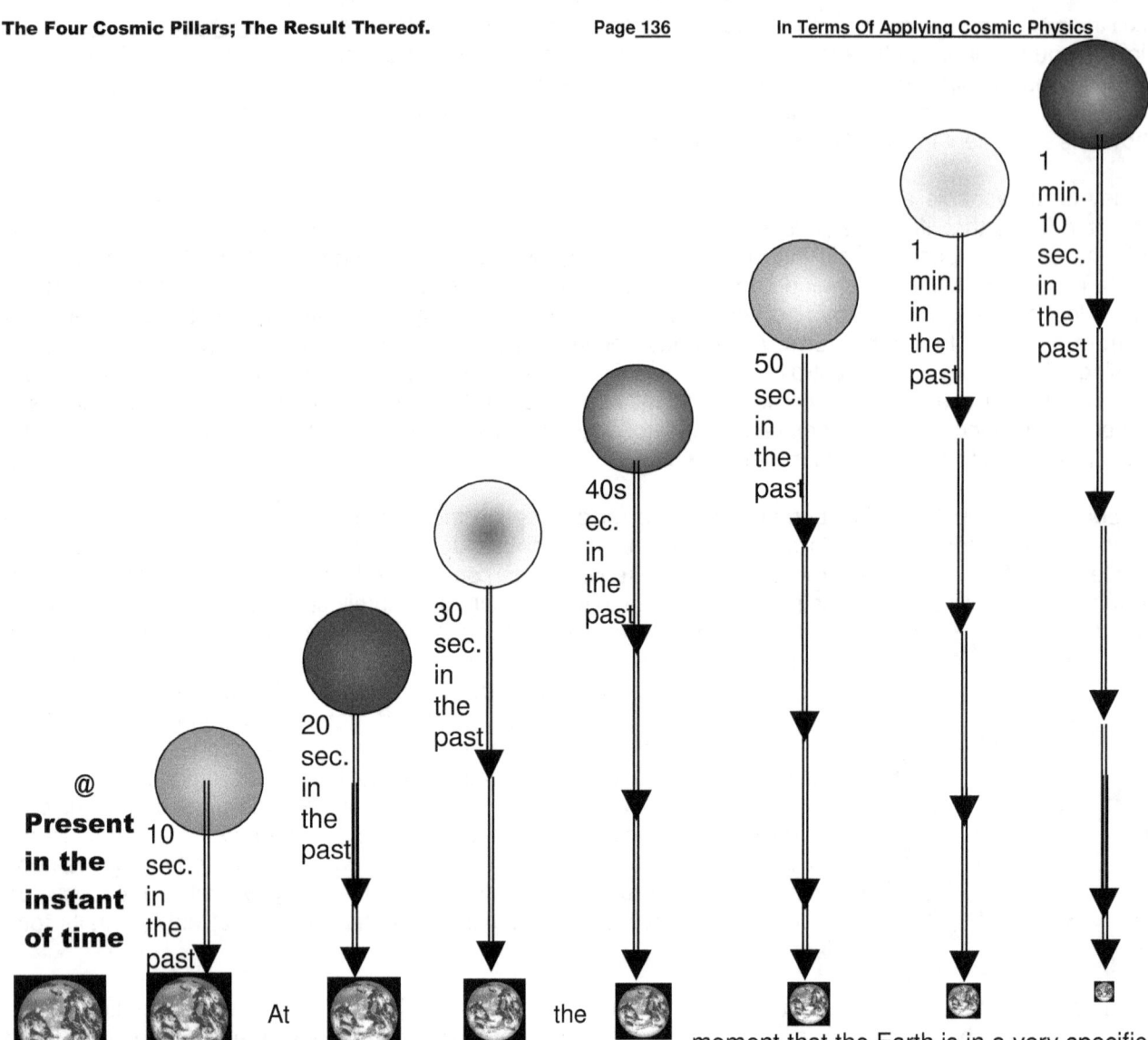

moment that the Earth is in a very specific position in time in the moment forming the present in singularity, time formed space as the past and in that relation the Earth will be in many various positions as it was in the past but would seem to be in the present in accordance with other planets being at a distant apart from the Earth by the distance separating them. Nothing and no one can go time travelling. Time is the spot where every dot is confirmed by time in the instant affirming the dot holding singularity. Then as time goes three dimensional time forms space by re-affirming every dot in relation to the movement of all other dots forming space in relation to time by allowing space Π to form as an extension of time Π^0.

The Earth would be in a very different position at the very instant we look at the Earth when viewed from 7 different planets when every planet is further than the previous planet in relation to the Earth. No one image we have would be actual-in-the-instant and not one image would be correct. Every image would merely focus the past of what time was when the image was the present. As time shifts the Earth it takes light time to flow from the Earth to the planet the Earth is being viewed from. From every planet the Earth would be in a different position and the Earth would also seem different as it would seem smaller. Time placed the image of the Earth in a different space in relation to the first space the Earth was in where the Earth is. However, what is seen is an image of space that time has left as a reminder of where time was when the space it left was valid in the instant of time. What we then see is a hallucination of what we think we see because it is not there where we see what we think we see. That is the cosmos...it is a hollow hallucination. The cosmos is a hallucination of mirage images that is not there but for our imagination placing it there in terms of where we see our selves being where time puts us in relation of what was the past followed by the present that confirms the future. What we see is an unreal left over of time that has gone by being the past.

Time is the instant in which singularity confirms points not representing space but forming space by committing to movement of the space. Space is what time then leaves behind when time moves on to form the next instant representing time through committing singularity whereby space is reaffirmed. As time is replaced by time in the instant the previous instant of time forms space as a reminder of what time left behind. Space is the history of time but space has no validity to be confirmed as substance. One cannot move through time in space to what we see being outside the solar system because what we see outside the solar system is no longer there and it would be most definitely not be there when we think we would arrive there. It would take us billions of years to reach the nearest star which is only 4.9 light years away, but we could never each it because if the gravity or the time that the Sun inflict on us.

When something is part of the Universe it can go nowhere but remain in the Universe. Time is the only thing that enters the Universe but as time can't leave because it became part of the Universe, time has to leave a reminder of what it was when it was where it was and the reminder it leaves we think of as space.

That which can only start by connecting to eternity

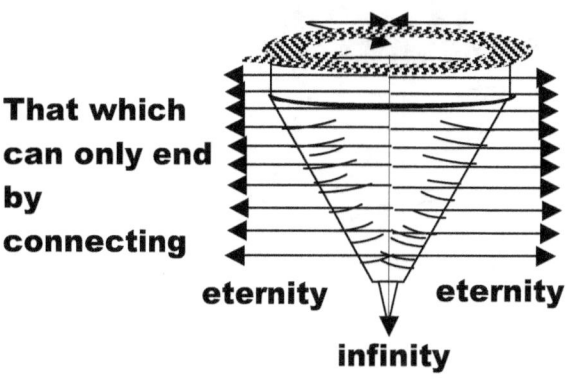

Every time in the instant when time form another presence time shifts whatever forms space in relation to the point representing time. Time is vested in the governing singularity Π^0 from where time forms space Πr^0 as the controlling singularity. Time moves in the instant Π^0 by becoming space Πr^0 as the governing singularity shifts to the controlling singularity $\Pi r^0 + r^0$ from where the controlling singularity shifts Π by the measure of r^0 or 1^0. However this happens on the condition that $\Pi^2 = \Pi^3 \div \Pi$ to form space-time. This applies when the Universe gore three-dimensional in time and space is affirmed in a new position.

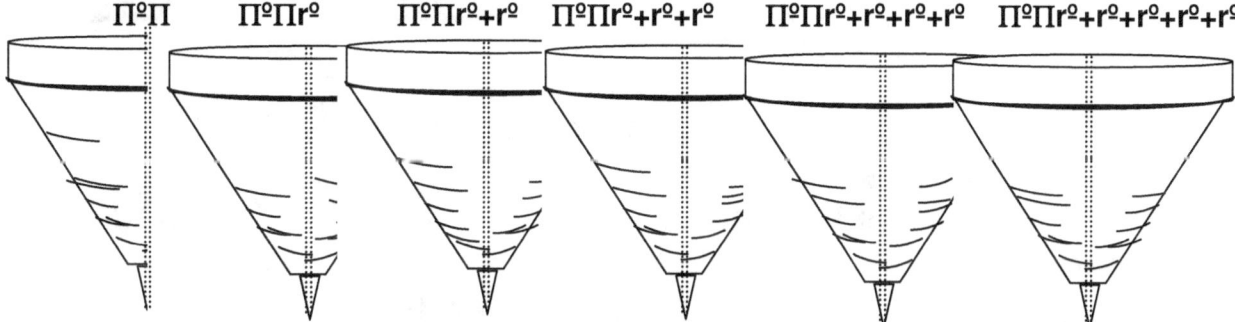

Then time goes singular. It is not like Einstein saw it where the entire Universe becomes flat. The three dimensional space never disappear even in the instant. We have to look at Π to find the practise of how time applies and there are many other books I go into detail such as **The Veracity Of Gravity** and **An Open Letter on Gravity Part 1 and 2**. We can see the one moves straight in relation to singularity $10 + 1.9991 + 10$ where the shift from 0.9991 to 1 is by the circle of 7. In that way we have time (going from 0.9991 to 1) moving the governing singularity straight 10, around the circle 7 and then straight again 10. At that point the shift time brings to space by controlling space through the governing singularity is complete.

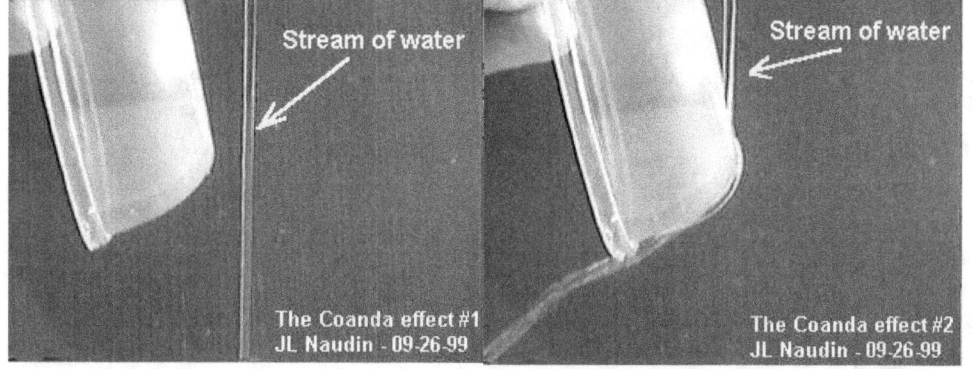

Lets put what I said in terms of an event that took place on Earth and by following the event we can see how time puts space in a location as a remembrance of time gone by. The electron is a solid pump spinning in order to pump liquids and that pumping action works on the Coanda principle, which is what I explained in much detail. The atom has the responsibility to contract space $\Pi = \Pi^2 \div \Pi^3$ by rotating $\Pi^2 = \Pi^3 \div \Pi$ solids inside liquids $\Pi = \Pi^3 \div \Pi^2$. Every atom is a future Black Hole.

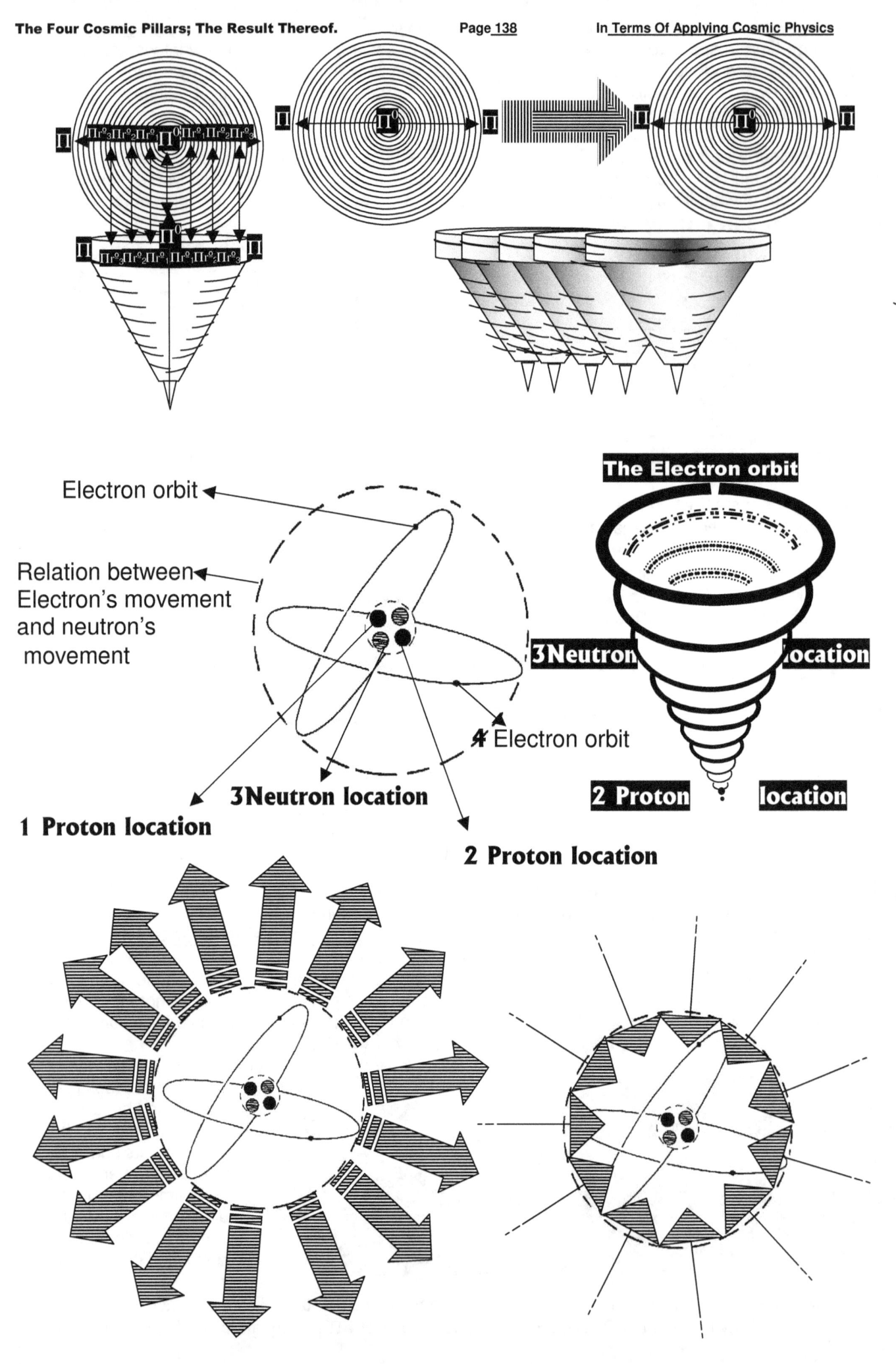

The Four Cosmic Pillars; The Result Thereof.

The atom is a little time capsule that determines the measure of applying time by setting the gravity standards and gravity is time. Gravity is what forms a relation between that which contracts and the centre whereto towards the contracting is directed. However, in the face of such contraction there is also growth of singularity applying because for every time that gravity or time reinstates the atom by shifting the centre to a new location, it also reinstates outer space by introducing one more space less dot and this introduction brings about movement by all the other dots already in place where they form the history of time. This replacing of dots by time all over the entire Universe is the Hubble constant. By contracting it compresses or condenses the liquid form of space surrounding the atom it condenses space and accelerate the movement. But in the same process it forms a more dense atom because the gravity determines the size $\Pi^2 = \Pi^3 \div \Pi$ of the applying atom that forms the gravity. The atoms form a collective effort to establish the applying gravity that establishes the applying space $\Pi^3 = \Pi^2\Pi$. The atom holding the relevancy between $\Pi^0\Pi$ is what determines the specific applying time or gravity within and around then star.

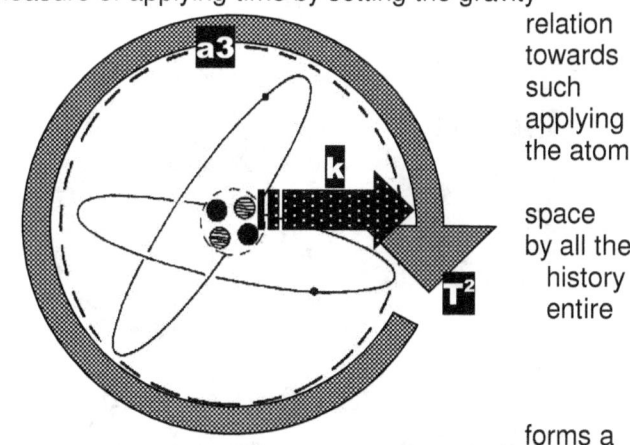

When the star gets smaller by the increase in the gravity applying within the star, the star gets denser and that makes the atoms forming the star get "smaller" or denser because whatever the star becomes, it then is because all the atoms within the star is what the star is. The increase in spin limits the relevancy, which in turn limits thee space, occupied by the atoms. The composition of the star gets denser and that makes everything within the star become more compact, denser and the increase in mass with that becomes exponentially more. Every controlling relevancy applying moves closer to the governing relevancy as gravity increases. This leaves us to realise what gravity is and that explaining I leave to the next books.

Even the atom being the substance forming the Universe will one day disappear into where it first came from and we all know everything came from singularity mismatching movement by control. The atom as all other smaller matter that combines to form the atom did, every particle no matter how small or big came from the relevancy time created when time parted infinity and eternity. As time increases that which holds no material so time decreases the space of that which is material. In the end the gravity will be so strong it will pull all material back into singularity as is the case in the present in the Black Hole and time again will be one eternal line as the three dimensional hallucination called space again hold infinity and eternity on one spot as it was when it began and everything will be one point. How this will come about is left for another book.

The faster the atom that is filling the star spins in relation to the Universe outside expanding or growing the less space it would have to occupy and the more space it would capture. By having less space it holds a denser relativity in relation to what applies outside and the shorter the relevancy between the governing singularity and the controlling singularity becomes, the faster would the relative spin of the atom be. Stars getting bigger is not getting bigger in the way Newton said it gains in size, but by getting bigger it gets smaller just as we can see by using Kepler's formula. The space outer space holds again is not getting bigger and neither is the space that the atom holds getting smaller but as the outer space is expanding it does so in terms of the star shrinking in size. Again it depends on the relativity applying. Again it depends on the relativity applying. Gravity and the speed light travels depend on how cold singularity is in terms of how hot outer space is.

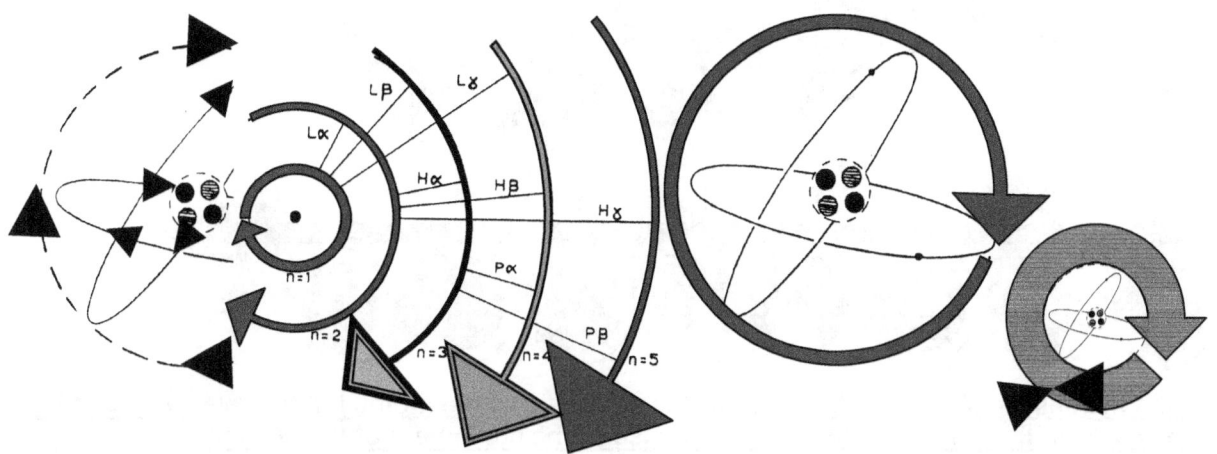

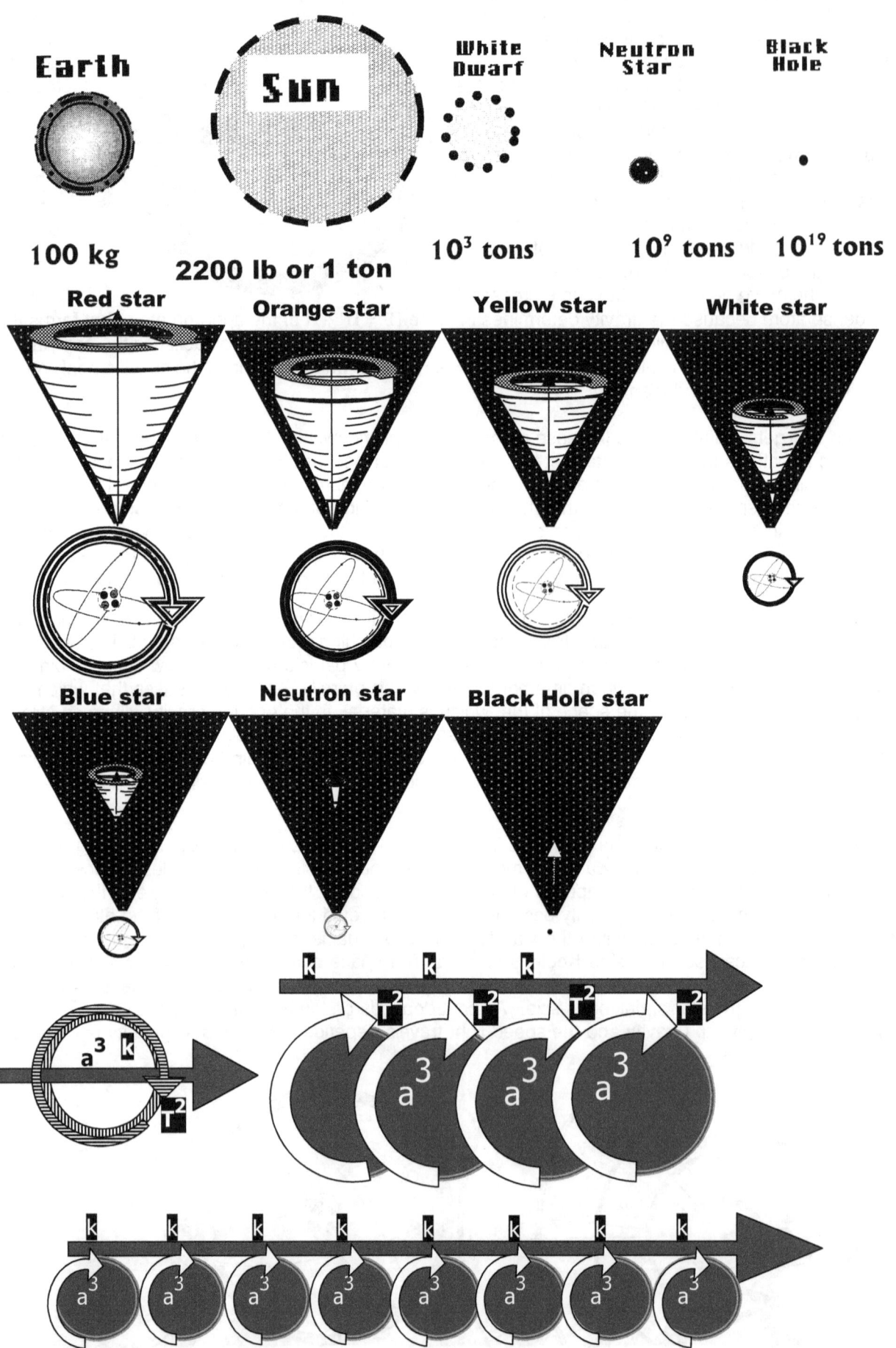

It is not the increase in size of the Universe or the shrinking of the atoms but the relevancy coming about delivering a discrepancy between the increase of the movement of the one enforcing a

The Four Cosmic Pillars; The Result Thereof.   Page 141   In <u>Terms Of Applying Cosmic Physics</u>

<u>relative decrease of the other that brings the increase in the gap but the movement and lack thereof that reduces matter as time increases in outer space. It is all about relevancies applying.</u>

Lets envisage a scenario where we have 7 plants all evenly apart located at intervals of say one second at the speed of light from the person playing a role on the Earth. At first we have the person going by the name of Sucker being bored and wondering how he would get through the day with nothing to do but sit around and wait for the market to open. The scene is in the present of the very instant time forms a presence that is directly connected to the instant that is directly connected with his body. In the case of Sucker it is very dangerous to be bored because Sucker thinks he can think but everyone else thinks he can't think. Lets, for argument, sake put the planets at an equal time-lapse time interval. That is, we put the light travelling from where Sucker is to every planet at that same time plus every one gets the accumulated distance plus the previous one added every time. The light coming to B would be the distance of B + A, and the light reaching C would be (A + B) + C. On every planet we have the American N.S.A. spying on everyone in the world and that will include Sucker.

Planet A sees the Earth with Sucker jobless and bored in Iraq, which is very, very dangerous.

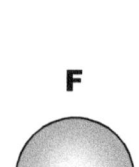
G

F

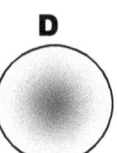
E

D

C

B

A
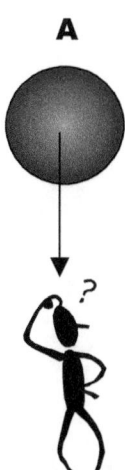

Then at the next time interval Planet B sees Sucker being bored while planet A see Sucker getting an idea as to bring some excitement to the people he share a city with, since as usual nothing is happening in Iraq except for Americans bombing the shit out of civilians and making mayhem. This is observed only by the NSA on Earth and in direct view of Sucker as well as the second scene being seen from the nearest planet that is infested by the NSA at the distance it is located.

Planet A sees the Earth with Sucker getting the idea while to persons on
Planet B sees the Earth with Sucker still being bored.

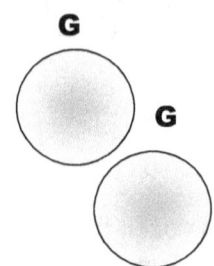

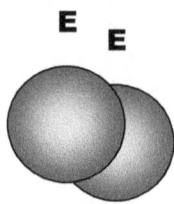
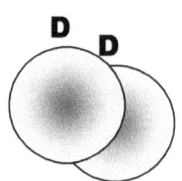
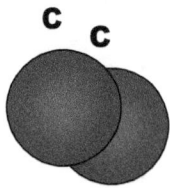

Then moving one time lapse on the reality shifted leaving the events unfolding while the past is soldering on at the speed of light further away in time forming space which is the prints time left in order to become space, which then space is the history or the past of time.

Where we then are at the next time interval at Planet C they see Sucker being bored while planet B see Sucker getting an idea as to bring some excitement to the people he share a city with and Planet A see Sucker reach for the biggest cracker the poor city ever saw.

Planet A sees the Earth with Sucker collecting the massive firecracker.
Planet B sees the Earth with Sucker getting the idea while
Planet C sees the Earth with Sucker still being bored.

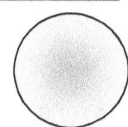
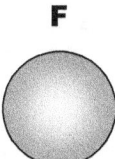

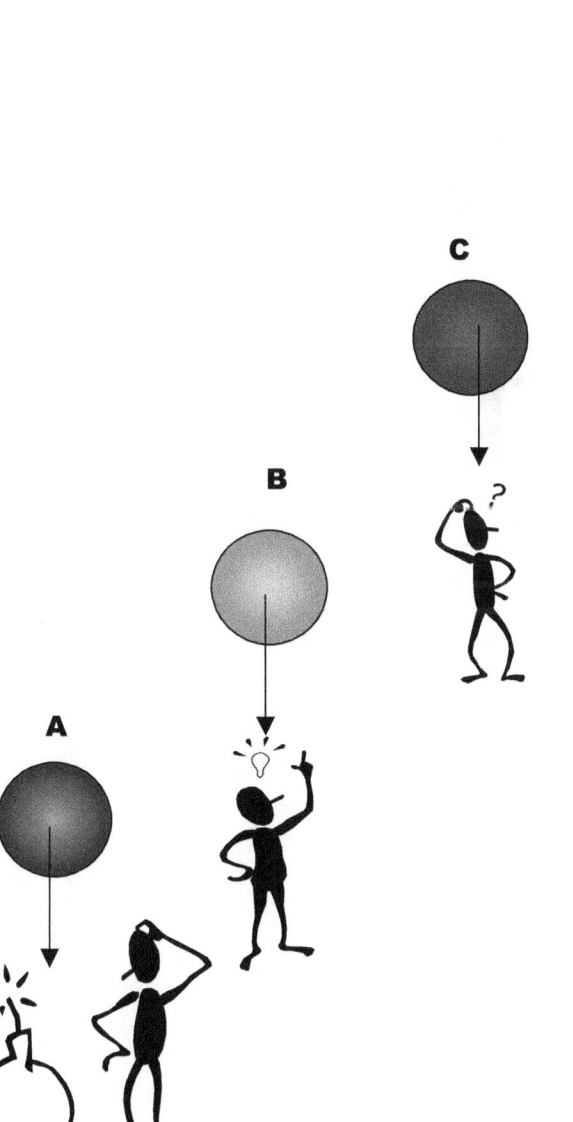
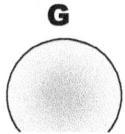

The next scene happening in the next time interval shows why Sucker having a low IQ could be life threatening to himself more than others especially when he is not taking care of the disease called mental inadequateness as one should when any person is suffering from stupidity. Sucker did not get a firecracker but instead collected an Iraqi home made bomb in a DIY kit, which he detonated and the result form that was he did not surprise everyone as he intended to but scared the life out of the lot, even the Americans that should by now be use to blowing people to fragmented bone and mince meat. At this point the Universe sees:
Planet A sees the Earth with Sucker detonating the massive firecracker.
Planet B sees the Earth with Sucker collecting the massive firecracker.
Planet C sees the Earth with Sucker getting the idea while
Planet D sees the Earth with Sucker still being bored.

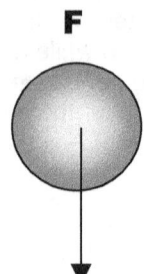

The Four Cosmic Pillars; The Result Thereof.   Page 145   In **Terms Of Applying Cosmic Physics**

What follows is what Sucker does best, because since his days as a child Sucker always runs from his stupidity while never realising his stupidity is running with him wherever he might go. Sucker thought everyone would be happy and laughing but now everyone is unhappy and frightful especially after the New York 911 and the 0707 London bombings not forgetting the 0311 Madrid train bombings because while the Americans are doing strategic bombing targeting as many civilian homes suspected (not proven) of hiding terrorists where such bombings of 500kg per shot and is maiming every Iraqi in the neighbourhood to the distaste and horror of everyone on Earth, The Americans bomb the daylight out of Iraqis to show the Iraqis that no one must do it to Americans.

Planet A sees the Earth with Sucker taking flight as a result of experiencing his personal stupidity
Planet B sees the Earth with Sucker detonating the massive firecracker.
Planet C sees the Earth with Sucker collecting the massive firecracker.
Planet D sees the Earth with Sucker getting the idea while
Planet E sees the Earth with Sucker still being bored.

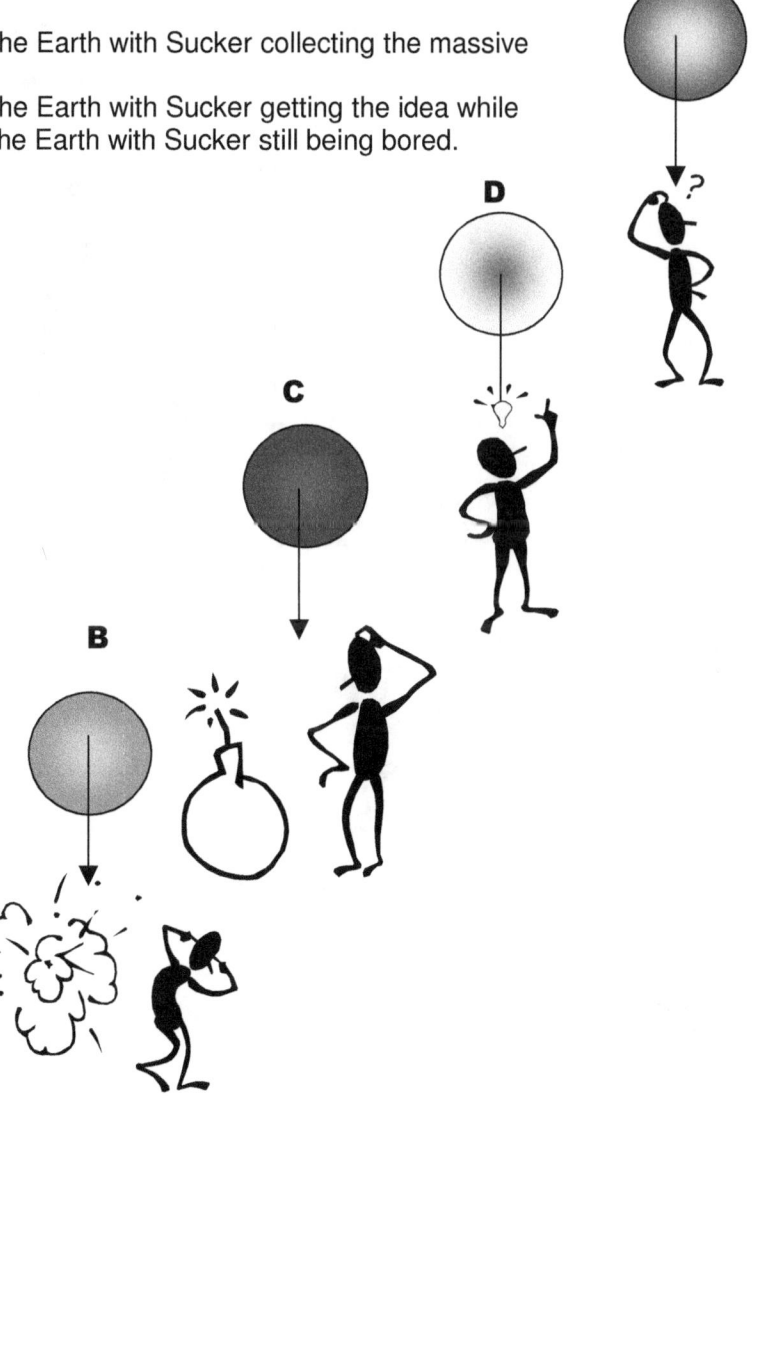

What follows is what Sucker does best. Sucker always run from his stupidity while never realising his stupidity is ruining with him wherever he might go. Sucker thought everyone would be happy and laughing but now everyone is unhappy and frightful especially after the 911 and the 0707 London bombings and the Madrid train bombings. This give the Americans more reasons to kill and maim Arabs in their own country.

Planet A has Sucker falling his freedom away for many years because everyone was surprised at his complete stupidity and chased him.
Planet B sees the Earth with Sucker taking flight.
Planet C sees the Earth with Sucker detonating the massive firecracker.
Planet D sees the Earth with Sucker collecting the massive firecracker.
Planet E sees the Earth with Sucker getting the idea while
Planet F sees the Earth with Sucker still being bored.

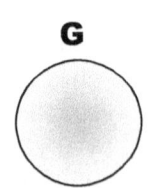

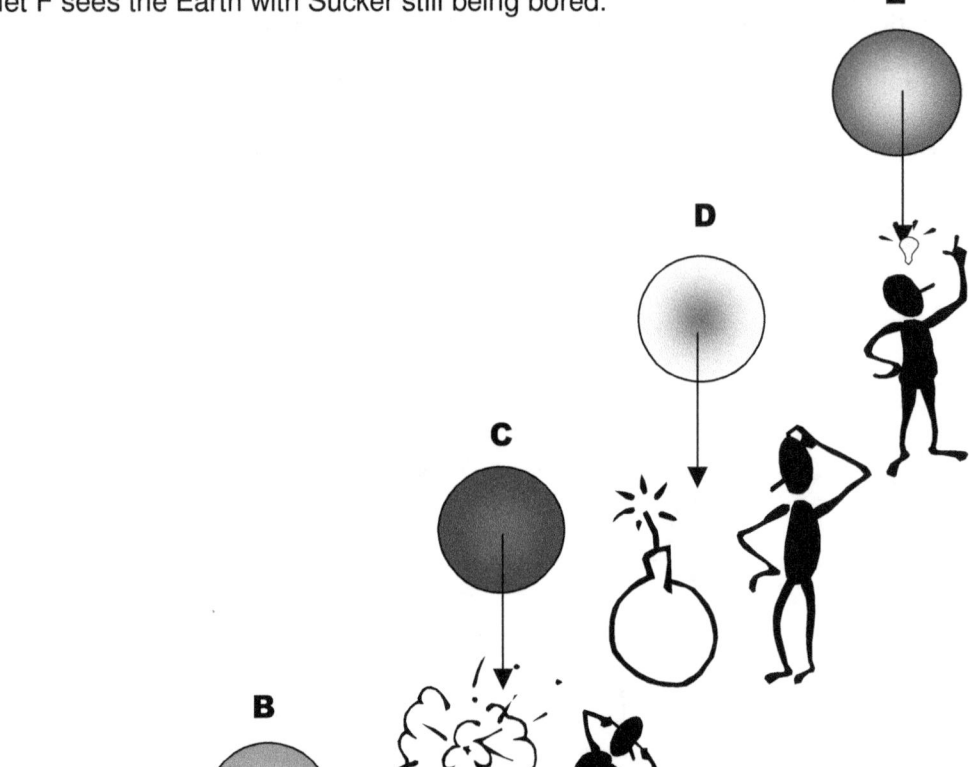

The Four Cosmic Pillars; The Result Thereof.   Page 147   In **Terms Of Applying Cosmic Physics**

When the town folk caught up with Sucker they insisted on life long incarceration and even after death his corpse should not be freed for burial because he acted as if he was an American, which he was not. Sucker committed an act of terrorism and such acts are exclusively reserved for George Bush and his executioners that has the privileged position to bomb Iraqis, destroy Arabic properties without compensation any Iraq civilian and then those that does not like the American destroying of their liberties and engage in American acts matching the American brutality George Bush labels those to be called terrorists because they have the audacity and tenacity to fight to have their country freed from American invaders, which according to Bush must be seen as American liberators.

Planet A sees Sucker lost his freedom and carries chains and shackles for many years to come.
Planet B sees the Earth with Sucker falling his freedom away.
Planet C sees the Earth with Sucker taking flight.
Planet D sees the Earth with Sucker detonating the massive firecracker.
Planet E sees the Earth with Sucker collecting the massive firecracker.
Planet F sees the Earth with Sucker getting the idea while
Planet G sees the Earth with Sucker still being bored.

This, Newtonians, is your Universe you measure and calculate and observe and dissect. It does not exist but for being some image seated within your imagination. This is the reality you wish to visit through time travel and in which you wish to establish space whirls and do all sorts of idiotic things that seems to make you feel special.

The Four Cosmic Pillars; The Result Thereof.   Page 148   In Terms Of Applying Cosmic Physics

So, viewed in relation too and from every different planet, Sucker has a different posture. Every planet has a different vantage from where such a planet would observe Sucker enjoying his daily life but in a different time slot formed as space. Sucker holds a different time frame in relation to reality at every point in space in time. Not one of the images are for real and everyone shows a certain perspective of what could be the reality while not one image is truly complying to the standard of being reality since every image is written in light as forming non-existing space. Sucker could be long gone dead. Sucker could have been a movie and all of his actions were just light flickered on film. In reality no one can touch Sucker in the position he is in or where they see him to be. Sucker is just an image with no degree of reality but forms a hollow hallucination in minds that is seeing thing and playing tricks on the observers.

Moreover, this is your Big Bang you try to chase down. This is the time you award to space forming the Universe you think you see while your awarding time shows how little Newtonians understand of things in the cosmos. Also this is your edge of your Universe that you find so many new things you are able to discover and what can claim as your personal achievements by awarding edges in space.

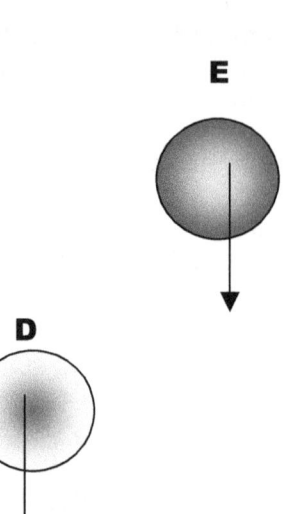

With the Master governing singularity that was present the instant the Big Bang took place and point that now is totally faded and split to every atom formed in the growth of the controlling singularity or forming time at present that is formed by every spot unseen and unaccounted for, the past formed in space can be seen but never again be experienced as it was back when the Big Bang was the news event of the day. The Universe you see is just a hollow hallucination of images not existing that is playing tricks on the minds of those incapable of understanding the cosmos.

The Universe we think we see does not exist. There is no Universe out there but there is some illusions we can associate with and by playing mind games we can value what we see for what we think they are worth by using our over excited imaginations, but in the end it is just an illusion, and that is the reality never to be forgotten.

This was the Absolute Relevancy of Singularity in terms of Applying Physics

When going onto to read the other three parts of the Absolute Relevancy of Singularity being

Book 2 The Absolute Relevancy of Singularity in terms of The Four Cosmic Phenomena
Book 3 The Absolute Relevancy of Singularity in terms of The Sound Barrier
Book 4 The Absolute Relevancy of Singularity in terms of The Cosmic

You will find out why the Universe is formed by …
Gravity that is The **Roche limit**,
   Gravity that is The **Lagrangian system**
      Gravity that is The **Titius Bode law**
         Gravity that is The **Coanda affect**

We as humans love to see the cosmos as spheres forming but having this view of the cosmos shows just how little thought we give to the idea of what the cosmos is!

While we use gravity the use of gravity makes us part of the Earth by mass forcing us onto the Earth as a semi unit with the Earth. Is that view truly portraying gravity?

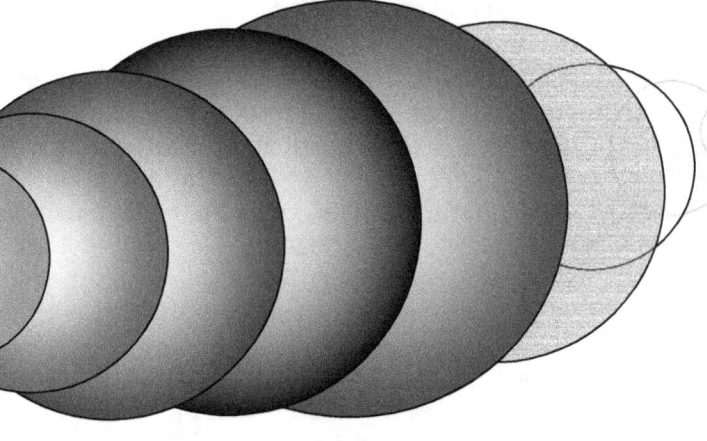

Kepler showed there is a growing Universe long before Hubble's birth

This view about the view we have thinking of the cosmos in terms of a sphere shows total arrogance we have and proves the misconceptions of Newtonian physics.

The cosmos spoke to Kepler about space-time coming from singularity

All the questions I put to you I answer in this book… and in the book I answer a lot more questions than what I ask here

Gravity being the Roche limit forms the principle in producing the sound barrier.

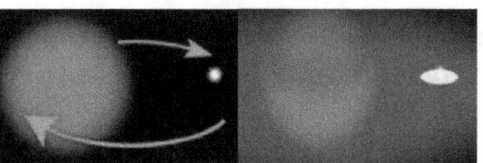

The books explain in the practical sense mathematically what the Roche limit is. We all know that Gravity is the sound barrier but the books explain how it is the sound barrier and how the sound Barrier implicates the Roche limit. It explains the sound barrier by indicating how gravity mathematically apply

These books explain what is motivating the expansion and the moving of the cosmos.
Therefore this book explains following phenomena:
It explains the principals **supporting** gravity
   The location and the value of singularity in relation to gravity
   Finding space-time, proving space-time in relation to gravity

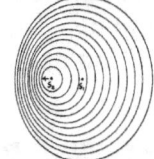

Gravity is **The Doppler affect: but it explains the Doppler effect in a new**

The Four Cosmic Pillars; The Result Thereof.   **In Terms Of Applying Cosmic Physics**

These books are about explaining the Roche limit as a derogative of singularity applying being one of the four cosmic laws that I prove

Using Kepler findings I can prove mathematically that:
Gravity is The Roche limit,
Gravity is The Lagrangian system
Gravity is The Titius Bode law
Gravity is The Coanda affect

'Some answers in this book are about telling the life story of the Universe: where did it all come from
...where is it going too
...and most of all...why is the cosmos going in that direction it is heading
...where did it begin in the time before the Big Bang.
All the answers came a result of studying Kepler.
The cosmos mathematically spoke to Kepler personally using mathematics as the medium that provided us information... ...about the cosmos.

It said Space $=a^3$ is equal to the motion T^2k of space a^3 = Time T^2k

There now is an explanation about the growth of space. There is an explanation that Kepler gave of why the expansion is occurring...and nobody since Kepler took any notice. The impression such as the above picture matches every logic view we all have about the universe, but does Science really provide the answers matching our modern logic, or are we filling in and compensating for science's shortfalls. Does what is out there really match official science? Does the Hubble expansion pictures match the explanations about how Creation all started... where it is heading...and where it will end?

Studying Kepler helped to understand why the phenomena are there to begin with and that enabled to explain in some way...
Why is the Universe depicted as a sphere...and why would that then be correct...
...How did everything become so much and so large...
...Why did it start small?
...Why does it grow from small to large?
...Why was the start so small?
...Why is it growing?   Gravity is
...Where is it going while it is growing ...
...why is it any specific size...
...What was everything before that?
And why in creation would it then reduce again!!!

The Coanda Principle
That I explain

Every one in science throughout many centuries ignored Johannes Kepler because all saw him as some derogative of Newton...until now. Kepler introduced space –time but nobody took the time to acknowledge Kepler's introduction. Kepler introduced space a^3 – time T^2k and showed that it is space a^3 – time T^2k that is performing gravity by relevance of k. Is our centuries long ignoring of Kepler truly the answer...Kepler introduced gravity by principle but no one in four hundred years took any notice of the manner in which Kepler brought gravity into human conception and understanding.

Kepler calculated that it is the motion of space a^3 during the time T^2k that forms the gravity that is keeping the sun and all the individual planets apart but moreover gravity is keeping the planets in orbit. While every one was surprised but now accepts there is a growth in the Universe by which the Universe is expanding...for four centuries Kepler said that and no one took notice. According to Kepler the expanding is the normal trend that the cosmos will follow...that he said four hundred years ago...yet in spite of Kepler findings...science still clings to the idea that what keeps the Universe secure is contracting the force by the mass value that creates an attracting in the distance between the objects. That is NOT what Kepler said. Gravity is the effort of independent objects to secure their position as the centre of the Universe by motion of space in space in relation to space by moving through space.

Let the other three books convince you to reconsider your verdict about what gravity is, because by you reading it, it most probably will...

The Four Cosmic Pillars; The Result Thereof. The Sound Barrier

© **KOSMOLOGIESE EN ASTRONOMIESE TEGNIKA**
WRITTEN BY PEET SCHUTTE   ISBN 978-0-9802725-3-6

EXPLAINING THE "SOUND BARRIER" IN TERMS OF PHYSICS BY PUTTING SINGULARITY IN ABSOLUTE

RELEVANCY

ISBN 978-0-9802725-3-6

All rights are reserved.
No part, parts or the entirety of this book may be reproduced by publishing, electronically copied, duplicated by whatever means that form reproduction or duplication of any description, without the prior written consent of the copy rite owner.

WRITTEN BY PEET SCHUTTE
© KOSMOLOGIESE EN ASTRONOMIESE TEGNIKA

APPLYING AS THE SOUND BARRIER

At $7(\Pi\Pi^2)\Pi^0$ the designated space in relation with the aircraft will not be sufficient to lift the aircraf of the ground. It would then require huge wings to increase the space the airplane holds in relation to the space the airplane requires to fly. The speed must increase to $7(3\Pi^2)1.25\Pi^0$ at least to become airborne and lift from the earth. The moment the aircraft is not touching the groun it loses the "mass" component science holds so dear.

At a speed ratio of $7(\Pi\Pi^2)1.25\Pi^0$ the plain exceeds the size to space ratio which the plain requires to remain at a level of $7(\Pi\Pi^2)1\Pi^0$ and it will according to the increase in space ratio lift from the ground and into the air. This is when the aircraft gets airborne. However this is where the craft loses the "mass" it had where "mass" is just an indication that by touching the ground it is part of the earth according to singularity.

The aircraft "increases" in space required and therefore moves to the next level of space above ground. Being above ground and not touching the earth changes the relevancy in movement from $7(\Pi\Pi^2)1.25\Pi^0$ to $7(3\Pi^2)1.25\Pi^0$ which places the aircraft in the liquid zone and forms no longer part of the earth's solid factor. Now the aircraft becomes space, which moves towards the earth. This linear movement has exceeded the downward gravity of $7(\Pi\Pi^2)$ or the craft will fall to the earth. It is the need for space that increases. Depending how you look at it the space the movement goes through increases and the size of the plain increases or the space remains the same and then the plain shrinks in size. This relevancy between that which moves and that which it moves through becomes so big that the space increases at low levels that it "breaks" the space or change the space to being so big sound does not exist

Specific density
This is $7(3\Pi^2)$ km / h

Static density
This is $7(\Pi\Pi^2)$ km / h

Virtual density
$7^0(3\Pi^0)(\Pi^2)(1.1\Pi^0)$
$7^0(3\Pi^0)(\Pi^2)(4\Pi^0)$
$7^0(3\Pi^0)(\Pi^2)(\Pi^2/2)$
$7^0(3\Pi^0)(\Pi^2)(\Pi^2/2)(\Pi^2/2)$

Variable density
From $(1.1\Pi^0)$ to $(5\Pi^0)$

Limitation density
$\Pi(7^0(3\Pi^0)(\Pi^2)(\Pi^2/2))$
$7^0(3\Pi^0)(\Pi^2)(\Pi^2/2)(\Pi^2/4)$

At $(\Pi^2/2)(\Pi^2/4)$ this is $7(3\Pi^2)$
At $(\Pi^2/2)$ this is $7(3\Pi^2)$
At $(4\Pi^0)$ this is $7(3\Pi^2)$ km / h Virtual density
At $(3\Pi^0)$ this is $7(3\Pi^2)$ km / h Virtual density
At $(2\Pi^0)$ this is $7(3\Pi^2)$ km / h Virtual density
This is $7(3\Pi^2)$ km / h Specific density
This is $7(\Pi\Pi^2)$ km / h Static density

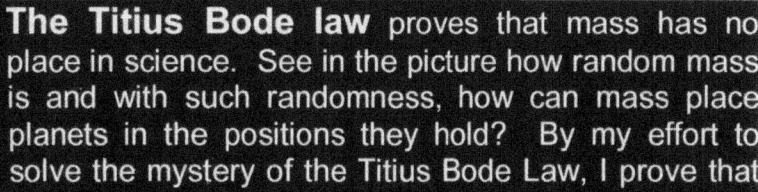

This is the Titius Bode law

The Titius Bode law proves that mass has no place in science. See in the picture how random mass is and with such randomness, how can mass place planets in the positions they hold? By my effort to solve the mystery of the Titius Bode Law, I prove that gravity forms not by mass but gravity forms by π forming in movement π². Solving the Titius Bode Law and proving from that how gravity works opens up a new view on the cosmos.

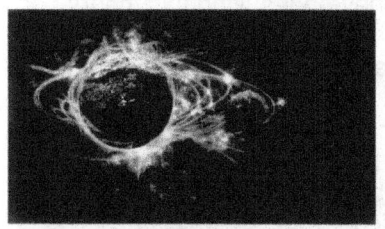

This is The Roche limit

The Roche limit has been around for centuries and with all the mathematical splendour available to apply in order to fathom concepts behind this phenomenon, still with all the computing ability of a machine all those physicists with all the mathematical superiority could not touch any understanding about the concept forming the background. Yet when using the truth about gravity in physics the answer is simple; it is that gravity is Π.

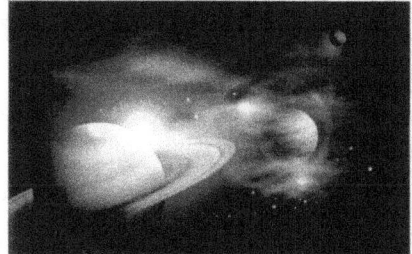

This is the Lagrangian points

The Lagrangian points have been known to science for centuries and with all the mathematical splendour available not one calculation could ever explain why this event is taking place. The satellites form precise locations positioned around the major planet and never comes closer while remaining in their positions.

This is the Coanda effect

The Coanda effect has powered turbine engines and aeroplanes in flight for almost a century and with all the mathematical splendour available to design the most terrific aircraft, not one engineer could mathematically compute one fact to show understanding why this takes place. How sad it is that those claiming of much superior intellect in physics remain just no more than having computing power. The understanding is not complex. I have to warn the readers that the topics are showing a very new approach with no quick answers. Understanding is in the proof and that does not come by reading just a few lines and then forming conclusions. The information is new but not hard to grasp. I did not put these phenomena in place and these phenomena nullifies Newton's correctness and the proof I bring goes beyond any doubt. I prove the Titius Bode law. Go to the internet and see how science doubt the Titius Bode Law and the correctness thereof while to solve the problem you add 3 plus 4 to get 7. That is if you want to find a solution. I have published the Titius Bode Law in four already published books but in this one I go deeper than the four already published. In each of the books I present I disclose how the Titius Bode Law forms gravity. These books are:

WHOM IT MAY CONCERN,

I do find much pride in my status as being Afrikaner and would like to have my names used by pronouncing it in the manner Afrikaans dictates...therefore I would sincerely appreciate the courtesy when readers will take note that my name and last name are pronounced in Afrikaans, which is originally from Dutch and must be pronounced that way. Peet one would pronounce "here" which is the closest English to the pronouncing of the "ee". The "Sch" in Schutte is pronounced exactly as school is where both actually are pronounced Skutte or "skool". By pronouncing my name in Afrikaans you do me the utmost courtesy any one can. Being an Afrikaner is what I am most proud of. Another point I wish to highlight is that I feel compiled to produce this work in a comic-like format. I have found that the more intellectual and the more educated Academics are, the less they understand the most primitive or classical mistakes in science as well as physics.

As I said my mother tongue is Afrikaans and my second language is English. I have per suiting this theory that I partly present in this book, of which the investigating research was done the past thirty years. Then I compiled my presentation thereof for the past nine years on full time basis whereby I was tying to introduce my findings to many academics without much joy. This past nine years saw me go without any income as I tried to get my theorem recognised. Going without a steady income left me almost destitute and in order to find a manner to get my theory across to the attention of influential readers, I decided to publish these books electronically as to try and get around the stranglehold of Newtonian bias controlling science at present worldwide. I decided to publish these articles through LULU.com which I saw as way the only manner whereby I could generate funding by which I would be able to have the twenty seven books I already wrote linguistically edited and then to have the books published on a Print-On-Demand basis. With my first language not being English and the books not linguistically checked by an expert there are bound to be language errors that readers will notice. In the past I tried to check my work myself but after checking say one hundred and fifty pages for language corrections, instead of having corrected work I ended instead having four hundred pages of new written information which is still not language corrected but holds a lot more information. This is because my priorities lie elsewhere. I aim to spend money on correcting the work as far as language goes, as I receive money and in the hope that I will receive money. I will have all my work including the one you are reading edited professionally and corrected as I find money to do so. . .

In the book that deals with gravity there are just too many and numerously wide ranging facts that form the complete picture as a whole, which leaves me unable to include a full introduction in a space as small as that which page will allow. The explaining include for instance those phenomena, which I call the four cosmic pillars, but wise as you are, you would not believe me at this point that I have cracked the coconut because I guess in your vast experience you have seen too many idle explanations in the past proving to be senseless and little impressive, therefore my mentioning my success would not matter much either way. The proof I bring is true about gravity being formed as a result of these phenomena, **1) the Lagrangian system, 2) the Roche limit 3) the Titius Bode law 4) the Coanda affect**, which I explain by delivering mathematical proof as to how they fit into the overall picture of gravity and which I mention just below. I prove the fact that every individual one of those phenomena is forming a unit that is in total being what we think of as gravity. The phenomena altogether constitutes a unit that forms the process working as gravity. Nevertheless my mentioning these facts will be just completely unbelievable to you without you reading the book, because I guess you have heard some attempt to explain the phenomena before but when I say you have not heard it in the context I put it, you might still be most sceptical because you have never heard it in the correct manner that I explain it and that poses the difference. Still you may not be convinced about my claims and although my explaining the phenomena is correct, does not change the fact that you don't believe me. The phenomena form an intergraded unit that results in gravity forming where each forms a part of gravity. You may still be you would be sceptical ...but convince yourself that I did manage to:

  1) Find the location, position of singularity as a factor forming space-time
  2) Finding space-time by dissecting Kepler's formula in relation to valuing singularity
  3) Finding and proving space-time and aligning space-time with gravity
  4) Find the working principals behind gravity as a cosmic occurrence.
  5) Find the reason for the Roche limit and explaining the resulting of gravity from that.
  6) Find out why the Lagrangian system, becomes the building form of the Universe.
  7) Find why the Titius Bode law mathematically provides the foundation of gravity
  By proving that the Coanda affect is gravity through activating space-time
  By using the above the four cosmic pillars, it enable me to present the proof where I now can explain what conditions bring on the sound barrier. By proving it is gravity that the individual structure generates motion above and beyond the gravity the Earth provide is what is producing individual motion that the independent object earned within the sphere of motion that the Earth's gravity provides where the independent and individual motion put the relevance that gravity has beyond the conserving means gravity has where **the space** that is serving the **independent object** is

independently in motion. The adding to the independence on top of the normal structural independence is creating more individualism by the independent motion of the individual structure being apart from the motion that the gravity of the Earth provides. The fact every one misses is that any structure that is not part of the Earth's crust has an independent gravity and the form this gravity applies is stronger than the Earth's gravity which is why the structure maintains its form and this provides the independent individuality the structure has giving the unique structural space. The gravity of the Earth strives to incorporate everything into the Earth's sphere and into the Earth's structure and therefore the fact that the object is not incorporated into the Earth shows defiance and individuality, which gives it, mass.

By applying individual motion on top of the structural individuality that increases by the motion that the Earth provides, the independence of the individual object is becoming further exaggerated by having independent motion, which is further defying the incorporation the Earth strives to achieve. As the motion of the independent object grows more independent by applying more excessive motion to such an extent where motion creates almost the ultimate independence that may free the individual object with independence from the motion the Earth creates is what is breaking the restraint gravity has on all objects with independence formed by their structure. The structure show independence at all times by not forming part of the structure of the Earth within the sphere of the Earth's gravity. Moving about shows even more reluctance on the part of the top when spinning allows the top to eventually become part of the Earth. Breaking the sound barrier is the motion in space duplicating space by crossing over gravity borders, which is the limit to what constraint the Earth may produce in accordance with what full independence would allow.

These are the definitions underwriting cosmology and while my work is that much ignored; let's see how far I stray from these definitions in comparison of how much Mainstream science underwrites these definitions by bringing indisputable proof in presenting unwavering hardcore facts.

Quoted directly from the Oxford dictionary of Astronomy the following:
The definition of space-time is as follows:
Space-time is a four dimensional position of the Universe where the position of an object is specified by three coordinates in space and one position in time. According to the theory of special relativity there is no absolute time, which can be measured independently of the observer, so events that are simultaneous as seen from one observer occur at different times when seen from a different place. Time must therefore be measured in a relative manner as are positions in three-dimensional Euclidean space, and this is achieved through the concept of space-time. The trajectory of an object in space-time is called world line. General relativity relates to curvature of space-time to the positions and motions of particles of matter.
The definition of singularity is as follows:
Singularity: a mathematical point at which certain physical quantities reach infinite values for example, according to the general relativity the curvature of space-time becomes infinite in a black hole. In the big bang theory the Universe was born from singularity in which the density and temperature of matter were infinite.
The Oxford dictionary of Astronomy defines gravitation as follows
Gravitation is the force of attraction that operates between all bodies. The size of the attraction depends on the masses of the bodies and the distance between them; gravitational force diminishes by the square of the distance apart according to the inverse square law. Gravitation is the weakest of the four fundamental forces in nature. I. Newton formulated the laws of gravitational attraction and showed that a body behaves as though all its mass were concentrated at its centre of gravity. Hence the gravitational force acts along a joining of the centres of gravity of the two masses. In the general theory of relativity gravitation is interpreted as the distortion of space. Gravitational forces are significant between large masses such as stars planets and satellites, and it is this force, which is responsible for holding together the major components of the Universe. However on the atomic scale the gravitational force is about 10^{40} times weaker than the force of electromagnetic attraction
I have to give potential readers this fair warning that *The Cosmic Code as the Absolute Relevancy of Singularity* requires a somewhat higher level of understanding and needs a greater degree of insight that the other books in this series does namely

1 Explaining Physics in terms of the Absolute Relevancy of Singularity,

2 Explaining the Sound Barrier in terms of The Absolute Relevancy of Singularity,

3 Explaining the Four Cosmic Phenomena in terms The Absolute Relevancy of Singularity and

4 Explaining the Cosmic Code in terms The Absolute Relevancy of Singularity
Which all are also available from Lulu.com.

Whilst recognising the work of Johannes Kepler, Mainstream science bluntly ignores the impact of his work, and in that they miss the full vastness of the wide influence of his work. Newton shrouded Kepler's work under a blanket of alterations which I show was most unwanted since Kepler's work needs no alterations or corrections and every one since then kept Kepler's work hostage under Newton's changes. It is therefore almost absolutely realistic to say that all information what you are about to read in this letter and article sent to you for your attention was never yet printed in the near or the distant past although Kepler's work has been with us for about four hundred years, during which time it went unnoticed. It seems to me that any research predating Newton never came into use or into practise. My investigation of Kepler's work brought about a conclusion that no one yet arrived at concerning them with the findings of Kepler because no one scrutinised Kepler's formula before. Everyone is satisfied with Newton's version notwithstanding the incorrectness of it. The world seems satisfied with the idea that Kepler found planets rotating around a centre formed by the Sun and because of that Newton saw a circle. Where Newton saw a mathematical circle and was unable to understand $a^3 = T^2k$, Newton added what he thought is mathematically required to indicate such a circle. Newton added a mathematical $4\Pi^2$ to the formula of Kepler and removed the distance symbolising measure that Kepler introduced using **k**. On the other side Newton changed the symbol of **k** by using the symbols G (m + m$_p$). This is just a longer and probably a more detailed manner of indicating **k** and better defining of **k** but it symbolises precisely to the point what **k** stands for nonetheless. I wish to draw your attention to the matter of Johannes Kepler's findings that Mainstream science considers as resolved and closed for many a century while it is not. My investigating Kepler helped me too resolve other unresolved matters but it was only possible by using Kepler's work. This brought about the idea that the Universe is in a state of contracting towards a centre of sorts where mass will form this contracting. This was prevailing until a man by the name of E. P. Hubble came to the forefront.

E. P. Hubble (1889-1953) confirmed an expansion through out the Universe, which contradicted all that science thought was known about our Universe. According to the accepted Newtonian cosmology everyone is of thought that the Universe is in a normal state of contraction because that is what $F = G \dfrac{M_1 M_2}{r^2}$ implies. Every person is very aware of the idea that the universal expansion would not last for ever, but has to start with some contracting effort at some point. Then all the heavenly bodies will collide and destruct, without any thought about any wavering on the matter and on the matters reliability there is evidently no doubt. When $F = G \dfrac{M_1 M_2}{r^2}$ apply, there should not be any force, which is able to keep the mass that is producing all the gravity that contains the Universe apart. Known for almost a century, science has failed to give any explanation about this cosmic phenomenon of a Universal expansion except for some silly notion about dark matter being dormant and not forming gravity, as it should. If the dark matter is present as is claimed, then why doesn't the mass form gravity as it should and contract? What does our ability to see or not to see or the luminosity that the dark matter does not have, got to do with the mass bringing about pulling power, that is if mass brings about any pulling power. If the mass is there, visible or not, then the dark mass has to pull because light has no standing in the forming of gravity and if mass does pull, it has to pull to form gravity. However Hubble's law contradicts this idea of a collective contracting Universe totally. This phenomenon about Hubble's constant finding the cosmos expanding should not occur with Newton's perception about gravity envisaging the contraction that must come by the force created by mass in $F = G \dfrac{M_1 M_2}{r^2}$. If the Universe is on a contracting as Newton said it has to, we have to first find proof about the location to where such contraction is pointing. In order to locate the contracting we have to locate the centre of the Universe, which means we have to locate singularity. With singularity eternal small, holding the place where the Universe started, we first have to differentiate between singularity and zero, should we wish to find singularity. In modern science the phenomenon we know as the Roche lobe comes more and more to the foreground, indicating an undeniable interaction between orbiting structures sharing a common axis.

That axis science at present does not recognise, notwithstanding the reality and undeniable proof there is behind all evidence. As apparent as it is to me, I went about divorcing $F = G \dfrac{M_1 M_2}{r^2}$ from all ideas forming cosmology and applying the roundness we have in Π to specific positions where one may locate singularity, which we have to locate if we wish to find gravity.

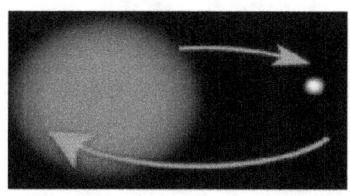 

The Roche limit in the practical sense

The formula $F = G \dfrac{M_1 M_2}{r^2}$ cannot explain the comic occurrence shown in the pictures above called the Roche limit, I should find some attention when I say I can explain what is occurring in this instance and this occurrence connects directly to the Roche limit, as explained above. Not only does the Roche limit explain this phenomenon, but also it ties directly to the Titius Bode principle, also being another inexplicable factor in light of the formula $F = G \dfrac{M_1 M_2}{r^2}$.

According to the formula of $F = G \dfrac{M_1 M_2}{r^2}$ all orbiting structures should collide with a bang, but instead they do the tango until one drop, but when dropping it still does not collide with the larger structure, as would the formula $F = G \dfrac{M_1 M_2}{r^2}$ suggest that is used by science. The position where the formula applies is most surprising. Where the formula $F = G \dfrac{M_1 M_2}{r^2}$ applies, one has to find singularity applying because the position of r is pointing to a specific pinpointing of space contracting.

This is not only limited to planets in our solar system. In the Universe, there are giant stars spinning around each other. These stars are binaries, which are also one form of double stars where double stars are another such a form. The difference between the types depends on the distance they remain apart. They keep a certain distance apart and do not collide. In the case of the sun and its planets, it could be a case that the systems might be to small, or they might be to apart. However, this is not the case with binary stars. They are close, they are big, and they spin around a mean axes called the Roche limit.

Bode's Law:

Planet	Mercury	Venus	Earth	Mars	Ceres	Jupiter	Saturn	Uranus
Bode's Law distance	4	7	10	16	28	52	100	196
Actual distance	3.9	7.2	10	15.2	28	52	95	192

Bode's Law:
A numerical sequence announced by J.E. Bode in 1772, which matches the distances from the Sun of the six planets then known. It is also known as the Titus-Bode law, as it was first pointed out by the German mathematician Johann Daniel Titius (1729-96) in 1766. It is formed from the sequence 0,3,6,12,24,48,96, and 192 by adding 4 to each number. The planets were seen to fit this sequence quite well – as did Uranus, discovered in 1781. However, Neptune and Pluto do not conform to the 'law'. Bode's Law stimulated the search for a planet orbiting between Mars and Jupiter that led to the discovery of the first asteroids. It is often said that the law has no theoretical basis, but it does show how orbital resonance can lead to commensurability. The importance that becomes known is the sequence the Ties – Bode law saw in the number arrangement of 3; 6; 12; 24; 48; 96 etc. The incorrect application of the Titus Bode law lies in subtracting the figure of 3 from 10 leaving 7. The other way of reasoning is to add four each time to the firs value of three starting with 3 and so on. The true significance of the Titus-Bode law is that it points directly

to a circular growth of 7 stages. The 7 relating to 10 is a precise derogative of the Roche limit or the Roche limit is a precise derogative of the Titius Bode principle because he two systems interlink.

The Roche limit is:

The region surrounding each star in a binary system, within which any material is gravitationally bound to that particular star. The boundary of the Roche lobes is an equipotential surface, and the lobes touch at the inner Lagrangian point, L_1, through which mass transfer may occur if one of the components expands to fill its lobe. It names after the French mathematician Edouard Albert Roche (1820-83).

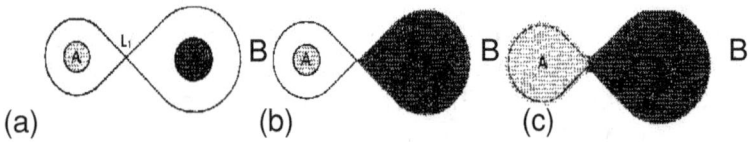

THE ROCHE LOBE: In a binary system, the Roche lobes of components A and B meet at the L_1 Lagrangian point. (a) In a detached system, neither star fills its Roche lobe. (b) In a semidetached system, one massive component, B, fills its Roche lobe. (c) In a contact binary, both components overfill their Roche lobes and share a common envelope.

LAGRANGIAN POINT:
The Lagrangian points are five equilibrium points in the orbit of one body around another, such as a planet around the Sun

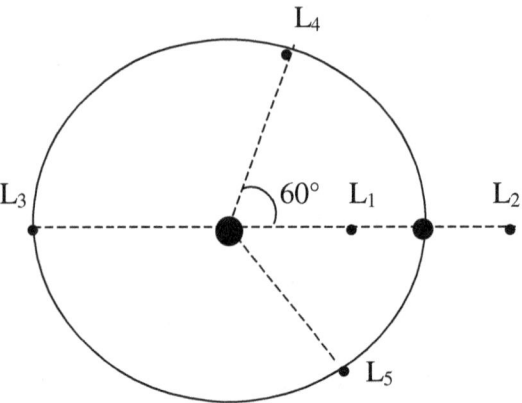

From singularity there comes three values each holding 180^0 and this fact science is familiar with. The straight line is always a potential triangle with on side apparent and the other side in infinity.

The Coanda effect

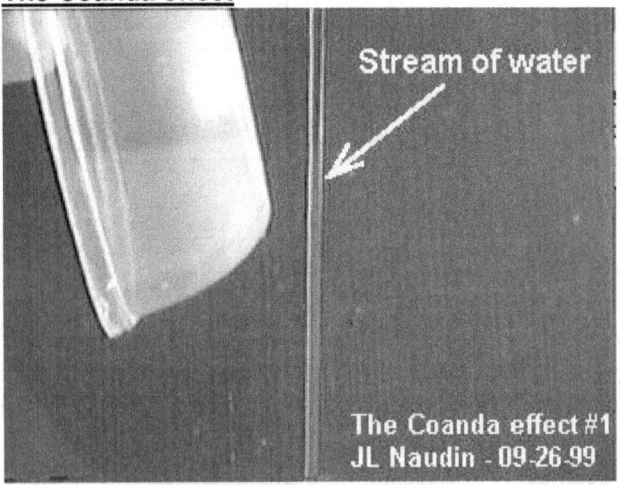

The Coanda effect where a liquid concentrates around the surface of a solid and by movement concentrates the density of the liquid to gather and compact while maintaining a relevance to the centre of such a round solid. I discard the idea that mass could be responsible for forming gravity because in almost four hundred years all evidence is indicating that the truth is to the contrary.

The condition for the presence of this singularity that forms everything, controls everything and is everything is centralised to a centre singularity $k^0 = a^3 / (T^2 k)$ that forms by movement $T^2 = a^3 / k$ of space $a^3 = k T^2$ placed in relevancy $k = a^3 / T^2$ that is centrifugally going both ways $k^1 = T^2 / a^3$ thereof (Newton's 3rd law). This explains the Coanda effect and the Coanda effect is gravity and gravity

"glues" the water to the glass by implementing Π to form singularity! *What is in the Universe is spinning. The entirety of everything forming the Universe is spinning inside the Universe*

I have no chance that what I state as my theory on **The Absolute Relevancy of Singularity** will be read, or much less that it will be seriously considered and I have not a snowballs hope in hell that it will be accepted by those with the authority to change physics principles. The theory I introduce here and now would never be accepted in my lifetime because science in the Newtonian way is bent on believing in the marvellous, the facts bordering the supernatural, the outrageously inconceivable and the magic of what can never be explained, although they claim to use facts. It is **the marvellous** to think that mass can create gravity. It is **bordering the supernatural** to think that with nothing between stars, yet by the magic of mass, mass has an unexplainable ability to attract another star many astronomical units away. It is the **outrageously inconceivable** to argue that life started on Mars, then overcame the Quite impossible to escape the gravity that Mars holds on all things, and after overcoming the unthinkable, then made a dive for the Earth just to come and evolve over here. Science think they my have the ability to create a Black Hole in a Manmade atom-accelerator because science thinks of the Black Hole as **the magic of what can never be explained** and therefore that proves that science has no idea of what a Black Hole is and I can prove what a Black Hole is. That fact that I can explain what a Black hole is, that the Wizards of Oz will never allow the explaining I present to be done in as simple manner as I am about to explain the Cosmic Code. However, when I prove what a Black Hole is I am going to destroy the fantasy world everyone makes believe as physics. To science a Black Hole is a world of magic where gravity has the ability to go mad and a Black Hole is something that man could manufacture by creating an atomic accelerator tunnel, or so science thinks. In other words the best science at present can do to explain the gravity in a Black hole is to give gravity a level of superior intellect and then take it away (by allowing gravity to go mad as it seemingly does in Super Novas and in Black Holes). Why can I prove what a Black Hole is…it is because I can prove what gravity is and believe me that is one thing science this far could never get around in proving. The facts they use is as much fiction as Little Red Riding Hood's talking wolf…when it comes to explaining the integrating details of how gravity comes about. In science, when following my theory, everything can be explained by using physics, but using my explanation will make all present science become fiction, make all present science look like a fairy tale and make all present science seem to be good bedtime stories deprived of truth…and the money spent on Newtonian fiction-science will never allow me to have success because that would be too costly for the industry money-wise. Why would I call science a fairy tale…well this is just one of many, many reasons. Science wishes to promote something as impossible as time travel, which I show, is impossible. Science believes in travelling at speeds unlimited that could exceed the speed of light. I prove all such thoughts are impossible because I show that gravity and time is the very same thing. No one can beat gravity because gravity as time maintains the structural integrity of the Universe. In beating gravity one wishes to beat the cosmos that hold us secured. That is why time can manifest as what is known as the Hubble constant. Time is the redeploying of space by extending the absolute relevancy of singularity and that is only one of several factors that serve as time. Every time I declare Newton was mistaken and therefore science is wrong in presenting the most basics of physics, the workings of gravity, I am barraged by rejection and silent ridicule. Every time I challenge the Members of science to either prove Newton correct or to prove me wrong, I am ignored…my challenge goes unmet, so please forgive me for showing much antagonism…it is a result of Mainstream Science rejecting my efforts unfairly for many years. What I write is undeniably and undisputedly correct, but the instant science admits to my work being correct, that admission demotes most of the work science has accepted in the past as correct to the level of science fiction. It will destroy the groundwork of mainstream science and demote what is accepted to become fairy tales, which is what most Newtonian based theories are. Let Newtonian science explain what the cosmic purpose or the function is of a star…of a galactica…of an atom…of gravity…they have no idea. By the time you have finished this book you would have found answers to all the above questions in detail.

Mainstream science has so little idea of what a Black Hole is or what could cause a Black Hole that they devised a "Mini Black Hole" to suit there marvellous misinterpretations of gravity. That is a form of fantasy that fairy tale writers can't compete with. Science is so misguided in understanding life that they put life in all places throughout the Universe without ever finding one shred of evidence of the presence of life. Yet they say they work only with proven facts alone. They hold the opinion that life could have come from Mars but fail miserably in explaining how it will be possible for life to escape the gravity of Mars and then fly all the way, ever so precisely guided; directly to the Earth. How would it be possible for life to escape the gravity of Mars without them when explaining such a possibility by employing realistic physics, going into so much fantasy it leaves the story of the three pigs and the blowing wolf seem real. Science has the explaining of the exploding Super Nova down to the last detail where they explain that a Super Nova is gravity that has gone mad without ever proving how gravity can go mad because the truth of the matter is that gravity has no intellect to "go mad" in any way. Mainstream science always places new object found where their findings prove that the newly found object is on "the edge of the Universe", meaning where the

Universe ends by forming an edge. This fantasy they dish up to anyone willing to believe them without ever telling what is beyond that edge. All they can see is an end of the Universe but in reality where there is an end there has to be a beginning of something else...this is physics. The Universe I show can't have an edge because I show where the point is that could never start and I show where the point is that could never end. I show that which can go no smaller and I show that which can go no bigger. I am about to introduce a Universe that mathematically can never start and the same Universe can mathematically never end.

I have been on a self-teaching mission that lasted thirty years and now that I have the answers and from which I have drawn the conclusions, I now find so much resistance from mainstream science in getting the findings my research uncovers out in the open. I offer tot academics many books in which I use diagrams, sketches, mathematical explanations and cosmic photos including other tools I employ to promote the required understanding needed to bring the ideas across that I wish to promote. However, publishing in this manner is very costly and money is one thing I do not have and therefore sending it to academics with no reply is an expense I cannot endure. Any academic feeling confronted by my accusation, please show how you prove $F = G \dfrac{M_1 M_2}{r^2}$ is applicable and is true. Show how the use of the formula could be applied meaningfully to present an answer worth of anything. Use the Newton's formula to show when the Moon is going to hit the Earth as the mass of the Earth pulls on the mass of the Moon. Better still, prove that mass does contract to create gravity and then explain how this is done...and please leave out the graviton because that is a joke! The idea that mass draws mass closer $F = G \dfrac{M_1 M_2}{r^2}$ is mathematically proven as an untruth, which means it is not true. What is the truth? ...when you have completed this introduction you will have had a peeping view, a tiny glimpse of the truth...but as little as you would gain from reading this introduction alone, when put in comparison to what any person can gain from reading all of my work in total, you will gain endlessly more than what science is to explain about the truth, because what you then have gained by reading this document is much more than what science know about the truth. What I try to convey is that there is a good reason why academics block any and all publishing of my work, and when finishing this book, in comparison to what I offer, you have not even opened a first page of what I offer as new information when judging what my other work uncovers. Still, your effort in reading this document allows you to discover so much more of true science than what previously was known If you think I am boasting I challenge you to show where any of my explaining gravity requires superior intellect to understand... however in my simplistic approach to gravity I prove everything I say by applying the simplest mathematics there is.

The effort that this book represents the informing about an entire new way of cosmic appreciation meant to show that there are grounds for concern in the way science thinks and this book does not even bring all such arguments indicating concern in full. That one can only find when reading the first ten letters forming books named as with a title beginning with **Open Letters...**and those titles are included as books which I mention on my website, having the same name as this book namely www.gravitysveracity.com.

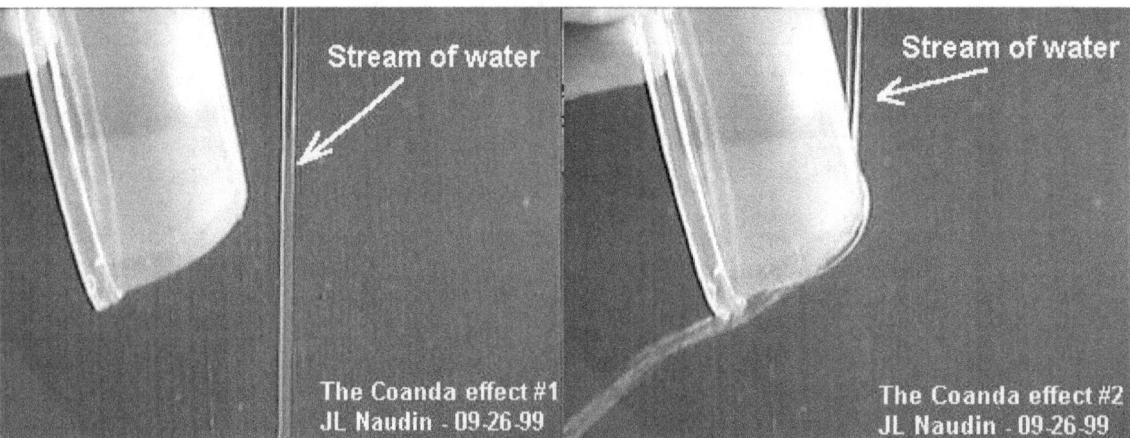

I am going to show, not only that this Phenomenon, which is called the Coanda effect is forming gravity, and not only why it is forming gravity, but how and why it forms the phenomenon science refers to as the "sound barrier".

I am going to prove that mass has nothing to do with gravity, but that mass is only the result of gravity forming as the Coanda effect.

What you are about to read is very new science and have never been written by any person ever in the past in any shape, suggestion or form.

I have no chance that what I state as my theory on **The Absolute Relevancy of Singularity** will be read, will be seriously considered and much less be accepted by those with the authority to change physics principles. The theory I introduce here and now would never be accepted in my lifetime because science in the Newtonian way is bent on believing in the marvellous, the outrageous and the magic of what can never be explained, although they claim to use facts as a basis. Science has no idea of what a Black Hole is and I can prove what a Black Hole is. However, when I prove what a Black Hole is I am going to destroy the fantasy world everyone makes believe as physics. To science a Black Hole is a world of magic where gravity has the ability to go mad and a Black Hole is something that man could manufacture by creating an atomic accelerator tunnel, or so science thinks. In other words the best science at present can do to explain the gravity in a Black hole is to give gravity a level of superior intellect and then take it away (by allowing gravity to go mad). Why can I prove what a Black Hole is…it is because I can prove what gravity is and believe me that is one thing science thus far could never get around in proving. The facts they use is as much fiction as Little Red Riding Hood's talking wolf…when it comes to explaining the integrating details of how gravity comes about. In science, when following my theory, everything can be explained by using physics, but using my explanation will make all present science become fiction, make all present science look like a fairy tale and make all present science seem to be good bedtime stories deprived of truth…and the money spent on Newtonian fiction-science will never allow me to have success because that would be too costly for the industry money-wise. Why would I call science a fairy tale…well this is just one of many, many reasons. Science wishes to promote something as impossible as time travel, which I show, is impossible. Science believes in travelling at speeds unlimited that could exceed the speed of light. I prove all such thoughts are impossible because I show that gravity and time is the very same thing. No one can beat gravity because gravity as time maintains the structural integrity of the Universe. In beating gravity one wishes to beat the cosmos that hold us secured. That is why time can manifest as what is known as the Hubble constant. Time is the redeploying of space by extending the absolute relevancy of singularity and that is only one of several factors that serve as time. Every time I declare Newton was mistaken and therefore science is wrong in presenting the most basics of physics, the workings of gravity, I am barraged by rejection and silent ridicule. Every time I challenge the Members of science to either prove Newton correct or to prove me wrong, I am ignored…my challenge goes unmet, so please forgive me for showing much antagonism…it is a result of Mainstream Science rejecting my efforts unfairly for many years.

I have been on a self-teaching mission that lasted thirty years and now that I have the answers and from which I have drawn the conclusions, I now find so much resistance from mainstream science in getting the findings my research uncovers out in the open. I offer tot academics many books in which I use diagrams, sketches, mathematical explanations and cosmic photos including other tools I employ to promote the required understanding needed to bring the ideas across that I wish to promote. However, publishing in this manner is very costly and money is one thing I do not have and therefore sending it to academics with no reply is an expense I cannot endure. My books carry this message as it uncovers much of the fraud hidden by Mainstream Physics. Yes, I say fraud because if one promotes untruths notwithstanding how well intentional it may seem, it remains fraud. The fraud is well hidden in physics. It is sand walled by those academics where they use the information they present as physics as if the facts they present are well proven principles and the manner of presenting these facts as such well proven facts became the teaching methods they use to fool the physics student about the truth. Any academic feeling confronted by my accusation, please show how you prove $F = G \dfrac{M_1 M_2}{r^2}$ is applicable and is true. Show how the use of the formula could be applied meaningfully to present an answer worth of anything. Use the Newton formula to show when the Moon is going to hit the Earth as the mass of the Earth pulls on the mass of the Moon. Better still, prove that mass does contract to create gravity and then explain how this is done…and please leave out the graviton because that is a joke! If you feel confronted by my accusations that science is committed to commit fraud by falsifying facts about gravity, then please prove how mass is able to create gravity. Do so before reading this book because if you complete this book you will know how gravity comes about and that gravity being there has nothing to do with mass being there although mass being there is the direct result of gravity being there. That mass draws mass closer $F = G \dfrac{M_1 M_2}{r^2}$ are mathematically proven as an untruth, which means it is not true. What is the truth? …when you have completed this introduction you will have had a peeping view, a tiny glimpse of the truth…but as little as you would gain from reading this introduction alone, when put in comparison to what any person can gain from reading all of my work in total, you will gain endlessly more than what science is to explain about the truth, because what you then have gained by reading this document is much more than what science know about the truth. What I try to convey is that there is a good reason why academics block any and all publishing of my work, and when finishing this book, in comparison to what I offer, you have not even opened a first page of what I offer as new information when judging what my other work uncovers. Still, your effort in reading this document allows you to discover so much more of true science than what previously was known If you think I am boasting I challenge you to show where any of my explaining gravity requires superior intellect to understand... however in my simplistic approach to gravity I prove everything I say by applying the simplest mathematics there is. That is why I have the courage to call science fraud. When anybody is telling some other person about what that person has to believe in while depending on the person believing while the something told to the second person can't be substantiated or the first person can't back what he tries to convince the second person to believe by giving solid evidence, the process of insisting on the person in believing in falsified facts boils down to something called brainwashing. I show how many facts that Newton claim is true can't be proven; therefore expecting students to repeatedly believe in the truthfulness of Newtonian science is the purist form of deliberate brainwashing ever devised by any group of persons that walked on Earth. If this then results in my work being out rightly rejected, then so be it.

The effort that this book represents the informing about an entire new way of cosmic appreciation meant to show that there are grounds for concern in the way science thinks and this book does not even bring all such arguments indicating concern in full. It aims to caution readers about the way academics teach "classical physics as science" and "classical mathematics" and whereby they intentionally or otherwise brainwash students to accept Newton's view on cosmology. I know the very second I say this all-academic interest disappear and let me add. With physics Newton is absolutely correct. Newton's failings came about when Newton started to dabble in the work of Kepler. In that, being the work of Kepler, Newton failed decimally and this book amongst may other of my books prove it. This effort aims only to warn students to look out because there are much more phoney science. My distress on this matter is so great that I am prepared to divulge this information for free as to warn students about the misconduct they are ruthlessly exposed to. However, telling about the new science as I uncover the fraud takes many volumes of writing since there are many volumes of fraud that has to be corrected. That one can only find when reading the first ten letters forming books named as with a title beginning with **Open Letters...**and those titles are included as books which I mention on my website, having the same name as this book namely www.gravitysveracity.com.

I am about to prove that gravity is **the Coanda effect** and gravity comes about from four cosmic phenomena never yet understood since it was never yet explained. Science doesn't believe there is something such as **the Titius Bode law** but science does believe that mass would generate gravity. Science has no clue about **the Roche limit** but science believes in spite of the Roche limit that big craters on Earth are reminders of massive asteroids that hit the Earth in giant collisions. With the Roche limit in place these crates are the result of something else because it can't be from asteroids colliding with the Earth. We all know how the bicycle rides and we all think we understand how the bicycle rides but having the bicycle ride on two wheels have little to do with balance and everything to do with the Coanda effect. The bicycle rides forward when peddled but also the bicycle rides downwards when peddled and the two are both linked to gravity. I am going to prove that the Coanda effect forms gravity. I am going to prove that the **Coanda effect** comes as a result of the **Titius Bode law**, **the Roche limit** and the **Lagrangian positioning system** but most of all how these are related to singularity. That means I am going to prove that mass has no effect on gravity but mass comes as a result of gravity. I am going to prove what singularity is and that there are two types of singularity that in the end is only one type of singularity.

Teaching ever since time began forms a pillar on which memory and remembering what you are taught is the most prevalent part of tutoring. One is expected to remember what those coming before and which are tutoring you, wish you to remember. The Tutor lays a foundation by ensuring that everything known and accepted coming from the past are well and truly founded in the mind of the student. In that there is no problem. The problem arises where the information studied is flawed and no one ever realised that. Fortunately this does not occur regularly, but if and when it does, notwithstanding the exceptional par it forms, ten becomes a major problem to deal with. Therefore what comes form the past are carried on into the future as unblemished truth and no person meddles with the thoughts called information given as study material. However, as unlikely as it could be, this did happen and it is part of the basis of physics. When the student is taught, the student is expected to accept without argument. What comes from the past are considered to be tested beyond suspicion of doubt! One can only start to think and through arguing set by reason new thoughts, after the learning by memory process is well established and it then forms a solid base for everything the student knows. This mostly takes about all the time one lifetime presents. Well, what happens when that everything that everybody believed in the present, inherited by all from the past, was totally flawed? It has happened to physics and no one in physics yet realised it. Then the mistakes will carry on forming the past, carried over as flaws into the future for as many generations as it takes to realise the mistake and could continue indefinitely, if there is no clear minds working the recognise and correct what needs to be corrected. I ask of you not to judge me for I fall short. Judge what I present to you, for then you will realise with all my shortcomings, I present you with a truth that exposes short fallings in the basics of physics.

Here and now and before the beginning of what this document may be to any potential reader, all parties reading take note that I state it emphatically that all members forming the community of science in physics judges me being not sufficiently educated and certainly not to the level where I am able to form any opinion on matters concerning Sir Isaac Newton or his physics. Any and all of my self-tutoring goes begging in their eyes notwithstanding and regardless of the fact that I did my private and individual studies by which I furthered my insight. That allowed me to show with clarity what destructive force Sir Isaac Newton released in order to corrupt the laws of mathematics, contaminating science along the way and mostly raping the work of a great man, Johannes Kepler and what Sir Isaac Newton did to derail the truth and disguise scientific correctness where such violation can only be expressed as being blatant criminal fraud. What his deeds amount to, is to corrupt the laws of mathematics, to render the laws of cosmology useless and to rubbish all of science. By your reading, you will learn what it is that those academics that are guarding science never wanted published and read by the public at large. What I say is don't run and hide from my attack and coward away from my confrontation as so many of the most intellectuals amongst the Physics Paternity did when I confronted their thinking. On every occasion where I confronted members of the Academic Paternity in the past, those I confronted acted in precisely such a manner, such as cowardly ending all reading by throwing the book down, and then pretending to show the utmost disgust in what I say.

Now show your academic worth and your educated dignity and accept the challenge I make to you and to all of your kind: I challenge one and all: **PROVE ME INCORRECT IN ANYTHING I SAY!**

What is it about gravity that I say which no one wants to know?

First you should decide what belongs to the gravity factor and what forms part of mass. Newtonian science is of the opinion that when a body is floating up in outer space the body has micro gravity…that just can't be the case. Newtonian scientists confuse the factors being responsible for mass and for gravity because if not, then please explain which is gravity, the part that tries to move the body to the centre of the Earth, or the preventing thereof? We have to see that mass is created by the pushing of an object onto the Earth and from the pushing (not pulling) comes mass while gravity is what is doing the pushing. While resisting further movement, mass comes into the picture and while moving towards the Earth or intending to move towards the centre of the Earth, that movement constitutes as gravity while stopping the movement leaves the object with having mass. Mainstream science loves to confuse the two issues because Mainstream science love to confuse everyone because Mainstream science is completely confused about the science they say they are the Masters of.

Is gravity that factor, which makes all bodies fall to the centre of the Earth, or is gravity that which prevents the further moving of bodies having gravity to fall further down to the centre of the Earth and then by restricting the movement, then forms weight or mass? By restricting movement towards the Earth a mass factor comes about which gives weight! It is presumed that the body has micro gravity because the body is weightless in outer space. This prompts me to ask the question underlying what has never been decided… what is gravity and what is mass. A body floating in outer space has maximum movement because when it moves slower, it starts to fall to the Earth. Being in outer space that body has little mass because it has a micro weight. At that point when floating in outer space the mass (measured as weight / kg) is indefinably small while the movement that is applying is at a maximum in maintaining orbit. However, that is speed measured by distance (meters) travelled in time (seconds). Mass has a value, which is measured in the same currency in which weight is, and then mass is weighed as much as weight is and therefore, undeniably and in contrast to the logic of mainstream science's confusion and frenzy trying to confuse what can't confuse any further, mass and weight is connected as the same way of measurement and using the same measuring tools while gravity is movement notwithstanding mainstream science trying to put mass and weight far apart. If mass was equal to movement as gravity is, then mass must be measured in meters / second. Instead mass has the value which is the same as weight which is measured in grams.

When a body stands on the Earth having mass it is a part of the Earth. The Earth with spin provides a governing singularity around which everything spins. The object having mass becomes part of the controlling singularity but presenting mass it forms part of the controlling singularity as it moves around the earth's axis forming a part of the controlling singularity.

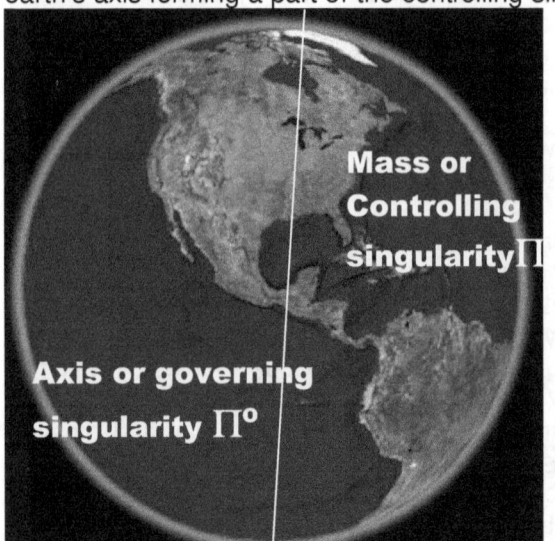

When an object stands on the Earth the object is in mass because the Earth holds control over the atoms in the object. All the atoms spin around their governing singularity that could be any place and all the atoms in the star connects to by bonding with the Earth's governing singularity. The Earth provides the centre around which the movement spins while the object with mass does the spinning as it then is part of the controlling singularity that the Earth provides. The governing singularity of the Earth establishes the worth of the controlling singularity of the Earth by providing gravity or time in relation to the space moving. When an object falls towards the Earth the governing singularity of the Earth captures the atoms of the falling object while trying to establish control over the object falling. In that manner the atoms within the Earth forms the Earth by becoming the controlling singularity in relation to the Earth's governing singularity in the centre.

This is the reason why Galileo said all things fall equal which dismisses mass being responsible for falling and when touching the Earth the object then receives mass as Newton said.

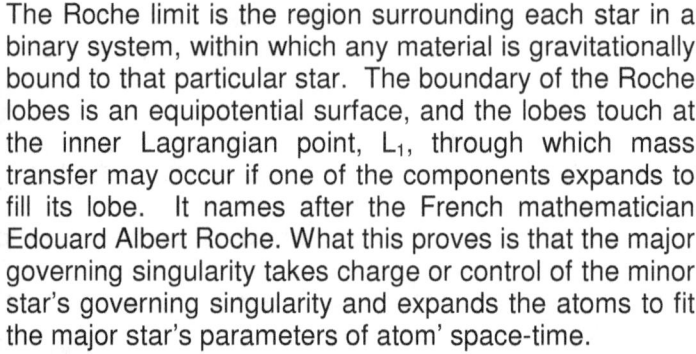

The Roche limit is the region surrounding each star in a binary system, within which any material is gravitationally bound to that particular star. The boundary of the Roche lobes is an equipotential surface, and the lobes touch at the inner Lagrangian point, L_1, through which mass transfer may occur if one of the components expands to fill its lobe. It names after the French mathematician Edouard Albert Roche. What this proves is that the major governing singularity takes charge or control of the minor star's governing singularity and expands the atoms to fit the major star's parameters of atom' space-time.

Then by overheating the minor star's governing singularity it forces the bonding between the minor governing singularity and the atom's forming the governing singularity to release. This is known as an explosion and what happens is that the governing singularity of the minor star no longer holds charge of the atoms forming the minor star's body. The movement of the atoms of the minor star charging the bonding between the atoms forming a single structure breaks and the atoms release its density provided by the spin of the minor star to produce the concentration of the star structure. As the star expands by the relevancy enlarging because the cooling of the inside of the star subsiding, this releases cosmic liquid containing atoms belonging to the minor star. However, the major star already took charge of the governing singularity of the lesser star's atoms and this hold of control remains in place. This proves that the star controls the size of the atom by the spin of the atom.

The governing singularity is controlling the atoms by means of placing the atoms in a relation where the atoms form the controlling singularity. In this expanding and the governing singularity losing control over the spin and therefore over the controlling singularity we have the Super Nova expanding the star's relevancy.

The Four Cosmic Pillars; The Result Thereof.   Page 167   APPLYING AS THE SOUND BARRIER

To understand what I just said we have to investigate singularity and where singularity hides. Singularity is Π producing Π° where this establishes Π and by movement Π² specifies space ending at Π³. In order to locate Π we have to locate Π° and this we can achieve by producing movement Π² that would allocate Π³.

Locating and finding the presence of singularity

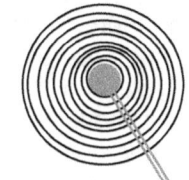

$k^0 = a^3 / T^2 k$ states that whatever is, is also spinning in order to be present.

What is in the Universe is spinning. In the **precise middle** of all **objects in rotation** is a precise centre dividing the object in sectors that will **start the spinning initiation** from that centre point. Thus, the spinning object **will have a middle point**, a very specific **centre point that does not spin** and only holds Π as a specific value because no radius can apply. But also the one value such a line **cannot have is zero** because the line **is there and holds contact** to the rest of the material bringing about that **zero does not start any** line and therefore the **value of the line must be infinite**, just as described in **accordance** and by **the definition of singularity**

As I am introducing a very new idea, I wish to explain in better detail what I try to convey.

While the toy top is spinning one will find singularity by moving the rotating line or radius progressively to the middle by reducing the length the line has from the edge to the middle. At one point all further reducing must end but the ending cannot include zero or nothing because the rest of the line still attach the rest of the top.

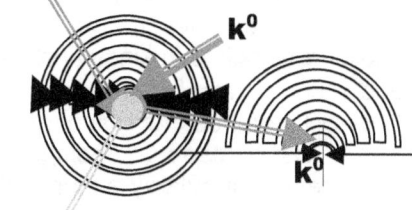

As the rotating direction moves inwards, the rings will become smaller and smaller.

That point albeit hypothetical, is also as much a reality none the less and is placed where that point **must be standing still** because every line **running from that point** in **opposing directions** is also **in opposing directional spin the other or opposing side.**

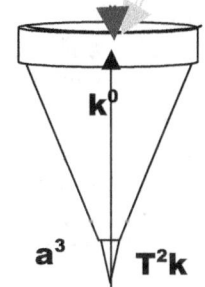

In considering the spinning motion in the fraction of time in the detailed instant every aspect of rotation will turn in every instant of change in time. Although the points had the same characteristics only one instant before, they oppose the characteristics it had just before and just after the very instant in which they are and to which they relate by similar points also in rotation. The fact of the graph proves my point in quarterly opposing dimensions and values.

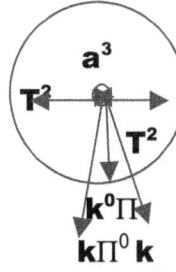

In dimensional terms, which I explain later on the value of **2k** relates to **T²**. That relation extends to the next value where **T²** relates to **k**, which relates to **T²**. The first space in the circle will then be **T² k**. From the centre being in infinity one can realise by applying mental power the single dimension factor not seen but present all the same. Extending that into the 3D comes six **k** and any one of the six will further extend to form a seventh point as **T²** All this is a multiplying of $k^0 = a^3 / (T^2 k) = 7$

In dimensional terms, which I explain later on the value of **2k** relates to **T²**. That relation extends to the next value where **T²** relates to **k**, which relates to **T²**. The first space in the circle will then be **T² k**. From the centre being in infinity one can realize by applying mental power the single dimension factor not seen but present all the same. Extending that into the 3D comes six **k** and any one of the six will further extend to form a seventh point as **T²** All this is a multiplying of $k^0 = a^3 / (T^2 k) = 7$

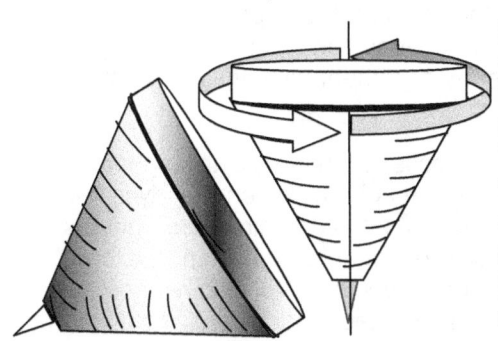

When the top is laying on the ground the atoms forming the top is part of the controlling singularity having the Earth spin that establishes the governing singularity in the centre of the Earth. The top with all the atoms forming the top still only remains a part of the mutual singularity f the Earth.

When the top starts to spin, such spin promotes the atoms within the top to form the individual singularity leaving the top to have the controlling singularity as the spin of the top charges a governing singularity within the top centre. Then the earth forms a part of the principle singularity and only holds the top hostage but no longer in control although it fights the independence to the very end.

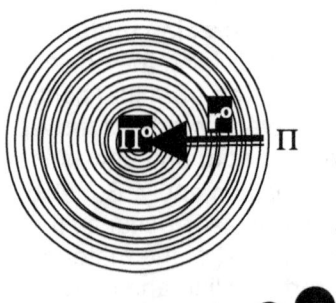

The circle can always reduce one step more to a point when the circle eliminates r completely by returning r to a point of singularity r^0, but the elimination of r as the factor reduces the major factor to the single dimension in Π^0. The point is not zero. If it was zero the point was not present and such a point is present.

By reducing the line we come to the end of the mathematical equation of the circle but the circle does not end there. That is what Newton did not recognise from the figures the cosmos represented to Kepler. The circle only secures the final cosmic figure and the value to singularity where all things have equal value. The movement of the circle splits singularity in two sectors. By forming Π the circle has to form Π^2 due to the movement coming about in securing the space Π^3.

That will not reduce the cosmos to zero, but it will only eliminate all potential lines r^0 to potential circles $\Pi^0\Pi r^0$ and from there the circle Πr^0 will come about by manifesting as a line but that manifesting can firstly only establish a circle Πr^2. The only value that singularity can have although the single dimension may host the entire Universe is Π^0. Pick a number and elevate it to the power of zero and in the process one may have established another point holding all points in singularity because that is the value of singularity. Only Π^0 or any other value holding one accompanied by zero as an exponential value can ever be the accurate value of singularity while singularity will then host the rest of all the possibilities in the Universe. This means that the entire Universe composes of and is made up of singularity... this much I am going to prove. Every point occupied or otherwise constitutes of singularity either under control by movement in a form we call atoms or being passive in a location we call outer space. This position one can derive from Kepler's formula $a^3 = T^2k$. It is just a question of how to fit this sensibly into Kepler's formula $a^3 = T^2k$ and find a way that will bring much understanding to cosmology and the way that singularity connects one Universe to form cosmology. The top spinning is what connects space to form the Universe. The top being still on the ground and not spinning holds singularity at a value of the dot forming Π^o while putting the relevancy on the Earth's roundness by Π. When the top spins the relevancy changes to the line from forming as a dot Π^o becoming a line Π. The line Π forms as a result of the top forming space Π^3, which is in place as a result of the movement that the top acquires Π^2. It is singularity without space so being a line or a dot makes no difference. The top no longer holds only a dot Π^o in the centre, but generates the relevance Π by forming $\Pi^o \Pi r^o$. The top, by moving adjusts Π to form space by movement which is $\Pi = \Pi^3 \div \Pi^2$. All of this is what makes gravity be what it is and all of that Newton missed and Newtonians never saw since all of that is covered by a blanket called mass being responsible for gravity.

Kepler chose to use different symbols too those being valid in my opinion, but the concept remain the same. Kepler said that $a^3 = T^2k$ while I show that $\Pi^3 = \Pi^2\Pi$. It still confirms that movement $\Pi^2 =$ is the forming Π^3 in relation with Π singularity Π^0.

At that point the half circle and the triangle and the line must start since all three having many different forms have equal value at 180^0. Only after that point does mathematics begin where all factors in 1 have the value of 1 being 1^0. In that conclusion one realises something must separate singularity from all other factors because singularity hosts all other factors but is by own initiative Π. That will be the spot of origin. That will hold the eternal spot...the smallest spot ever because all spots that ever can be was secured in a position in the centre of that spot. Because of the progress singularity follows from the single dimension singularity only allow mathematics a start at Π^0 progressing further too $\Pi\Pi^0$ and from there the line is born as $\Pi\Pi^0\Pi^0$ or $\Pi^2\Pi^0$ $\Pi 3\Pi^0$ $\Pi 4\Pi^0$ $\Pi 5\Pi^0$ where Π^0 then may form the concept and value of r. But the line starts at $\Pi^0 = r^0$. Because cosmology is singularity based and the value is $\Pi\Pi^0$. This escaped the attention of the greatest mathematician about the work of the greatest cosmologist ever because Newton incorrectly introduced $4\Pi^2$. The introduction of $4\Pi^2$ exaggerated the value of time and removed space / time from the concept. Mathematics in cosmology does not apply pi, pi is the root value of all concepts in cosmology. The factor pi impersonates as much as it represents singularity. This is my argument with which I support my claim that I made

The fact of form proves that the sphere captured all sides that can possibly influence the sphere. The sphere therefore holds $k^0 = a^3 / T^2k$ within the boundaries designated to the sphere. When a body is placed in a location on the outside of such spherical borders that object seems to float in any direction. There is no control one can establish which will secure movement in any specific direction of preference except by releasing heat to counter act the required motion in a specific direction of choice. We all have seen what happens to any object that comes into the border area of a sphere. The object suddenly is motivated by motion to follow a specific designated direction and the motion leads the object to move towards the centre of the sphere. It is as if the support of the six opposing sides has lost one side where the sphere took over the control and movement starts in the direction of the Earth centre. The support of one side is literally removed by the centre of the earth where Einstein claimed the strongest gravity is and the motion of the object starts in that direction. There is no pulling on the object but there is removing of space by the centre of that specific point leading the object and the space it is in as well as the space it carries to move to the centre spot. In the sphere the borders the sphere holds are deliberate and very distinctly placed edges forming a specific distance from the centre. The centre is also proven beyond any debating. The centre of any sphere has to be at the very point where space completely falls away. That will put that space at that point in the single dimension and centre is the single dimension.

The claim becomes obvious when observing the connection between the half circle, the straight line and the triangle, which could also promote all the qualities lurking behind the pyramid. Consider the connection between 180^0 sharing three different forms all part of mathematics where each is different in form, but equal in value and then one may realise in considering the very basic in mathematics being the Law of Pythagoras on which all mathematics are focused. The triangle stands in for one factor represented by one at a value of 180^0. So does the straight line become a factor of one and the half circle also becomes one where the factor of one equals all 180^0. All three are most seriously part of shapes in the cosmos. Revalue any one form to zero and the rest too must follow and share the same value.

When you walk outside and look at the vastness of the blue sky or at night at the black night sky, you are physically part of singularity in the part of 1^0, the part that moved away from 1^1. You are within the part 1^0 that has no end because it has only one side, which is the inside. It is 1^0 going nowhere. It is the part that I named the spot that had the dot 1^1 moved away from. In our Universe it all seems the same but where singularity and only singularity matters there is differences to be seen mathematically. At the border falling outside the realm of our cosmos there is a Universe applying which we can only detect by applying intelligence, much the same way as one can detect a Creator that one cannot see but one can gauge just by being more intellectual than the animals of the field are. This placing fall outside the parameter of the Universe we know.

This is the dot that has no start. It is 1^1, the part that released from the spot 1^0 when motion parted singularity. It came apart when motion unleashed the dot 1^1 that has no start from the spot 1^0 that has no end. It is the Universe born from motion that was driven by heat. It is still there because once anything is part of the Universe and forms a principle within the Universe it has nowhere to go but to remain within the Universe. Walk outside at any time day or night and you are a witness to the result that there is a 1^1 and there is 1^0.

The spinning established a value between the object's movements in relation to the movement of the electrons spinning around the atoms. The evidence that proves the statement is the fact that the minor star relinquishes density when the major star takes control of the minor star's governing singularity and thus takes control of the movement making the control of the atoms dependent on the major star's governing singularity.

The movement of the atom is interlinked with the movement the star establishes which we call gravity. This movement depends on the spin but also depends on the movement of the structure in its rotation of the secondary controlling object as the Sun is in the case of the Earth. This is in relation to time applying to the atoms versus the space that is either dismissed or displaced. The movement is as all movement is, space – time incorporated.

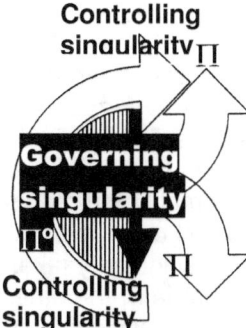

In the spinning top we find that singularity Π^0 can be generated by motion. But singularity Π^0 has no motion within the dimensions we find allocated to the Universe in which we live. Since the singularity found in the centre of the spinning top is in truth just a mathematical point, which means in mathematical terms the point with no sides cannot even be calculated as a factor since the measure thereof goes beyond what mathematics ever can calculate. Mathematics has a use within the 3D Universe but singularity that keeps the spinning top attached to singularity governing the gravity of the Earth, that singularity is truly single dimensional and beyond mathematical measure. It is singularity Π^0

If we put this in terms of singularity (Π^0) we find the Earth (Π^3) is in relation as viewed from Alfa Centauri (Π) four point six years (Π^2) while moving in that space that is time that has gone by. That secures the three dimensional status the Earth has (Π^3) in terms of a present (Π^0) that depends on a location (Π) secured by a future (Π^2) that will come by movement where the future ($\Pi = \Pi^3 \div \Pi^2$) moving forward that also doubles as a past ($\Pi^{-1} = \Pi^2 \div \Pi^3$) by the light coming from and thereby confirming the past. That is space formed three dimensionally by keeping time in infinity apart from time in eternity. The relevance (Π) that forms in relation to the present (Π^0) will relate to movement (Π^2) and the movement is circular which ensures that the relevancy forming is circular (Π) by securing that the movement is circular (Π^2) in terms of one specific point (Π^0) in infinity which then secures a roundness (Π^3) that forms an everlasting eternity ($\Pi\Pi^2$) which validates a never ending circle Π^3. In this time in infinity (Π^0) that secures that there is an everlasting eternity ($\Pi\Pi^2$) in space (Π^3), it is not the space that is everlasting but the movement of time by the line ($\Pi\Pi^2$) that is everlasting. The **governing singularity** (Π^0) holds a **positional validity** (Π^3) of three dimensions Π^3 =($\Pi\Pi^2$) in terms of any **relevance** (Π) formed by the **controlling singularity** ($\Pi\Pi^2$) thus mathematically it equates to $\Pi^0 = \Pi^3 \div (\Pi\Pi^2)$. If a **relevance** ($\Pi$) did not validate a **positional validity** (Π^3) securing a **governing singularity** (Π^0) in terms of movement formed by **the gravity** (Π^2) that produces the **controlling singularity** ($\Pi\Pi^2$) in space, with a three dimensional status Π^3, then space (Π^3) would not be obtained and thereby the Universe would not be secured. That is why space-time is $\Pi^0 = \Pi^3 \div (\Pi\Pi^2)$. However this must be seen where it applies. It applies where singularity as time meets space, which means it applies at a point in the Universe where time still grows and that is at the position that predates the Big Bang. It is where material forms before material forms. It is where the visual will never come. It is where singularity Π^0 forms space Π^3 by singularity (Π) moving (Π^2). This means there is a time space delay validating a connection where any connection is (to us) absent. Stars in an axis dispute are not yanking each other around because mass is pulling one another because of some magical medieval territorial dispute. There is control over atomic control coming as a result of singularity charged by and charging movement. By virtue of a controlling singularity effectively in place due to a governing singularity much the same way as a top spinning tries to surge into the air when spinning too fast or try to fight of the control the Earth takes on the top just before the top is grounded by movement deprivation coming from the Earth.

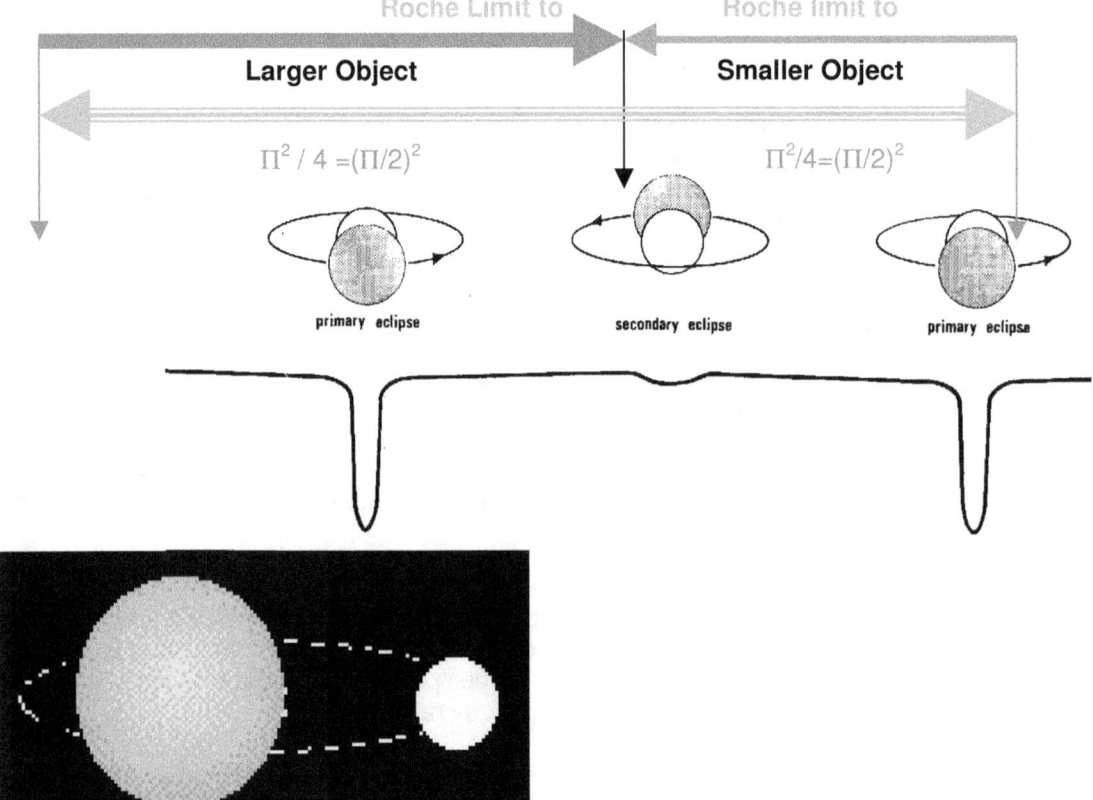

This is the reason why stars would form massive binaries, where they share a common combined circular displacement or a controlling singularity in terms of a governing singularity, separated only by each stars Roche limit with no linear value. As the Titius – Bode Principle comes into effect the linear displacement would once again grow, or the common spin value will be to grate for either one, or both, and their structural composition will collapse, forming smaller structures with less space to occupy the time in which they are. This proves that atomic movement shows loyalty to the control of a governing singularity that forms a president in movement and a governing singularity takes charge of the entire atomic movement. The governing singularity of the one star takes charge of the other stars movement and thereby the controlling singularity which is also taking charge of the mutual singularity and in that it takes charge of the individual atomic singularity as the star deprives the other star's atoms of any independence adhering to the atomic governing singularity in the star. It is this evidence by which we can gauge that singularity cross refers and establishes space-time worth in not only the star forming the mutual singularity but also the star of which it took charge of the controlling singularity.

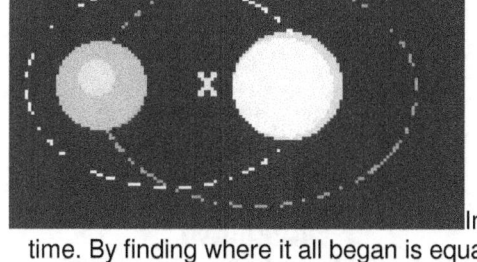

Important to note is in the case of Binary stars the two stars "lock out" space-time. By finding where it all began is equal to finding where the line began we have to trace the line in order to trace the development of the entire Universe. As seen the first development went beyond where mathematics may take us. The Universe did not become more but only focussed better on detail. What is was present because nothing can be new to the Universe that started out to be what it presently is.

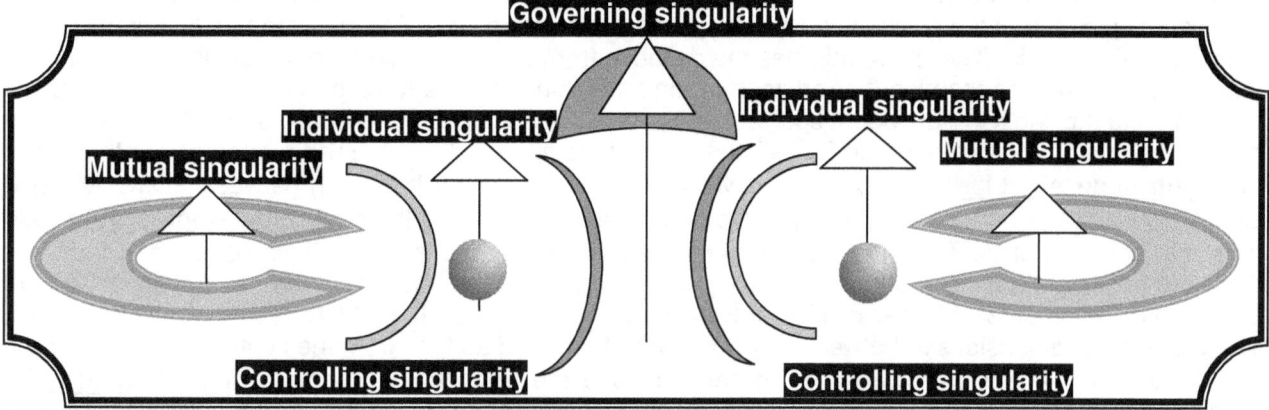

An object can rotate in outer space as long as it can maintain a speed that will keep the object rotating in that orbit. No line can go straight because part of all lines in the cosmos is the curve. There is always some relation between the factor of how much **k** influences or how much **T²** influences and the combined unit determines **a³**.

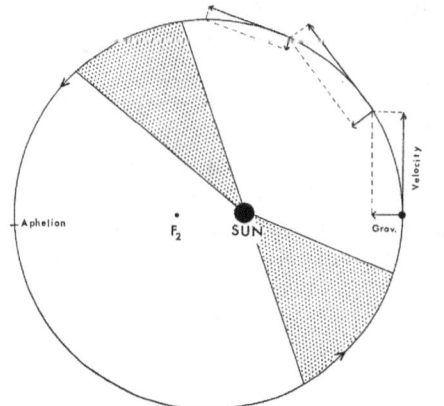

That reason too has its footing in the Titius Bode law because 7/10 is half of 10/7 and by going seven the factor also has to go 10. The compliment of this is gravity at Π^2

One of the four most important values in the Universe is the Roche limit. On this and the Titius –Bode connecting to the Coanda effect by means of the Lagrangian points and on the inter connecting of the four principles rests the growth of the universe, as space relate to time. The sound barrier is the four cosmic principles applying to form the relevancy between the governing singularity and the controlling singularity.

The Roche limit comes into effect when the linear displacement factor reaches a value of one and part of the circular displacement value. In this is the value $\Pi^3/\Pi^2 = \Pi$. When two stars are at the Roche limit, the linear displacement reaches a value of the lesser one, and excites in an electrical sense the singularity within the lesser one charging its atmosphere to extend the atmospheric level to that of the major one.

$$\Pi^3/\Pi^2 = \Pi$$

$$(\Pi/2)^2$$

$\Pi^3/\Pi^2 = \Pi$ ←

The circular displacement reaches its full complement of half Π^2 which is the Roche limit. The Roche limit forms a centre by the value of the atoms spinning inside the structure and the atoms all forming a governing singularity secures independence while the structure holds a unity by measure of the governing singularity that the star controls the atoms by.

Atoms spinning inside the structure independently form the individual singularity

- Atoms inside forming the total structure is the mutual singularity
- Inside centre forms the Governing • singularity
- Curvature formation spinning as a structure entity forms the Controlling singularity
- Spinning around the Sun produces the principle singularity

This indicates four factors forming singularity that absolutely dictates the cosmos in terms of movement. Holding that in mind, I therefore had to name the four positions that equally form singularity by dictating gravity. To argue this concept of singularity guiding movement, let's take the Sun that provides a centre k^0 for the Earth a^3 forming a centre where k points a line that forms the orbital circle T^2 wherefrom the edge of the line k is pointing at the position of whichever planet a^3 forms a circle T^2 in relation to a line coming from a centre of the Sun k^0. The line k indicates the distance from the Sun's centre to the planet that orbits and this forms the circle as the planet a^3 orbits T^2 around the Sun. The line k will provide a line from the Sun's centre k^0 and the line k will provide a spot where T^2 produces a circle holding space a^3 in a located position by running around the centre of the Sun k^0. In this view the space a^3 of the Earth rotates and in that forms the **controlling singularity** that holds the value as Π indicated by k forming between k and k^0 being singularity Π^o. The Sun holds singularity in the centre, which is forming the **governing singularity** Π^o and from that point the circle T^2 comes that forms the orbit Π^2. That means every single point that k indicates there are positions forming space a^3 implicating sides of a double dimension. In the same manner is k not limited to distance or is T^2 lesser by size. If Kepler said $a^3 = T^2 k$ then $k = a^3 / T^2$ is also what Kepler said. There are three dimensions a^3 between any two points T^2 flowing as time from the centre of the Sun, which is indicated by the line k. However in the next scenario the Earth holds the **governing singularity** Π^o running from the centre k^0 to k forming the edge while the circling rotation T^2 then forms the **controlling singularity** Π indicating the point in rotation. There are also two other points holding **the mutual singularity** and **the primary singularity**, both which I do not explain in this presentation but without which the four phenomena would not form gravity.

The value of k is not to be put in place as a measured value, but is there to bring a reference to the location of singularity $k^0 = a^3/(T^2 k)$ applying as to place a specific singularity in as the **governing singularity** and acknowledge the position of another singularity in place as the **controlling singularity** because there always has to be a **controlling singularity** determining the orbit while there has to be a **governing singularity** determining the spin of the body in relevance performing as the space a^3 in question in the formula $a^3 = T^2 k$ where in that formula k determines the relevance of k^0 as in $k^0 = a^3/(T^2 k)$. However, this burdens k forever with the responsibility of forming a line and a line is what places the Universe in place while the circle T^2 is forming the Universe a^3 at the same time. Every space a^3 in question puts singularity k^0 in position by the motion T^2 in relation k to the position allocated to k in the Universe a^3. Nothing in the Universe can move without moving straight k that is also going in a circle T^2 to form space a^3 in relation to a centre k^0 while in orbit around another centre k^0. In this point k^0 time forms space and space develops as the history of time running from k^0.

a^3 symbolises in a mathematical interpretation of implicating the three-dimensional space holding a specific centre in relation to another specific centre indicated by k that could apply to either centre points in question. This is always a straight-line k representing the position of the **controlling singularity** moving in a circle T^2. The space forming a^3 is a **positional validity** of the space indicated by $k^0 = a^3 / (T^2 k)$.

T^2 is representing the circle that goes around the **governing singularity** k^0 or Π^o that forms in relation to the line k pointing to the controlling singularity or Π in reference to the centre k^0. The space that forms holds the orbiting planet a^3 in direct circular contact with the space in relation to a very specific centre k^0 moving from point T_1 to T_2 that then forms Π^2 in relation to a precisely placed centre k^0. The circle coming about from T^2 is the **controlling singularity** Π, which is always a circle Π relating to the centre Π^o that is positioned by the line k in relation to the centre k^0 and by forming a circle Π it holds reference to the **governing singularity** Π^o. Where **the governing singularity** is the centre of a spinning object such as the Earth, the centre of every atom holds **mutual singularity** Π^3 that collectively puts a mutual value of all the atoms' singularity as a combined equal to the **governing singularity** Π^o. The solar system will provide a **primary singularity** $\Pi^3 = \Pi\Pi^2$. The one would represent T^2 the other forms k that then produces the third singularity forming space a^3.

k indicates <u>**controlling singularity**</u> from the centre **k⁰** ending at the line **k**. This line shows the location around which a planet circles. The specific value about the centre is most important because from the specific centre gravity indicates a positional worth. The line forming **k** is pointing the circle or the <u>**governing singularity**</u> formed from a line that ends at a circle **T²** running from the centre **k⁰** to where the space **a³** is indicated.

The turning **T²** of any circle holding space **a³** is valid only if forming a reference **k** to a centre **k⁰**. $k^0 = a^3 / (T^2 k)$. This depicts a position the domineering singularity **k⁰** fills in relation to another point serving subordinate singularity **k**. There are always a dominant and a serving singularity interacting. If **k** indicates the centre of the Earth then **T²** rotates initiating the <u>**governing singularity k⁰**</u> where then the centre of the Sun **k** will form the <u>**controlling singularity.**</u> When the Sun rotates, the Sun's centre **k⁰** forms the <u>**governing singularity**</u> giving the Earth in orbit **k** holds the <u>**controlling singularity**</u>. The measure of **k** is not a specific value but serves only as an indicator to which space rotates or applies by the space rotating in a circle. This role of singularity being **controlling** or **governing** is playing part in movement of gravity forming and is very important when trying to understand the role that the four phenomena play in forming gravity. It is important to understand what happens in the event of an object going through the "sound barrier" or when escaping from the Earth's atmosphere. Where the object is standing still holding a position that allows the object to have mass, the object is part of the Earth while the Earth has the <u>**governing singularity**</u> and the Sun has the <u>**controlling singularity**</u>. As soon as any object moves on Earth, the movement switches singularity by allowing the object to obtain the <u>**governing singularity**</u> while the Earth then for fills the directional circular control in forming the <u>**controlling singularity.**</u> All four phenomena interacts in a manner forming this role where for instance in the solar system the Sun holds the <u>**controlling singularity**</u> and Milky Way forms the <u>**governing singularity.**</u> To find validity in my argument one must draw this statement of motion back to the point where singularity is getting sides or said mathematically Π^0 is going Π. Π is the <u>**controlling singularity**</u> and Π forming Π^2 is in relation to the <u>**governing singularity**</u> Π^0. When there is singularity there can be no sides. The one forming singularity Π^0 by measure fills no space while form Π develops Π^2 into space. The space that even the dot fills being Πr^0 does not really exist in the manner we humans see space to exist. It is a spot that is there without being there. It does not visually exist because it is not filling any substance and it cannot be recognised since it is not three-dimensional. The spot and the dot have no dimensional worth of any measure but holds relevance. This Universe I am addressing has never been unveiled by any one since this is the flat Universe. This Universe holds a line in time made up of dots and spots forming no space but holds a Universe relevant.

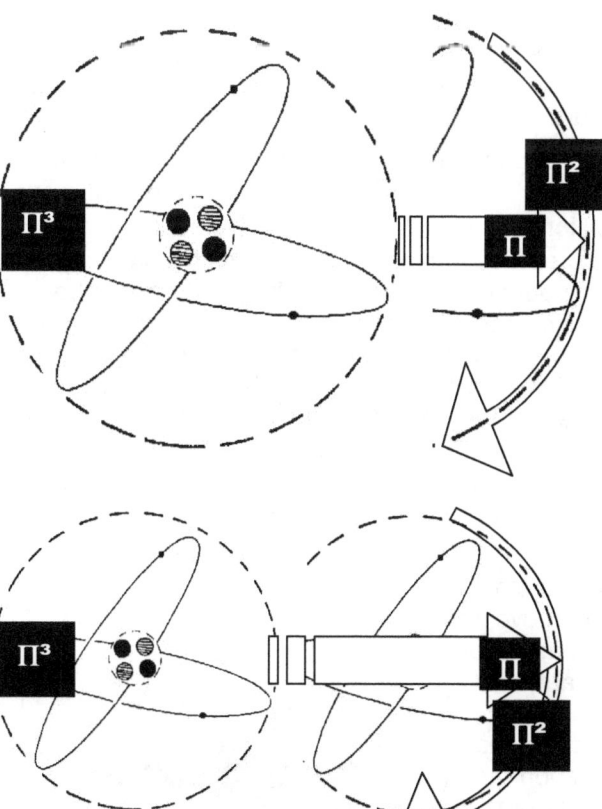

The movement of the atom $k^0 = a^3 / (T^2 k)$ is also widely known as gravity and also as time.

This is where the simplicity in the approach of Newtonian science renders the understanding about science that Newtonians have to a joke. Newtonians dump all on mass and with that they whish to produce the most elaborate mathematics to derail all concepts that validates science.

When an electron rotates around an atom it takes time to do so. Notwithstanding the mathematics Newtonians apply, the speed of light can't be 1 because we know that in computers the speed of light is slowed down to a point in silicon where light stands still and yet time moves on all around the silicon chip. In massive stars the gravity exceed the speed of light and therefore if light was the same as time the time in those stars would run backwards and leave the cosmos approaching the Big Bang and not developing away from the point of connection. The electron is in time when spinning around an atom. It takes time to move around the atom. When the atom moves forward, the movement has to reduce the electron circle because the time it takes to spin around the atom is reduced by the forward movement of the atom as such. The forward movement will take away electron time when moving forward because it will reduce the time the electron has to complete the circle. The atom's space **a³** is depending on both factors that determines sizes such as the circular rotation **T²** and the linear movement **k**.

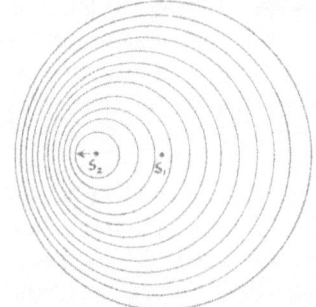

More evidence of movement influencing the rotation is evident in the Mach principle where the moving train pushed the sound of the train moving in a specific direction forward by concentrating the lines and to the back it expanded the circles. This shows the effect of the movement of the atom $k^0 = a^3 / (T^2 k)$ that circles T^2 while moving forward.

In every case scenario as gravity intensifies the atomic space becomes smaller as the density factor rises in every case and the mutual singularity becomes more compact because of a firmer controlling singularity being the product of a more intense governing singularity.

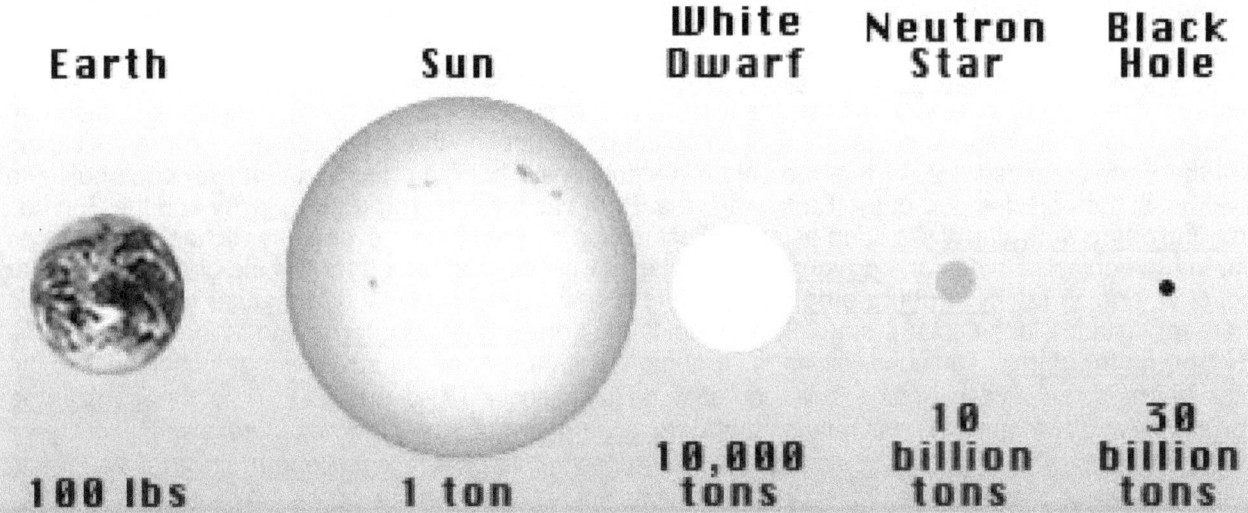

Earth	Sun	White Dwarf	Neutron Star	Black Hole
100 lbs	1 ton	10,000 tons	10 billion tons	30 billion tons

This understanding of the reality of science coupled with the fact that one has to understand science and not dump everything on magic and mathematics makes the understanding of the sound barrier possible. I would not have bothered that much with the explaining if the phenomenon was not at the heart of cosmic development that tiny stars use when the come out of the heat cocoon in the centre of galactic insides. I explain this occurrence in the Cosmic Code.

The Roche principle is a viable phenomenon and unlike the implementation of mass that has no proof, this cosmic phenomenon is widely applied. Movement puts a body of atoms in a different relevance and the movement applies different circumstances to the object moving by shrinking the object's entire structure ever so slightly up to all consuming shrink such as what happens when objects enter the Earth atmosphere.

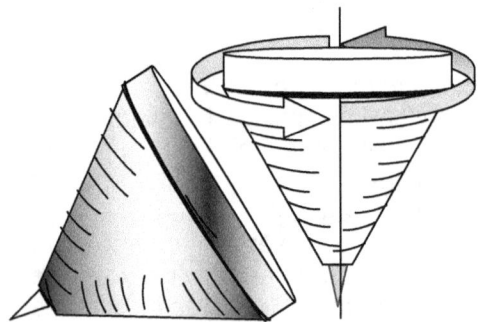

The big lesson the spinning top teaches science is that the top complies with Kepler's formula of $\Pi^0 = \Pi^3 \div (\Pi^2 \Pi^1)$. What this says is that a line Π^0 forms in singularity when movement in spin Π^2 confirms Π the presence of a body Π^3 of atoms that is in a structure that is independent by the spin from the Earth. The body is in a formed space that holds its atoms secluded from that of the Earth by the movement the moving body applies. The movement of the top places the top in a secluded area and it is the relevance that this secluded area brings that allows the top to maintain an erect stance. The top is in a "bubble" of its own making by the movement in the air of the Earth. It is this "bubble" that we have to trace when finding the "sound barrier".

The circular distance, however remain valet as one can see from the "gravitational pull" that has nothing to do with jerking each other around by inflicting mass on one another.

When the object in this case an aeroplane, starts to move the movement establishes a controlling singularity by the measure of the movement of all the atoms forming a governing singularity that is without space and could therefore be anywhere. The movement of the Earth remains the most dominant factor and it is around this that the rest of the movement comes to value. As soon as movement starts the "bubble" takes on the movement value of the Earth's gravity. This "bubble carries" a value of $7(\Pi\Pi^2)\Pi^0$, but enough said since I do not wish to jump the gun so I will leave it at that and later on in this book more will be said about that.

Every moving object forms a container by providing a centre that by he movement forms a governing singularity. This movement forming a governing singularity takes control over the space that form by the movement of the containing structure. The Earth is relegated to the position of forming the principle singularity and while the object is stationary the object holds mass the object then forms part of the mutual singularity of the Earth.

This is the very reason why objects like "falling stars" burn out when they enter the Earth's atmosphere.

The fastest speed a falling object can reach is the rotational speed that the Earth provides. This is the limit to the linear displacement. Past that speed, a body needs additional and independent drive to exceed the linear to circular balance.

Between Π and Π^2, the body finds itself in its circular displacement, which is the Earth's linear displacement value and therefore as it moves with the linear movement of the Earth this brings about an additional movement of Π^o.

When an object exceeds the Earth's movement or displacement value, it is refer to as Mach 1, or the speed of sound. However the movement is always still subject to the earth and although the moving object may reach total independence in movement, such movement is still subject to the movement of the earth and hence the Roche limit value of Π^2 but that is cut in half by depending on the movement the Earth provides and therefore mach one would come at $\Pi^2/2$.

The linear displacement value surpasses the circular value, therefore it is $\Pi^2/2$ and this still carries the measure of kilometres per hour, because the distance (kilometres) and the time (the rotational speed of the Earth) applies directly to calculate any movement concerning the Earth.

This will come into affect when the projectile reaches Mach 1. I have to stress the fact that the Doppler effect is, once again, merely a co-incidental but duly related by product.

BETWEEN *POSITIONS ONE* **AND POSITION TWO** it is $\Pi^3 / \Pi^2 = \Pi \times \Pi^o$ reaching $\Pi^3 / \Pi^2 = \Pi \times 2\Pi^o$
ATMOSPHERIC CIRCULER DISPLACEMENT Π^2 **and the linear displacement** Π.

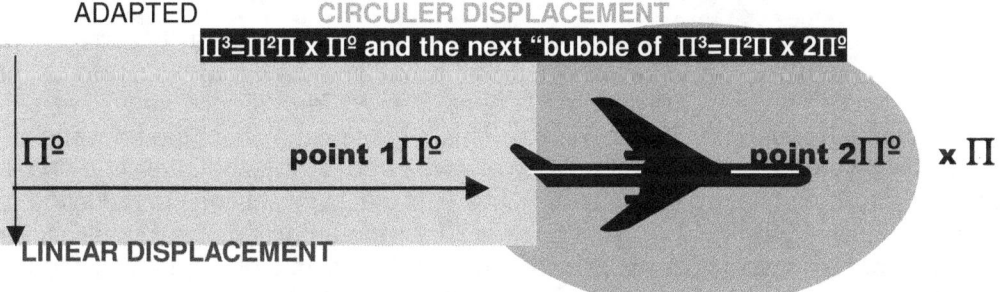

BETWEEN POSITIONS THREE AND POSITION FOUR

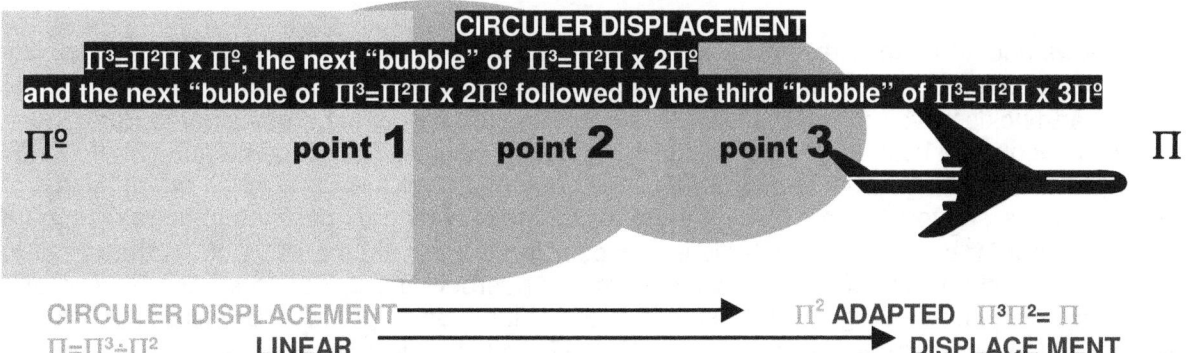

Doppler's effect has nothing to do with the process, and neither has sound anything to do with the entire affair but is simply a derogative of the principle that has deep roots in other cosmic development. Sound is merely an innocent bystander and Doppler was an onlooker that knew nothing about what he witnessed. However, the Doppler effect, as such, plays no part in these phenomena, or in the outcome of the application. The "Newtonian scientific mistake" in this case is the connection with Titius-Bode law that has gone unnoticed for almost a century and a half as well as how to couple this with the Lagrangian positioning. In this is the reason why a spacecraft when launching into outer space does not "break the sound barrier". The circular space-time is growing faster than the craft can apply linear displacement. The higher up in space, the less the prevailing circular space-time would be therefore the more linear space-time will be required to penetrate the circular value. Through the four cosmic pillars creation came about and by not recognising the Titius Bode law The Roche limit and the Lagrangian atomic structure we shall never understand the cosmos. The importers of these systems reach way beyond the solar system to the very heart of what ever forms the totality of the Universe.

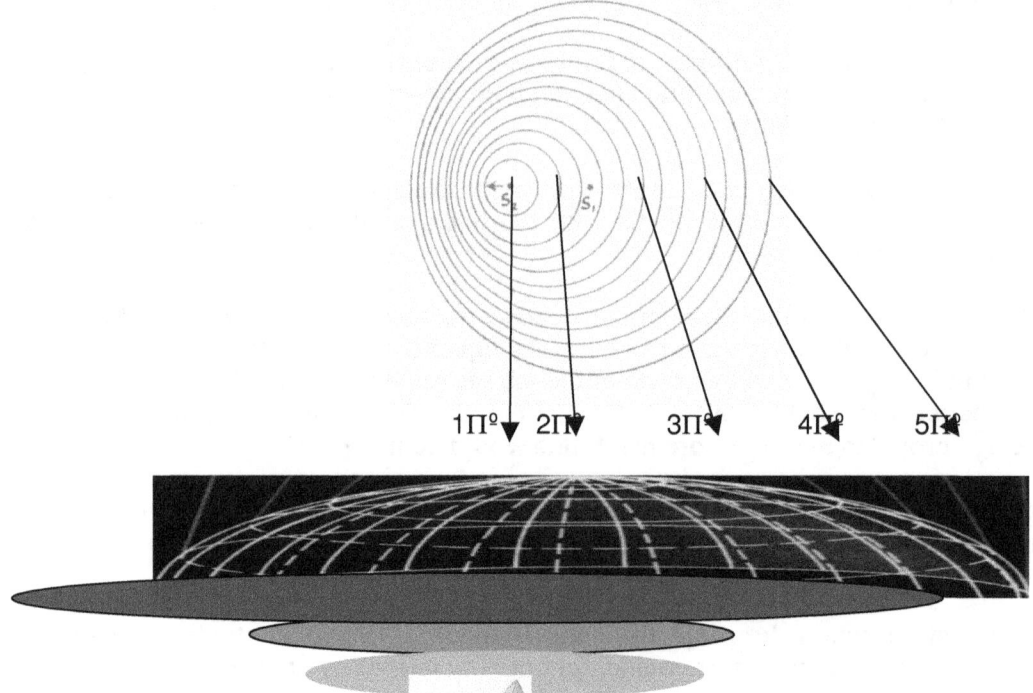

THESE TWO COMPONENTS FORM **SPACE-TIME** ($\Pi^3 / \Pi^2 = \Pi$), THE DIMENSION WE FIND OURSELVES IN, AND WHICH IS NOT FLAT.

It is the Coanda effect that produces gravity and it is the Coanda effect that is keeping the Universe together. Let us take the "Sound Barrier" from a point where we see phenomena apply laws that matter complies with. The sound barrier is a prime example of the relevancies, which I suggest, takes place. There are two points in singularity always referring to each other and one is expanding while the other is contracting. In nature there is the Roche limit placing a limit on the reduction of space and the inflow of heat to sustain proton cooling. At a point of $(\Pi/2)^2$ the reduction of space disallows any object to immediately compromise space claimed in time because the cosmic object cannot reduce space while an entry to its area demands such a time reducing in space claimed by the material. The first question that one can ask is why would there be the value of $(\Pi/2)^2$ between orbiting structures positioning themselves in a time relation to space.

Every person knows about the entry restriction an orbiting spacecraft finds that forces the craft to comply with. The entry maximum is 21,991 and the minimum entry is 7. This is without doubt, the number of Π (21,99/7). The Earth holds its value to $4\Pi^2$ and when an object is not part of the surface of the Earth, even say a mountain; it becomes a holding value of 7. Later in this part I explain the sound barrier in more detail and the 7 will then become better understood. At this point one must see the Earth in the proton status of $(\Pi^2+\Pi^2)$ while acting as an atom. In this relation the atmosphere including all particles in the atmosphere will in relation be either Π^2 or Π. When water is in a vapour form, it will have a value of 3Π, having heat separating the water to the factor of 3. By dislodging the thunderbolt, the 3 receive a square value and displaces to the Earth in the linear light to time stance of 3^2. With heat (3) grouping by initial spin value, it will remove from space leaving the water to the value (Π to Π) and this will then give the water a relevancy of Π^2. The factor of Π^2 places the water no longer amongst heat as gas, but heat as a liquid (rain) or solid (hail). The relevancy of the water will change from 3Π to Π^2 placing the water's position from space (3) to liquid or solid (Π^2). Where does the Π that one find in the Roche limit $(\Pi/2)^2$ and the vapour (3Π) finds its relevancy to gravity? Every particle that enjoys space-time outside the Earth's structure ($\Pi^2 + \Pi^2$) will hold a neutron position of ($\Pi^2\Pi$). The Π^2 ends will be at the point where heat passes through the object directly to the Earth and this position of space-time relates to the neutron time link of Π^2. The space link of the neutron will then form the Π link. The value of the Π link we find to be $(Π/2)^2$, but the explaining to why it is $(Π/2)^2$ is rather more complicated.

The different values can represent different motion, i.e. the atom where the electron's space-time displacement is as follows:

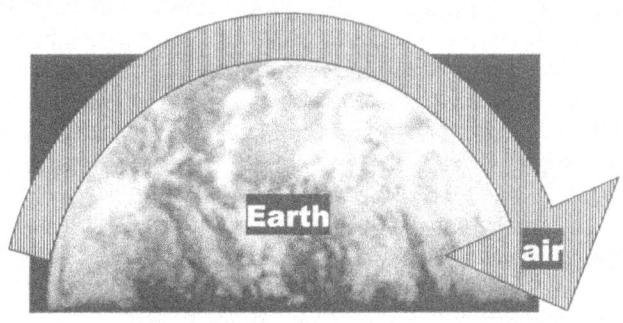

Explaining the "sound barrier" the quickest would be as follows: the Earth spins at a rate of $7ΠΠ^2$. Since the Earth is just another centrifugal pump, it spins the liquid around the Earth we call air at a rate of $7ΠΠ^2$. The absolute top speed that any object could fall at, will therefore be a rate of $7ΠΠ^2$.

All the liquid flowing around the Earth, which I repeat is the liquid we call air forming the atmosphere, will have to flow at $7ΠΠ^2$ at the highest speed it possibly could when compressing as the Earth contracts the air. This is not wind speed, but is "atmospheric compressing" that forms "atmospheric pressure", which are all terminology used that applies to the same phenomenon that goes by the name of gravity. If any object is on the Earth, but that moves within the liquid, where such an object then moves faster than what the liquid moves, then there are certain criteria applying as laws of physics.

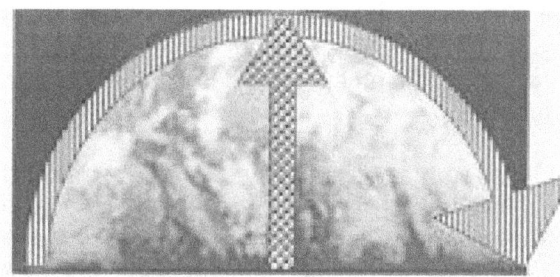

When any object is in a state of having mass the object has to be standing still and being secured in a position on the Earth at that point of having mass. The object has to be in a position of absolute rest while it is on the Earth. At a point of standing still in relation to the Earth while excepting only the movement the Earth allows any object to form mass and it is where at that point that the object with mass is resting while all the rotational movement is equal to the movement the Earth delivers where the Earth is rotating. Rotating at the speed the Earth dictates form the factor science call mass. When the object leaves the surface of the Earth such an object will have to move much faster than the Earth moves or have less density than is required to maintain a steady position on the Earth.

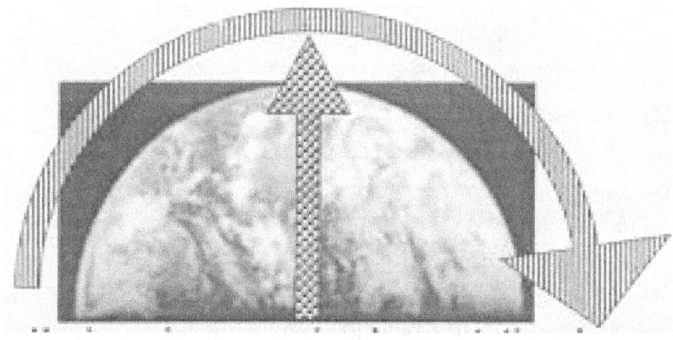

When any object is standing still in mass on the surface of the Earth, an object has micro gravity because the individual gravity left to the object in mass is infinitive small and is left to become an indication of attempting further movement towards the centre of the Earth while the Earth's material blocks the micro gravity to move and hence apply mass in doing so. Mass is not something inherent of the object but is the annexing of the object given mass by the Earth to secure the position of the object to ensure the object becomes part of the Earth structure. Having micro mass (not micro gravity) is where the body in rotational movement extends beyond the limit at the point where the Earth surface would award a mass factor. The movement speed goes beyond the speed required by the Earth at the Earth limit where rotation velocity secures mass as a factor. By exceeding the rotational velocity at a higher rate, such movement would exceed the movement or gravity of the Earth that is required in order to grant a mass value.

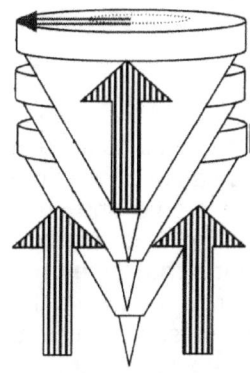

When the top is spinning it is this line that urges the top to excel from the limitation of the gravity of the Earth and extend up into the air and away from the ground. It is this centre line holding singularity that drives the top to lift up from the ground and fight the mass that the Earth inflicts as to retain the top with the limitation of enforcing the top into a state of mass where the mass holds the top onto the ground. The top is fighting the Earth's effort in restraining the top with mass by producing gravity that lifts the top into the air. It is the top's spinning that is producing anti gravity to fight off the gravity of the Earth. I have heard so many scientists refer to man discovering anti gravity as if such a discovery of a force of anti gravity will give humans the power only God can have. It is this mindset that I refer to as science wanting the marvellous, the magical and the unexplained. To the masterminds of science having anti gravity would come to the same as unlocking all the witches' forces and opening the Pandora's Box of forces while anti gravity is simply jumping in the air. If gravity is what is pulling you down, then anti gravity must be something lifting you up. The top wishes to leave the position of $7\Pi\Pi^2$ to then become $73\Pi^2$ and that is finding release from Π or then having "mass".

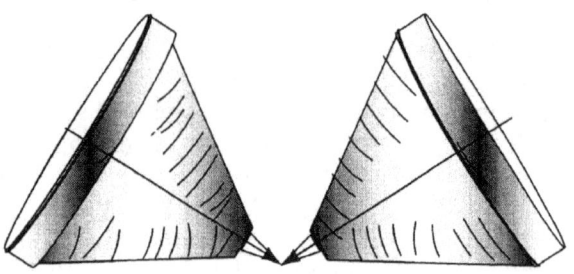

When the spinning has died down so much it will arrive at a point where the gravity of the Earth will reduce the spin of the top to lying still while the Earth secures the top with going into having a state of mass or having no independent movement which is having mass. It is at that point that the centre line within the top that is securing gravity by the spinning of the top will seize to be and the top will once more come to rest in a state of mass. The gravity of the Earth is fighting the gravity of the top which is equal to the singularity of the Earth is fighting to destroy the singularity of the top which is the movement of the Earth is fighting to destroy the movement of the top and all of these relevancies are all the same. It then has mass or is in $7\Pi\Pi^2$.

This issue is of cardinal importance and could deliberately be altered to hide the misinterpretation science wishes to connect to mass in order to hide the fact that mass does not bring on gravity but it is gravity that brings on mass. Mass is achieved when the object is resting motionless on the surface of the Earth while it is gravity that is still attempting to obtain movement as to try and move the object down to the centre of the Earth. This movement consists of two parts where one part is following the curve of the Earth while the Earth is rotating and the other part of the same movement is the thrusting of the object educing the object to move to the centre of the Earth. Mass is the result of gravity and not the other way around. Gravity brings on mass and mass depends on gravity to have any value or function.

A person that acquired the skills of peddling while staying upright on the bicycle has achieved the method of rearranging gravity within singularity. Without motion the bicycle falls on the spot it holds. When the bicycle is put in motion the bicycle can maintain the upright stance as long as the motion applies. When the motion stops the bicycle drops. To introduce motion to the bicycle the motion brings about a stable unsupported upright stance where balance can result from the motion the Earth enforces to the balance coming about by the bicycle using independence gained from motion of the space holding the bicycle. The space that the stationary bicycle holds is the direct result of the Earth providing the motion.

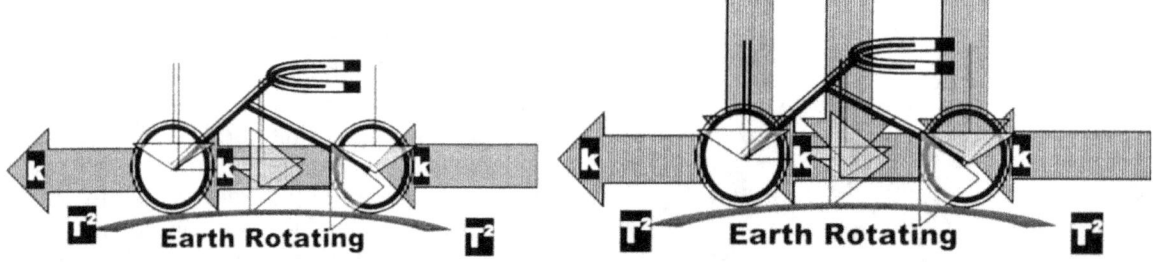

That is why it then will adhere to the gravity or motion that the Earth will enforce. The motion restricts the static bicycle to one allocated position that the Earth supplies to the bicycle. When the bicycle starts to move, the bicycle gains a cosmic independence. The gravity effecting the redirecting of the Earth gravity response comes about as the result of additional motion that is introduced to the bicycle. This is the very same process that the aircraft need to get air born because it replaces or repositions the singularity the Earth holds to the singularity the bicycle develop in motion. The aircraft only takes the change in direction of what the gravity is insisting on through changing direction in motion through faze one and into faze two. It all is still part of the Coanda effect. With more motion contributing to acceleration the bicycle will become airborne on condition that it is also given the advantage of a set of wings to increase the effect of creating space-time to the advantage of the motion requiring the change in singularity direction.

I specifically chose to use a bicycle in my explaining because the bicycle is the object that rely the most on singularity achieving the required balance in which to operate. It is singularity, which puts space in balance of the time the space uses to duplicate. The singularity create space-time and such space-time results in a balance of space and time $a^3 = T^2k$. **It is through the Coanda effect that marries the motion to the space that gives the balance** that keeps the bicycle up right while it is singularity that allows the bicycle to move or duplicate the material by relocation through time. It is an act of balancing singularity that gets the bicycle as a machine working properly. It is also the next best thing to illustrate how singularity by

The Four Cosmic Pillars; The Result Thereof.   APPLYING AS THE SOUND BARRIER

motion provides gravity in addition to that which the Earth already produces. In the Coanda principle there are two factors where one is motion and the other is space and the two provide both duplication as well as contraction of space-time.

We think of the bicycle moving in a mono dimension where the bicycle as one unit moves from one position to the next position.

That is not true because accidents show otherwise as the bicycle fragment when having collisions.

The bicycle as it moves by atom much part part of the a new forms the movement re-affirms its position changes by displacing the body atom and the movement is as of the bicycle as the bicycle is cosmos. Every movement is association with whatever bicycle's surrounding and the is as much part of gravity as the bicycle moving independent. It is very important to see the bicycle moving as not forming part of the cosmos but as a result of life manipulating the cosmos by forcing the bicycle to move. Without the direct

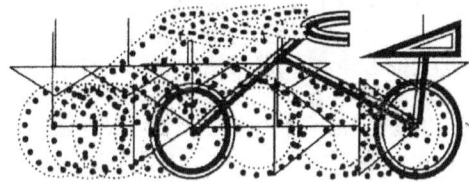

involvement of life the movement of the bicycle will never come about and therefore the bicycle receives independence as a cosmic structure not through the cosmos energising the bicycle but because life intervenes and life is not a natural part of the cosmos. The bicycle stays upright and riding because life puts it there.

In the time of Newton steam was the Rocket science of the day...and that was literal because the first steam engine was named "The Rocket" and knowing anything about steam engines meant knowing everything about the steam engine called the Rocket. The concept of anything being drawn or pulled or powered by any source outside life physically pulling such as man or donkey or horse or whatever, was still in a concept form and to place much emphasis on movement or to think about what drove what whereto was in its infancy. Everything was either wind driven or animal driven or powered by slaves...yes slaves was still part of the daily practise of getting things done. That is how far back Newton's ideas go.

Downward thrust

In gravity there is never just one movement but there are always two movements combining as gravity.

Taking the bicycle as an example we have the downwards thrust of gravity as well as the forward moving of the peddling cyclist that bring about movement. Under normal conditions and without acrobatics the forward peddling is required

$$k = a^3 \div T^2$$

$$T^2 = a^3 \div k$$

to keep the bicycle upright but just as much is the downward thrust required to keep the bicycle peddling. Try as anyone may, the cyclist will not be able to keep the bicycle upright in outer space where the downward thrust is not present. Using Kepler's formula of $a^3 = T^2 k$, it is the body of the bicycle filling the space the bicycle claims a^3, that will move forward T^2 on condition that the bicycle is thrust down

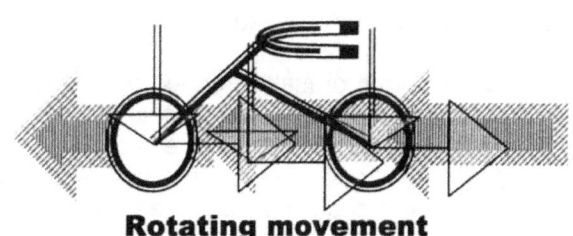

Rotating movement

k. The downward thrust can only help if the forward movement T^2 grants the space the bicycle holds a^3 independence of the Earth making the moving singularity the bicycle then claims become the governing singularity while the movement demotes the Earth's normal governing singularity to become the controlling singularity. Space to grant a body independence from the Earth requires two directional movements acting as one movement.

The body can only be independent a^3 in space if the movement is backed = by rotational movement T^2 of the wheels as well as having **k** become the directional

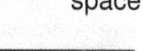
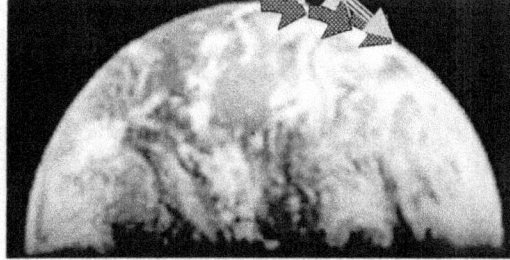

movement. The value of **k** is in the reference the object holds in relation to the direction of movement.

Let us take it from a point where the Sun provides a centre as one starting edge of **k** giving **k** a directional relevancy, then that centre **k** will provide a line from the centre and the line **k** will provide three spots in a formation that produces a structure by the square T^2 of the dimension where T^2 becomes the spin of the Earth. Not once did Kepler indicate size as a contributing factor to a^3. That means every single point that **k** indicates there are three positions a^3 implicating sides of a double dimension. In the same manner is **k** not limited to distance or does T^2 lesser by size declining. $a^3 = T^2 k$...That is what Kepler said. There are three dimensions forming space a^3 by movement where two is between any two points T^2 flowing as time from the centre of the sun, which is indicated by the line the relevancy factor **k** indicates.

From the sun there are three points moving between two points from one point to two other points giving the six dimensions we find in space. It is space in time or space converting space through the movement of time. It is a location of a point in the third dimension a^3 that will move according to the second dimension T^2 that will implicate **k** as a reference in the first dimension. It is about dimensions in reference to one another.

When an object falls to the Earth the body moves in the direction of the Earth **k**. However at that point the Earth is also rotating away from the point where the falling object is heading to the Earth, which is indicated by the symbol T^2. The space in which the body is a^3 is falling **k** to the spinning T^2 Earth $a^3 = T^2 k$ all the while the body is descending (falling) $k = a^3 / T^2$ as the Earth is rotating $T^2 = a^3 / k$. It is in this that Kepler's formula comes to prominence. As the body moves towards the Earth k^{-1} the Earth shifts T^2 and the falling body is re-aligning with the Earth by associating with the position the Earth has that then changes. The falling body is declining in space represented as $k^{-1} = T^2 \div a^3$ while the Earth is rotating $T^2 = a^3 \div k$. At this time **k** shifts from straight down to slanted because the reference point relocated the reference position.

Direction of descending $k^{-1} = T^2 \div a^3$ by 7°

Directional re-alignment

Direction of Earth spinning $T^2 = a^3 \div k^{-1}$ by 7°

Having to re-align its position of reference as the Earth changes position in rotation, the object changes direction by implementing the triangle of Pythagoras because the Earth moves to the side by 7° and the object fall by 7° and in that the triangle that forms a right angle triangle (both sides equal to a change of 7° and therefore equal forms establishing a hypotenuse where the hypotenuse forms gravity. That forms the basis of what all that gravity is about and gravity is about Π coming into space-time by movement Π^2. Mass comes into question when the body falling has no further space in which to fall but are then obliged to stay still on the Earth surface and form a part (mass) with the Earth. While it is falling, it is in the space surrounding the Earth that contracts by the margin of Pythagoras and the body that s filling the space is also contracting with the space. The body is merely moving down with the space in which it is and it is the space in which the falling body is that is moving down with all the surrounding space also moving down. Being solid the body restrains the contracting that reduces the size of the body but still the space surrounding the body is becoming denser and compacter. In the Universe as in cosmology there is no possibility of the presence of nothing and therefore outer space cannot be "*nothing*" but is cosmic fluid that has the ability to become denser by contracting.

There are two forms of matter formed by one cosmic substance, which is singularity. There is singularity formed as a liquid and there is singularity that is forming material. That is it...there is no possibility of anything else and all substances are composed of singularity being controlled by movement (material or matter) positioned in relation to singularity not controlled by movement or forming a liquid. Everything is heat that is forming space but some heat is controlled by movement, which we call elements and the rest of space is heat forming cosmic space, which is uncontrolled and is therefore not dense but totally expanded. We think of this substance as outer space. There are elements and there are heat covering and surrounding the elements as space forming a liquid in which elements of all sizes float. That is the cosmos. That is material floating in cosmic liquid having no mass because it has buoyancy by movement. That is why not one of the planets indicates any positional arrangement by virtue of mass because the buoyancy of any cosmic object excludes mass as forming a factor.

It is assumed that a person uses oxygen to breathe or that fire uses oxygen to burn. If people were to use oxygen to breathe or that fire does consume heat to burn, the oxygen on Earth would by now be exasperated. If it was oxygen that burned, there will be no oxygen left on Earth to burn. It is the relation oxygen has with heat that makes oxygen the carriers of heat and oxygen transports heat from point to point. Oxygen only delivers the heat to a fire or to the body and it is this heat that then forms the aging process in all living bodies. Atoms grow in size when consuming heat and that forms the expanding of the Universe where the unoccupied space-time transforms to the occupied space to maintain heat balances.

In all of nature there is no **NATURAL GAS** as a natural element as much as there is no **NATURAL SOLID** as a natural element.

No element is either a gas or is a fluid but all of the elements forming material are a solid. This solidness comes about because the atoms spin and the spin provide a density that cosmic liquid or heat lacks. We arrange the elements in such a manner, but that is only applying to the situation the Earth grants the elements a status to be thought of as a "natural fluid" or a "natural gas" where in fact even the hydrogen atom is a "natural solid" that boils at a very low temperature. In outer space all elements will classify as a solid because all elements freeze under those (to our thinking) "extreme conditions"

Always being a solid	Becomes a fluid	Becomes a gas
Hydrogen 1	melts at -259^0 C,	boils at -252^0 C,
Helium 2	melts at $-269\ ^0$ C	boils at -268.9^0 C
LITHIUM 3	melts 180^0 C	boils at 1300^0
BERYLLIUM 4	melts at 1287°C	boils at 2770°C
BORON 5	melts at 2030^0 C	boils 2550^0 C
Carbon 6	melts at $804\ ^0$C	boils at 3470^0 C
Nitrogen 7	melts at -210^0C	boils at -195.8^0 C
Oxygen 8	melts at $-218.8\ ^0$C	boils at -183^0 C
Fluorine 9	melts at -219.6^0 C	boils at -188.2^0 C
Neon 10	melts at -248.59^0 C	boils at -246^0 C
Sodium 11	melts at 97.85^0 C	boils at 892^0 C
Magnesium 12	melts at 650^0 C	boils at 1107^0
Aluminum 13	melts at 660^0 C	boils at 2450^0

When an element freezes it is solid notwithstanding… because then there is much less heat in between the solids. The ratio of cosmic liquid to solids favours the solid overwhelmingly.
When an element melts it becomes a liquid and that means there is just more heat in between the solids
When an element boils it is a gas again notwithstanding…and a lot of heat (cosmic liquid) is added where the ratio of cosmic liquid to solids favours the liquid ratio overwhelmingly.
In the cosmos at all times everything composes of singularity that holds form either as a solid restricted by movement, or as a liquid, which is unrestricted by movement. That is the cosmos.

Stars are balls of heat mixed with particles we call atomic elements. There is heat and the there are elements structurally held within atomic capsules. One of the main issues that I wish to protest by my writing this is my argument that if the Universe can be compressed back to the size it had at the point of 10^{-38} seconds after the Big Bang, at that point the daily outdoor temperatures will return once more to 10^{27} K. The expansion was the result of compressed space, which then formed into heat and in turn resulted in finding a Universe with all the insufficiency of space- less-ness prevailing throughout and wherever space was needed. Elements can't compress… And therefore it must be liquid that compresses as air does when the Earth compresses outer space into atmospheric air. By that it forced space-time to come into being as solid as light particles are. Space-time came about at the time without space availability, which brought about the period of the Big Bang forming space and converted massive heat into massive space. If the Universe was in a vacuum as big as being available now, then what was the temperature of the vacuum while it was empty before material filled it later? The Universe has no outside therefore it can have no unfilled vacuum. If the Universe then employed the space of say one atom, the impression comes through that from edge to edge and from Universal border to border the space occupied was the same as one atom will claim in our present day and age. The Universe has no outside therefore it can have no "from edge to edge". It is accepted that normal gravity started at 10^{-43} seconds. The Universe was the size of a neutron. The Big Bang began and GUT, or the grand unified theory, produced the attempt to describe the strong and weak nuclear forces and electromagnetism in one single mathematical theory. Somewhere before 10^{-12} seconds of counting the Universe cooled to about 10^{15} K the electromagnetic and the weak interactions acted as one single physical force. Science reckons that unification may come about at temperatures of 10^{27} K, which was the temperature of the day at 10^{-38} seconds after the Big Bang. This statement echoes my viewpoint but one has to look carefully for that to surface. In the suggestion the presumption claims that all the space that the Universe made available at that time was the total space one atom might take up today. If that might be the case then where was the rest of the space that now fills the Universe? Or was the rest of the space we now find in the Universe and what is now explained away as the vacuum, also available back then. Did the Universe only have that one tiny hot spot it filled with huge volumes of heat? Was the rest of the space vacant out there all along during all the time running to the present date filled with emptiness standing around as a big vacuum with nothing better to do than suck on the Universe while the Universe was exploding at the speed of light. I say everything is singularity forming the entirety.

The Four Cosmic Pillars; The Result Thereof.

APPLYING AS THE SOUND BARRIER

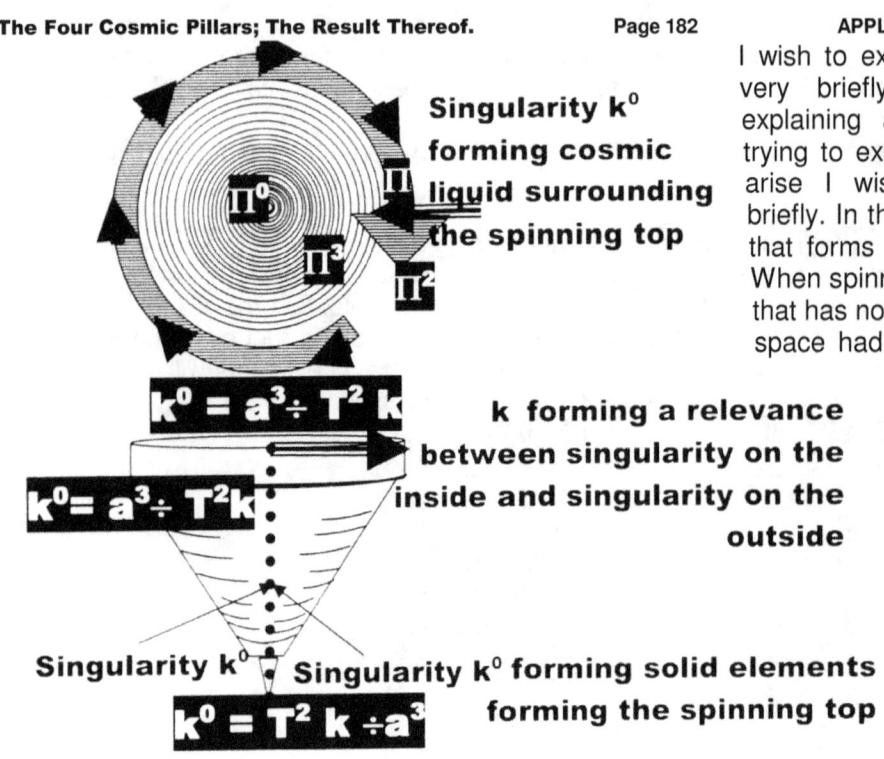

I wish to explain the location of singularity very briefly since shall return to the explaining about singularity later on but trying to exclude much confusion that may arise I wish to explain singularity very briefly. In the centre of all things are a spot that forms a centre, which has no space. When spinning that centre spot forms a line that has no space. If the line had space, the space had to choose sides and since the line forms the divide between that which spins to opposing sides, the line in place can't have space. It is the movement that brings about the relevancy k and with k in place the singularity inside the centre forms the space, which becomes the top. Putting the relevancy k in place brings about the validity of the space a^3 by the movement of the spin T^2. By starting to spin the line forming singularity in the centre $k^0 = a^3 \div kT^2$ and that is what mathematical significance Kepler's formula indicates. As to why things spin has to do with heat and cooling but I will deal with that argument as the book progresses.

Locating and finding Singularity

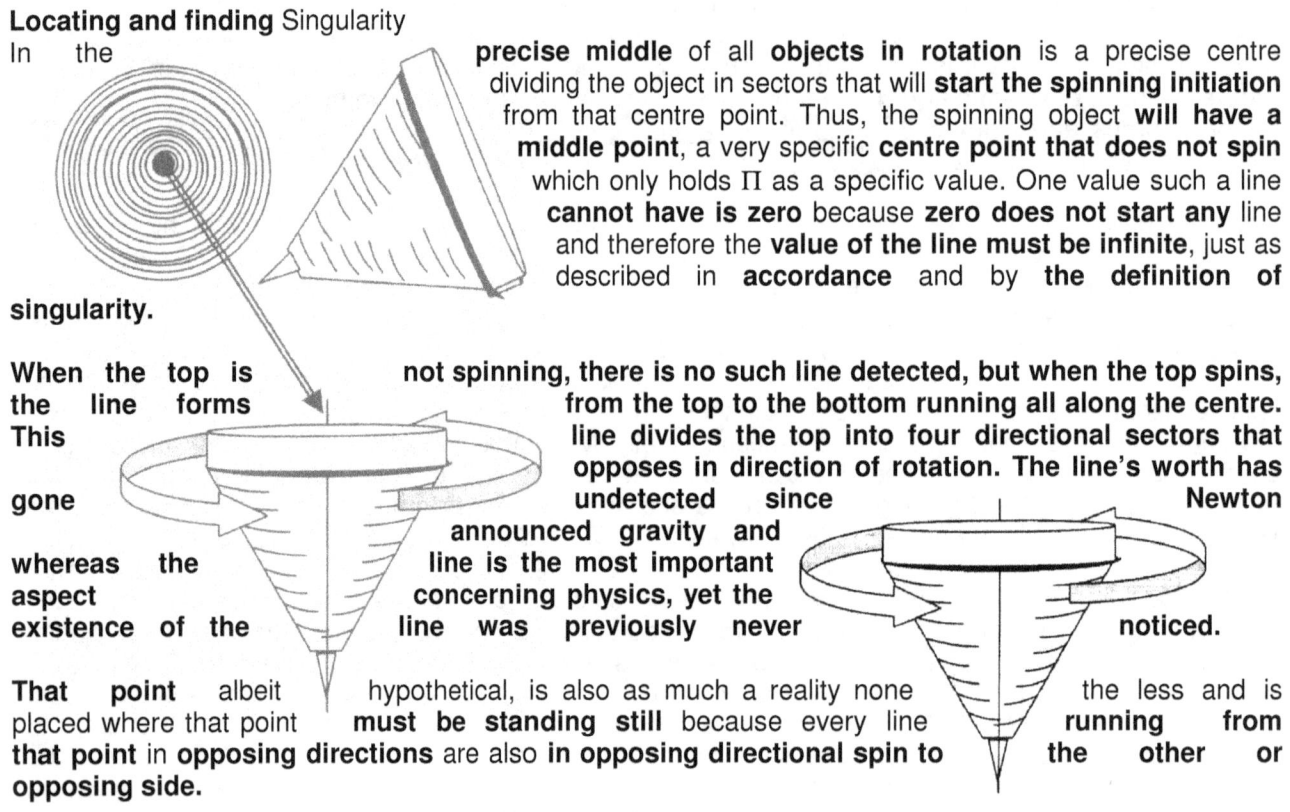

In the precise middle of all **objects in rotation** is a precise centre dividing the object in sectors that will **start the spinning initiation** from that centre point. Thus, the spinning object **will have a middle point**, a very specific **centre point that does not spin** which only holds Π as a specific value. One value such a line **cannot have is zero** because **zero does not start any** line and therefore the **value of the line must be infinite**, just as described in **accordance** and by **the definition of singularity.**

When the top is not spinning, there is no such line detected, but when the top spins, **the line forms** from the top to the bottom running all along the centre. **This** line divides the top into four directional sectors that opposes in direction of rotation. The line's worth has **gone** undetected since Newton announced gravity and **whereas the** line is the most important **aspect** concerning physics, yet the **existence of the** line was previously never noticed.

That point albeit hypothetical, is also as much a reality none the less and is placed where that point **must be standing still** because every line **running from that point** in **opposing directions** are also **in opposing directional spin to** the other or opposing side.

Spinning or movement inside the line would **be zero,** but the line, although **being without space,** also **can't be zero** since the line is there for all to see. The movement in the line **is zero** and the space the line uses **is zero** and the line holding a value in size might **be zero,** but the line as **a cosmic reality** just can't **be zero** since the line controls all the spinning taking place.

From this centre line that is only theoretical definable, but is still there all the same, an opposing value always form from a previous turning position to the next turning position that becomes real and distinct when rotating, but loses its distinction when not rotating because then all traces of the line that is not there is lost as the line disappears.

As the line disappears the value of the line not being there changes from being noticeable to zero, and as the line removes from having a notice ability in securing a value by spin to then when not spinning have a value of zero, where this zero value then replaces the most original value it had. When not rotating, zero removes the line from a position it never held in space previously. When rotation begins, the line forms and is only backed in value by having only a hypothetical position claiming zero in spin and in space but not in presence. Being without space doesn't make that the line is not less distinct but the line is more distinct than any other part that in reality does hold space and therefore participate in spin because from that point every rotating piece of what ever is, then will spin around this line that is not there to start with and such spinning will clearly carry from where the line only has a distinction value in the singularity to carry on with a value of Π implicating rotation. The line forming holds 1 in singularity and from where such a line ends, only there does the circle value of Π start.

If the spinning top is all the evidence any one needs to come to such a conclusion that will bring any proof that the singularity governing the top connects to anything anyway, we will then find it when studying the spinning behaviour the top represents. Placing singularity in a location not being present in the Universe is fair and fine, but what will the evidence be in proving its activeness as part of the creation at large?

There are solids that form elements and by the control of singularity producing movement that confirms the structural integrity and discipline of singularity and is contracted to solidity by movement. That then is a solid forming various atoms of all sorts that we know as elements or as material. The elements in atomic cocoons are formed by singularity but the immense fast spinning contracts the singularity to a solid substance. Then we have liquids that are singularity that is unattached and are loosely connected and will accommodate any solid spinning where the forms of solids are in need of occupying that space for any duration of time. The liquids are able to accommodate or house the solids without being affected in way. There are cosmic liquid accommodating cosmic solids where the solids Newtonian science does recognise but the cosmic liquid Newtonian science fail to recognise. Newtonian science calls the solids elements but the cosmic liquid they call "nothing" and then give "nothing" a measurable value. Lately Newtonian science came to think about "dark matter" that has to keep the Universe in tact and I suppose in a way this forms the "dark" or invisible matter Newtonian science so desperately needs. The only viable conclusion about what keeps the Universe in tact would be the idea of a liquid in relation to a solid. Should there be any one that disagrees with my statement about the cosmos formed by liquid, then please tell yourself why would the atmosphere be a liquid with a density of about six hundred time less than water and where does the atmosphere (not density differences) stop and the nothing filling the atmosphere starts.

Elements are solid and that which house elements can't be nothing but has to be a fluid / gas/ call it what you like. Hydrogen is as much a liquid as iron is a gas and neon is just as much a solid. In fact all material (atomic compositions) is a solid and the ratio of heat in between the atoms determines whether it forms a liquid or a gas. It depends on the element relating to the space/heat in the circumstances surrounding the substance at that very precise instant in time. We have to stop telling the cosmos to show us what we wish to find and start accepting what the cosmos is telling us to find. The culture that I am referring to is all about **nothing.** At present we find that there is something we think of as nothing in outer space. Because nothing is what we wish to find and nothing is precisely what we are getting because we think of outer space as nothing. If you accept the cosmos to be nothing, then please define nothing to yourself and find the definition in the cosmos. What we think of as forming a gas / liquid is when the mixture of cosmic liquid becomes more in ratio than what the solid (atomic element) is and when the substance "freezes" there is less of the cosmic liquid than there is when the mixture turns to gas or liquid. We confuse water with what is a liquid since water tends to mimic liquid because water is very adaptable to changes in form when being in a fluid state. However, being fluid like does not change the substance of water since water forms with the combining values of material and material is a solid notwithstanding human connections to the idea.

This brings us back to the importance of Kepler's relevancy, which Newton got rid of so easily. The value of Kepler's space he indicated as a third dimension a^3 does depend on indicating a structure a^3 that is in rotation T^2 but also needs one position having a constant of some sorts in relevancy to singularity. Any point where **k** may indicate a position one will find a value matching a^3 and the matching location will fit T^2 at that point on the condition that T^2 forms the margins of the specifics of **k**. That is the relation there is in the solar system between all planets and the Sun. The Sun always indicates the centre and the planets always indicate the rotation. But $a^3 = T^2 k$ is only producing a relevancy of three dimensions that is equal to two plus one dimension. That indicates the space a^3 is in place by the movement T^2 thereof in relation k to singularity k^0.

In order to argue this idea that outer space is zero let us return to the sketch and take it from a point where the Sun provides a centre as one starting edge of k then that centre k will provide a line from the centre and the line k will provide three spots in a formation that produces a structure by the square T^2 of the dimension. Not once did Kepler indicate size as a contributing factor to a^3. That means every single point that k indicates there are three positions a^3 implicating sides of a double dimension. In the same manner is k not limited to distance or is T^2 lesser by size. $k = a^3 / T^2$ That is what Kepler said. There are three dimensions a^3 between any two points T^2 flowing as time from the centre of the sun, which is indicated by the line k.

The value of k is not to put a measured value in place, but to bring a reference to singularity $k^0 = a^3 / (T^2 k)$ applying as to place a specific singularity in as the **governing singularity** and another in place as the **controlling singularity** because there always has to be a controlling singularity determining the orbit while there has to be a governing singularity determining the spin of the body in relevance performing as the space a^3 in question in the formula $a^3 = T^2 k$ where in that formula k determines the relevance of k^0 as in $k^0 = a^3 / (T^2 k)$. However, this burden k forever with the responsibility of forming a line and a line is what places the Universe in place. Every space a^3 in question puts singularity k^0 in position by the motion T^2 in relation k the position allocated in the Universe.

The implication of the relevancy produced by the use of the formula $k = a^3 / T^2$ brings about that when dividing T^2 into a^3 there is k left. The fact is that a^3 is a three dimension (^{3}) of single k (^{1}) showing one or T^2 is two dimensions of k being the one dimension it means that k is a part of space a^3 or T^2 which is time. It is the same thing in a double dimension or space being a triple of k then k is one factor and k cannot show a position of zero. If $k = 0$ then there is no possibility of $k = a^3 / T^2$ because $k = 0$ then $0^3 / 0^2 = 0$. That does not make sense. Mathematically space cannot be zero because those being of the opinion of space could be zero or nothing must first prove mathematically that space is zero. Moreover they then must prove mathematically how zero grows through the Hubble constant. By translating Newton's vision of the circle in completing a cycle would become zero through rotation...well that does not count because in the use of the formula when calculating a^3 please replace any factor of $a^3 = 4/3 \Pi r^3$ with zero and calculate the end result. If k cannot be zero then k could not start from zero. With $k = a^3 / T^2$ no point can be zero because k shows space $a^3 = k T^2$ is no reference to the volumetric mathematical formula used to calculate $a^3 = 4/3 \Pi r^3$. Nor does it show the use of the circle in the second dimension being $a^2 = \Pi r^2$. In the case of the Kepler formula the circle factor becomes the square as indicated by the duration of the time T^2. The factor standing in for the line which normally would be r and then be the square value is in the case of Kepler's formula not the value indicating the square. That means Kepler never indicated a circle of mathematical procedure but said mathematically the distance of the planet from the Sun k holds space a^3 in relation to time T^2

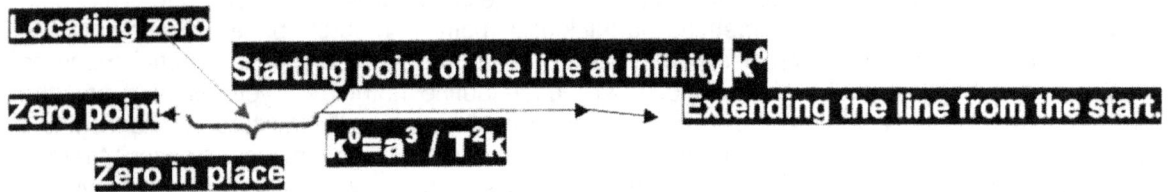

Lines mathematically cannot start at zero because there is no evidence of zero as a factor in mathematics. **Should you disagree with my statement** the question in need of answering is this: **What will the length of the shortest hypothetical line imaginable be and moreover, what would the total overall length be in that case?**

The shortest line that can be valid is a line having the start of the line and the end of the line holding the same spot. That points to singularity forming 1. There can be no line shorter than that and a line having zero as a start must have no start in order to qualify as being zero. The Universe is lines that form from every angle possible and with zero removed from the Universe we find that the Universe starts with singularity 1^0, 1^1, 1^2, 1^3, 1^4 and so on. With zero excluded from mathematics we can return to gravity.

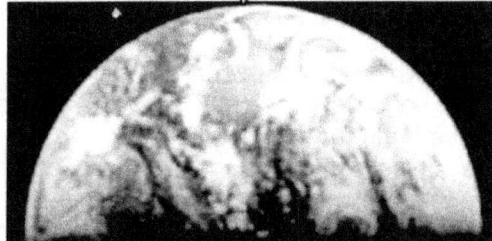

Nothing stands still to anything else in the cosmos

When having "mass" the body is stranding still in relation to the Earth

The Earth is taking on all the moving responsibilities and the body "with mass" is taking on the moving speed and density of the moving Earth. The Earth moves in two ways

When an object has gravity the objects is following a line that will form a direction of movement. Every conceivable object in the entire Universe is spinning while also moving in a specific direction, which will result in forming a circle. Everything in the Universe is submitted to movement and all that holds space moves. When an object is standing still on Earth with mass, that object is moving. This is because time is the movement of all things in relation to specific point.

Gravity is a product of movement and not a product of the influence of mass. By orbiting at a specific distance, the distance from the Earth is determined by the rotational speed the object encounters. When the object reduces the orbital rotation (circular velocity), the gravity by slowing down will bring the object to start moving towards the Earth, which is falling and which is what everyone knows is to be gravity. One then must accept that mass is having an object being in a point of only moving with the Earth while gravity is the movement or inclination it shows to produce what is required to further move towards the centre of the Earth or the inclination of forming movement towards the centre of the Earth. The inclination to move to the centre of the Earth is gravity while stopping such movement is forming mass. That is the difference science never finally concluded…gravity is movement or having the inclination to move while mass comes into play when that which moves is standing still in reference to the Earth while having the Earth move.

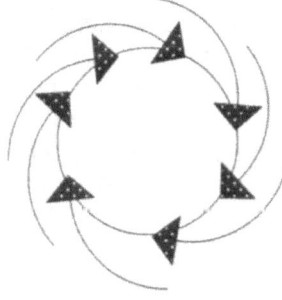

Anything that is spinning, by spinning shows contraction in moving surrounding space towards the centre of that which spins by the centrifugal forces pulling space to start to decline. Everything is moving towards the centre and I have indicated that resting in the centre we will locate singularity. The spin will comprise of two parts where the movement the spin holds will form one part and the next part will be the movement of space going towards the centre. This is exactly the features gravity holds as Kepler's formula indicates in $a^3 = k\ T^2$. The linear k movement shows the direction in which the space a^3 moves as the space compacts and reduces while the rotation T^2 is coming from the object spinning around its centre k^0 or forming its axis by spinning around its singularity. All hydraulic pumps work on this principle and I show that all stars are hydraulic pumps pumping cosmic liquid into such stars. The planet we call Earth is just another very poorly developed star on its way to become a star in the far, far future. The entire Universe is working on the principle of hydraulic power, which in fact, is the most powerful source of drive thee is. Electricity is just more hydraulic drive that implies cosmic fluid as a liquid source…and that too I am going to prove mathematically as the book progresses.

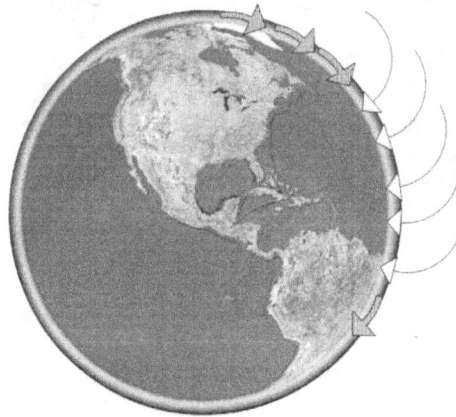

The space that is drawn and is forever becoming more concentrated and denser as it contracts towards the Earth we named as being the atmosphere. This movement towards a centre by a rotating body drawing space is as much physics as mass is not natural physics used by the cosmos. If any object spins, such an object will influence the surrounding space holding whatever in that space to move towards the centre. Having gravity is forming movement that is inclined to move towards the centre of the Earth. That is called centrifugal force and therefore one could call gravity a centrifugal force. It is movement drawing all the space that is around it towards the centre. The Earth, when spinning has to show centrifugal force because it is part of the spin the Earth shows. All space surrounding the Earth will move towards the Earth and thereby become more compact, or compressed or denser, no matter what name one attach to the process. Mass don't even need to be mentioned in this process!

I am going to explain in this book why everything closer than 2.4674 times the radius of a star will dissolve into liquid and become a fluid that a star incorporates into its structure. This is called the Roche limit.

As the Earth spins the spinning of the Earth is engulfed with space and the space surrounding of the Earth, the Earth draws from the outside towards the centre. There are so many layers named different names of atmosphere where everyone has a different density and each holds a name. In terms of science naming names to layers the real issue of why it is there in the first place as well as the fact of why the layers form is left to gravity, according to science, which is left to mass, which is left to magic. The contraction of the space immediately around the Earth becomes dense and hot while the further it expands space towards outer space, the fewer particles the space holds and therefore the colder it gets. I put this incorrect view to the test in other work and show that what we think of as being hot is in fact cold and what we think of as cold is extremely hot. However, that argument I leave for another opportunity because at this point we look at gravity in its most basic form. This comes about because there are two forms of substance forming the Universe. There is a liquid holding a solid and it is movement that makes the solid secure. By spinning within the liquid the solid brings about gravity. Every atom is a centrifugal pump within a liquid forming gravity.

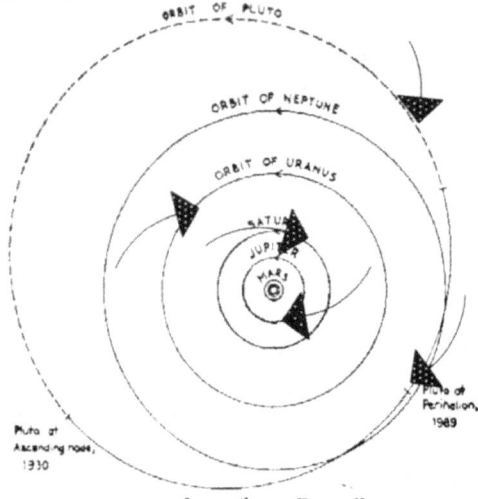

There is one BIG centre pump pumping space-time towards the centre. There are nine smaller pumps, pumping space-time towards each one's individual centre and this is aligning according to the Titius Bode law of positioning the allocated position according to the specific requirements that the Titius Bode law prescribes.

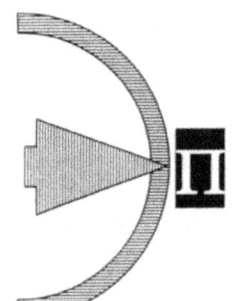

The main factor required has to be Π since everything so far is relying on the mathematical factor of Π. Everything is spinning and when it spins it involves a circle and a circle can only be a circle by using Π. In that sense when searching for the four phenomena we have to connect the use of Π to every one. I will explain the Π connection in all four of the Phenomena called the **Titius Bode law**, the **Lagrangian positions**, the **Roche limit** and **the Coanda** effect and I will not use any force because forces belongs to witchcraft and in physics that is very absent.

Being a circle requires two factors and both those factors Newton dismissed in his search for gravity. More important is the fact that modern science are so well equipped with the skills of mathematics and yet for hundred years after the fact not one in science came to a conclusion about Π having to be involved as well as with the Earth being round therefore the Earth having to have a diameter when dealing with a circle...this requirement fits any circle and the Earth is just another multi dimensional circle.

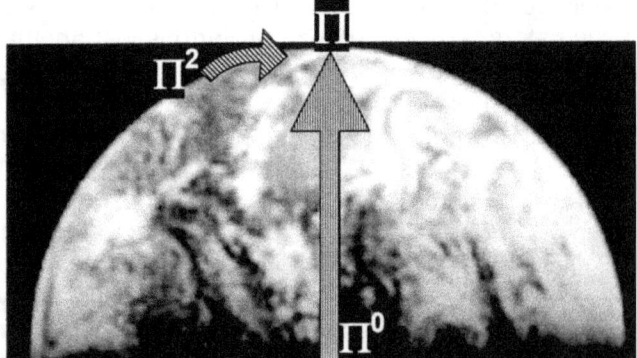

If r is the diameter, then the position science so feverishly award to mass has a point that actually holds Π as reference in the laws of mathematics.

Nothing stands still to anything else in the cosmos. When having "mass" the body is standing still in relation to the Earth. The Earth is taking on all the moving responsibilities and the body "with mass" is taking on the moving speed and density of the moving Earth. Everyone accepts gravity is taking a body "straight" to the centre but it is not. The Earth moves in two ways and this seems to be Π which is the point ending the circle and Π^2 where Π moves = Π^2.

Modern science still supports the Neanderthal idea that a body fall straight towards the Earth as "mass" draws the body directly to the centre. This is as outdated as any view science may have on gravity where they hold the informed opinion that mass is having a pulling power that can pull by force where such pulling is aided by magical powers and forces.

Ever since Newton everyone has been cherishing the idea that gravity moves a body straight down without ever taking into consideration that the earth is rotating while the body dropping is falling straight down. The makes the body moving straight down falling at an angle. Even in having mass the body is still circling as the Earth rotates. This oversight by science is most crucial a mistake as any Newtonian mistake can be.

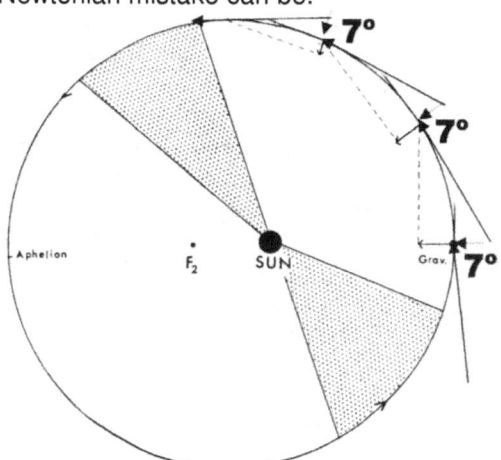

The Earth is going around the Sun by 7°
The Earth is going around its axis by 7° in a cyclic rotation

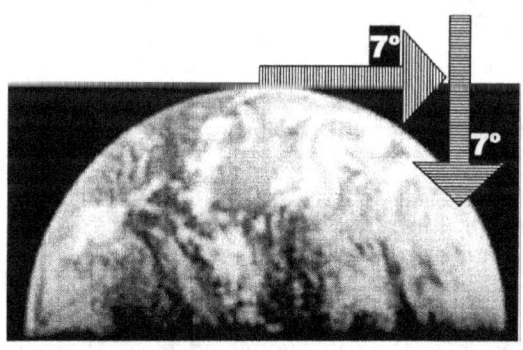

Although being in a state of having a "mass-attack" such "pulling" on your body moves the body to the centre of the Earth. By the Earth rotating the Earth is moving notwithstanding your body being in the state of having no motion, still moving with the Earth has your body falling by 7° as it circles with the Earth around the axis of the Earth. The Earth falls by 7° when rotating and therefore your body is falling with the Earth being connected to the Earth by 7°.

Not only is the Earth falling by 7° as it revolves around its axis, but also it is circling the centre of the Sun and by doing that the Earth is falling another 7° by rotation. This is pivotal in understanding gravity as a mathematical fact.

That puts the falling of the object completely in relation to the speed that the object holds and that places gravity by falling in direct relation to gravity by orbiting.

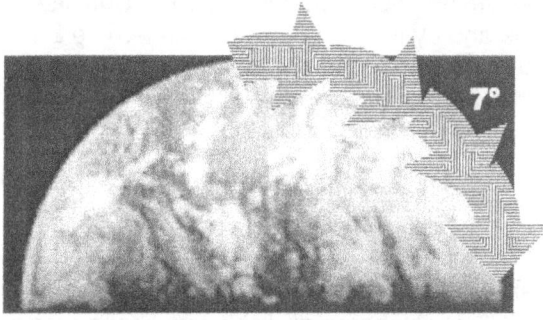

When a body falls there is no mass involved because all objects fall equal and this was accepted long before Newton started fantasizing about his mass involvement in gravity applying. That is what Galileo proved eighty years or so prior to Newton. The distance the object orbits measured from the centre of the Earth and the orbit circle holds a direct link to the speed or time in relation to space that the object rotates. If the speed in revolving declines, then the orbit circle declines and this reduces the distance the orbit circle is from the Earth centre and the orbiting diameter reduces. If T^2 reduces then space in which the orbiting object is a^3 declines from the centre k^0 and k the relevancy factor depreciates. It is all in Kepler's formula. The orbit circle T^2 is directly associated with the distance k the orbit takes place a^3 measured from the centre of the Earth k^0 in a ratio of time taken versus space travelled through. This has to do with speed or movement and applies to all objects equally holding no specific relation to size or mass. It is a relation between the orbit circle (circumference) and the distance from the Earth centre (circle radius) and if that is the case, then gravity forms by Π having some sort of involvement and that throws any idea of mass playing a part in forming gravity out of the window where I hope it takes all of Newton's ideas of mass-forming-gravity with when going out the window. In forming gravity the centre line (diameter) holds a specific value to the orbit (circle) and with that being the case then we have to search for the part Π plays in the function gravity has and when doing that we can leave mass out of the frame because big or small, all things fall equal. Galileo was the one that proved that.

So you think that it is much simpler to maintain the argument that gravity is the force created by mass pushing the object onto the Earth only when the object moves at the same pace as the Earth rotates...and mass is always present and not only as I say when the object is on the ground and finds mass or weight!

So you still think that explaining gravity remains as simple as putting gravity in a connotation with a force fed to measure by having mass attracting whatever is attracted and this then allows the simplicity of the Newtonian concept to deal with the confusing part of the entire issue!

This is how I prove mathematically how gravity works. There is no pulling of mass or by mass or even that having mass plays a part in forming gravity. On the contrary, it is the forming of gravity that establishes mass when the space can no longer reduce and the reduced space locks whatever then has mass onto the solid surface of the Earth.

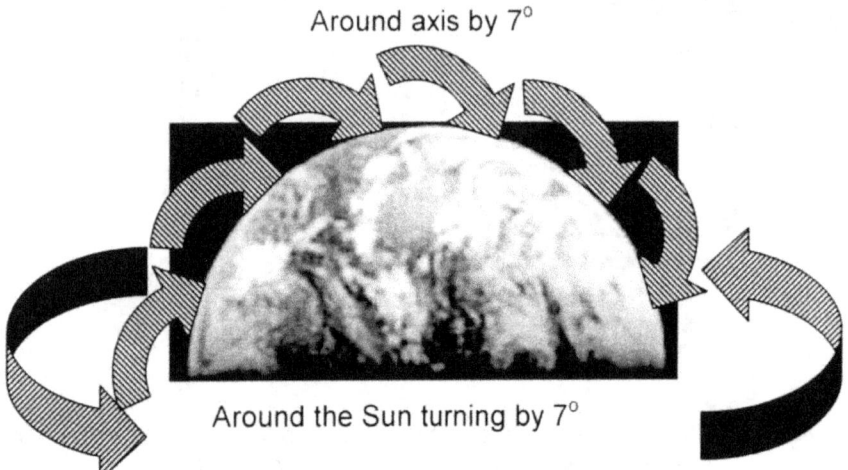

When the object moves while being in space or in contact (in relevance) with the spinning Earth, the object wishes to continue moving straight ahead while the Earth also moves straight ahead by turning 7°. Therefore, the Earth by spinning is falling away by turning 7°. That clears space or compresses space by the margin of 7° declining (compressing) of air / space. The Earth pins around its axis by 7° and turns around the axis the Sun provides by 7°.

The Earth is moving, constantly spinning and in this is contracting space by compression (we call this contracting of space in air the atmosphere) and while the air is getting more compact, it takes whatever is filling with space towards the Earth constantly at a rate of 7°. By the Earth rotating, it is compressing space and with space compressing it is moving objects in the direction of the Earth. That is why objects that is falling, has no mass and only the stupidity of the simple Newtonian mind will force scholars to accept that it is mass that is pulling gravity.

The entire idea of gravity is secured in movement of everything in relation to everything else. There is nothing in the Universe that ever could remain still because everything cosmic that is filling the Universe is spinning while it is also at the same time moving in a straight line and by that is following a circle. The Earth is only moving straight ahead because the Sun is spinning and while the Sun is spinning, it is compressing space, which allows the Earth and all other rotating objects to spin around the Sun in a perfect synchronised fashion. This process is going on throughout the entire Universe.

This places Pythagoras in the pivotal role of gravity by forming a calculated value of Π when dissecting Π mathematically. Gravity is Π using the law of Pythagoras.

There are three movement involving $Π^2$ from which gravity takes effect. At this point however, I will only introduce the two of the three ways in which the value of gravity $Π^2$ takes charge of displacing space-time in a process of dismissing space-time by diminishing space through the flow of time. It might sound the same thing but it is processes putting the Universe in dimensional positioning.

In finding the double diverting of 7° from the straight line it could be seen as the Earth spinning around the Earth axis and the atoms within the Earth spinning around the atomic axis forming the controlling singularity or it could be as I said the Earth spinning around the Earth axis and the Earth spinning around the Sun's axis in orbit forming the controlling singularity because it is in relevancy with singularity that this applies and singularity is equal as well as the very same everywhere.

The Four Cosmic Pillars; The Result Thereof.   **APPLYING AS THE SOUND BARRIER**

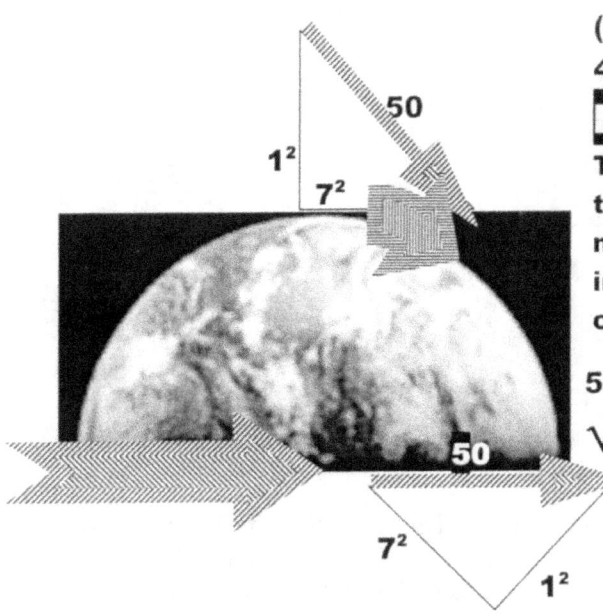

$(7^2 + 1^2) = 50$

$49 + 1 = 50$

Pythagoras This applies twice in one movement (going in a circle and circling the Sun)

$50 + 50 = 100$

$\sqrt{100} = 10$

Since the movement involves two equal phases that acts as one, therefore the double value of 7 in relation to forming ten becomes what forms Π. We have the movement of seven forming one direction standing in relation to singularity which is the square of 1…According to Pythagoras that will bring about fifty. Since singularity is equal a one and seven is combined with singularity, the equality of singularity brings about the seven uses the same attached 1 in the square making the fifty a combination of another fifty and from putting a double fifty in the square as Pythagoras demands we have ten as a result in relation to seven. In Pythagoras's square the one side, let's say the adjacent side of the triangle forms a square in seven bringing on forty-nine. The other side let's say the opposite side of the triangle uses the square of one (singularity), which also remains one. The sum total of the two forms fifty, which is the measured value of the hypogenous. Since gravity is always applying in the double movement of seven the total of the hypogenous then is fifty plus fifty which is one hundred. The square of one hundred is ten and that brings about the value of one side of Π. When explaining the total worth of Π such explaining in detail requires a lot more information which will claim about as much space that which this article in full would allow. I complete this explaining as well as the explaining of how Π comes about through the forming of gravity much later on in **THE VERACITY OF GRAVITRY.**

That is what gravity is. Gravity is space moving or changing position in time and when an object can retreat no further towards the Earth centre, it only then forms a solid that aligns with the spinning solid material and with that then receives mass… Gravity is the movement of space in regard to any one specific point…and that is also precisely what time is. Nothing is standing still in the entire Universe. There is not one fragment of a sub-atomic particle standing still in relation to any other particle through out the entire Universe that is standing still. Having mass is when one object is standing still in relation to the Earth forming a part of the Earth while the Earth does all the moving on behalf of the particle having mass as well as the Earth and only happens when through having mass the object becomes part of the rotating Earth.

What does this all of this controversy mean…it means the way say the Sun and all the planets formed is total rubbish. The way say the Universe came about is hogwash. The age gives the Universe is proof of their total incompetence understanding cosmic principles (Newtonians can't even understand or explain any of the four cosmic principles I named the cosmic pillars) and total by lacking such fundamental understanding shows complete ignorance on the side of Newtonian concepts. The Universe is something can't dream to fathom…or begin to understand and then wish us to consider their positions they hold in society as the wise experts that can explain it all, while all along they can't even explain gravity. …

This explaining abandons Newton's idea that gravity is being formed by mass that through some form of magical intervention is pulling on other mass and this is forming gravitational contraction, which is madness. I hope this idea is finally going down the toilet. If one takes the Kepler formula $\underline{a^3 = kT^2}$ that Kepler introduced, which Kepler received from no less than the cosmos at large, one find the **space a^3** is equal to the movement of the defined space in a **straight line k** as well as a **circle T^2.** In the cosmos no line can go straight without circling as well and no circle can go on without going straight at the same time. That mathematically explains the Coanda principle in detail.

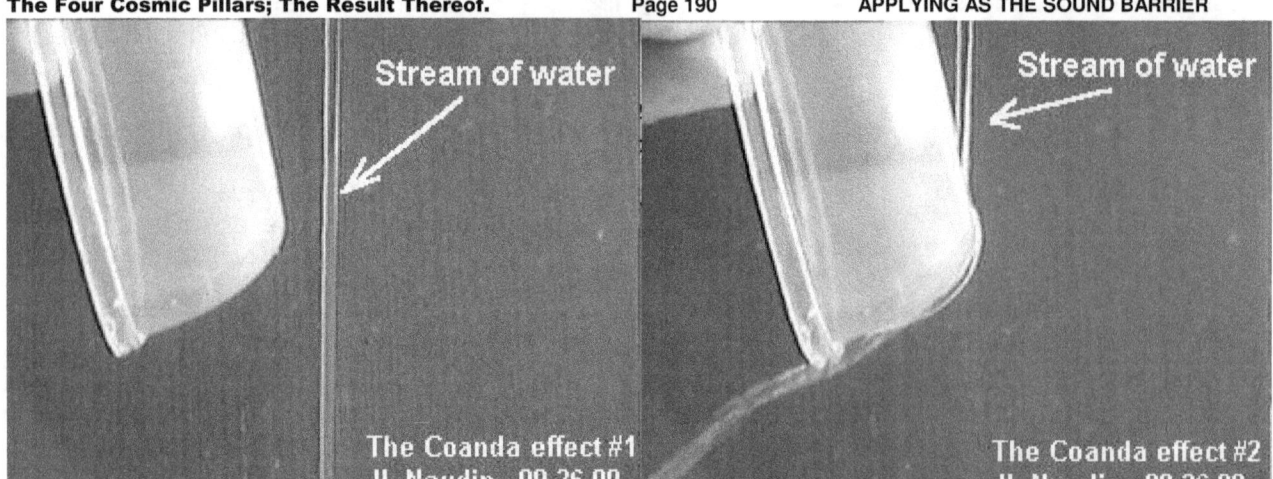

In the books **The Dissertation on Gravity** or the more informing **The Veracity of Gravity** I explain the process in much detail. Gravity is the Coanda effect as the above picture indicates. Should you wish to find more information on **The Dissertation on Gravity** or **The Veracity of Gravity** please visit the web site called **www.gravitysveracity.com.**

In gravity there is a circle turning. Where there is a circle turning we have Π involved. Also we have a radius involved. The radius or distance from the centre Kepler called **k** and Newton classified this as zero ($\frac{dJ}{dt} = 0$), which is an insult to mathematical principles. Where a circle formed while spinning in a cube, three factors are involved, namely the radius, the spinning speed and most of all there is Π. Without Π there can be no circle and the Earth is a circle, even Newton should have been aware of this. Therefore there can be no gravity without having Π as a contributing factor to what forms gravity.

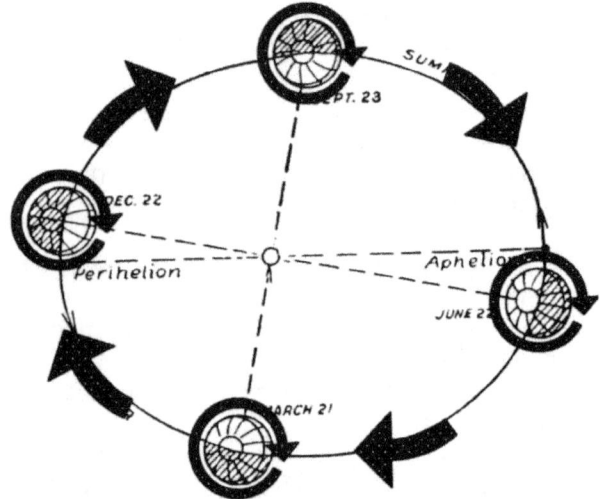

Since gravity is tightly interconnected with a circle formed as a sphere and is spinning around an axis, the main issue of research has to start with finding the factor Π. There is no connection with a circle and mass but for sure the circle will find in form an end serving in a measured value as Π. It is common knowledge that in calculating a circle the formula used is $Π2r^2$ or $Πd^2$. Would it not be mathematical plausible then to start looking for mathematics in gravity while leaving Newton's magical mass out of the picture since there might just be some common sense to be found in this. In science and in mathematics we have to see where true mathematics fit and what role every factor has in playing a part. We can't keep on dumping all findings on mass since mass has no part in mathematics.

If we wish to award mass a value as a factor we then must see the role we give mass. Giving mass a value is issuing a body a value as it would have when being part of the Earth. We take a cube of water and we award the water a measured size. One meter by one meter by one meter of water would give a thousand litres of water which would leave a thousand kilograms of weight and that would leave a mass value of one metric ton or a thousand kilograms. The awarding of mass is giving the object a relevancy of being part of the Earth. In Newton's time the Earth was the Universe because people were starting to get used to thinking that the Universe was not spinning around the Earth but the Earth was only a small part of the Solar system and that was a small part of the Milky Way which was a small part of whatever was a small part of another small part of another small part of something getting all-the-while into a bigger picture.

There is also a movement that should go straight but is in fact going in a circle and by never going straight but always circling the Universe becomes eternal on the one end and infinite at the other end. The infinite point I am going to explain in due time. Everything eventually is going in rotation and in that we find the measure of outer space being eternity. In Kepler we find $a^3 = k\ T^2$ where T^2 is a circle but on the other hand **k** is also forming a bigger circle and everything a^3 that is going away is going to return someday.

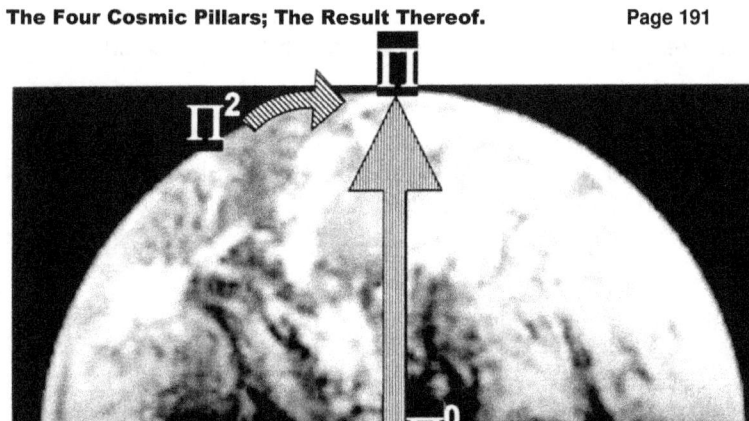

APPLYING AS THE SOUND BARRIER

Science always awarded the position an object holds standing on the centre of the Earth with a measurable value of mass. This is not incorrect in normal physics but as far as astrophysics goes there is no mass factor present anywhere. In normally practised physics we are allowed to award mass as a usable tool because on Earth all things adhere to the singularity holding the Earth, however that is as far as the use of mass could go!
The value of such a position should be in the relevancy of the movement that the Earth holds as

This is the process whereby gravity forms. Mass has no influence on gravity except for resulting from gravity compressing the Earth.

At this point so far after all my numerous attempts in trying to establish some contact with academics world wide I wrote seven books in a combination I titled **"Matters Time In Space: The Thesis"** covering the entire issue of my work plus the mentioned books wherein I combine all the various letters I wrote to academics through out the eight years of ardent trying to establish some line of communication. The last letter I addressed to academics I include as part of the content of my web page called www.sirnewtonsfraud.com for your insight and which forms part of this and other of my books where I join and elaborate on the letters that I combine to form a unit as a book.

To all those that feel disgusted by me accusing the greatest name in science that ever lived being of fraud, please go on and prove me wrong! That will be easy for prove the following formula correct. $F = G \frac{M_1 M_2}{r^2}$ This is the formula Newton used with which Newton proved gravity. Now prove gravity by using this formula. Do the following to prove me wrong. To find the force of gravity one has to multiply the mass of the Earth (M_1) with your personal mass (M_2) and then divide the distance there is between you and the Earth (r^2). Using these factors by multiplying (M_1) and (M_2) and dividing with (r^2) should present gravity coming from mass. But science uses a fixed value to calculate gravity. Now, convince your mind about my correctness. Do the simple calculations.

Take the mass of the Earth (M_1). Multiply the Earth mass by your personal mass that any scale should indicate (M_2). After multiplying the two mass factors, then proceed to the following step by dividing the multiplied mass factors with the square of the radius there is between your feet and the Earth (r^2), which should not amount to more than a few billionth of a millimetre. If the answer in front of you is not 9.81 Nm/s^2 then there is something very wrong. The incorrectness has to be either one of two possibilities presented: The measured value of gravity is not 9.81 Nm/s^2 as science uses it, or Sir Isaac Newton's formula suggested as $F = G \frac{M_1 M_2}{r^2}$ is complete fraud...Now which is it...you can decide...the force of gravity that the world of physics uses to do measurements is 9.81 Nm/s^2. If the answer you have in calculating your force of gravity is not 9.81 Nm/s^2, then it is either this measuring value of gravity that is wrong or it is Newton's formula that is wrong because by the calculation you did, the calculated answer you got could not possibly have delivered a measured value of 9.81 Nm/s^2. After all, science maintains it is the pulling of the combined mass in relation to the boosting that the radius would present to the force created that delivers the force of gravity! If by using the factors of mass and the radius does not accumulate to 9.81 Nm/s^2, then how can mass deliver gravity? Multiplying the mass of the Earth with the mass of a person and then bringing this answer in relation to the radius by dividing must be 9.81 Nm/s^2. If not, something is wrong with either the prescribed value science puts in place or Newton's suggestions.

To teach students that $F = G \frac{M_1 M_2}{r^2}$ are the measuring formula in determining gravity, while knowing very well it is not totalling gravity at 9.81 Nm/s^2, then doing that to students while enforcing a thinking pattern in the minds of a student is committing brainwashing because by forcing examinations on students, expecting them to confirm the falsified statements used that the tutors present as correct, is brainwashing, a way of enforcing mind control and it is manipulating the thinking process of students. If you can't prove that my manner of thinking is incorrect and you keep surmising that science is correct then recalculate the formula or start reading the rest of their fraud.

Gravity is a constant of 9.81 Nm/s². This is used in all cases of scientific calculations. Mass is an individual factor that is different on anything on which it is applied as a measuring factor. How could something as different as mass that is never constant even on Earth form a constant such as the force of gravity and still be the same in all cases?

How does M_1 connect to m_2 forming $\frac{G}{r^2}$ when the connecting medium constitutes of nothing that then has the value of zero as the Cosmologists wish to value outer space?

In these following books forming a series of four and parts of this series about gravity there are the books entitled:

This Work disputes the correctness of the formula $F = G \frac{M_1 M_2}{r^2}$

Using the formula above as Newton did does not imply a suggestion or carry an idea across as a thought but must be seen to be acting as confirmation about a fact because one cannot suggest anything mathematically, one can only confirm a fact mathematically. There is no mere suggesting of any possible movement in a specific direction of any suspected behaviour by an object moving from and to a point as suggested, but this formula says the gravity of the Earth measured in mass at it's totality is colliding with the falling body's measured mass as the two factor's diminish the radius from both ends. This mathematical formula as it stands is no mere suggestion, but in its use it must back up or prove a fact!

Let us briefly investigate the way mainstream science advocates the process of how the solar system started in a random selection of material. Is it not rather a bit simple to leave explaining of complicated issues to mass pulling mass without any more investigating or elaborating explaining?

Let's investigate Newton's first concept of gravity before fame got the better of his senses. Newton stated that $F = \frac{r^2}{M_1 M_2}$ which is mathematically expresses as a force that will destroy the square of the radius between two falling objects where the tempo of destroying of the radius by the square holds the mass of the two objects in product to achieve that. The mass factor combines by multiplication in order to diminish the radius between the Earth and the object at a distance from the Earth. The mass factor makes the bodies conjunct and it depends on the mass of the bodies how the bodies will fall. That is Newton's personal view. My contradicting this so many times lead to me being rejected by the scientific institution.

I say this to indicate that am fully aware of all the resentment I am about to release and what I have already released thus far. I say what I have to say in full realising what resentment any physics administrator feel at this moment towards me. I say this remark to inform that as much as I am fully aware about the damage that I am doing too my chances of having my work accepted because I realise that he who does not praise Newton is condemned. I still am adamant in continuing with such relation damage that I evoke between physics academics and I. While academics are denouncing me at this moment please concern yourself with the following. When I fall down a cliff within a waterfall, I fall at a steady pace. It is the same pace that I would have when I fall down a waterfall holding a cup in my hand. Should there be water in the cup the water in the cup will not stay behind or spill faster than any other water or I. The water in the cup will not spill by emptying the content going up or down faster that any of the rest. The cup will not fill with more water adding to the content just because the waterfall has more water thus more mass. If I had to fill the cup with water while falling I will have to supply upward motion to the cup in order to gain water in the cup. It will require an increase in upward motion to the motion with which we fall. The water and I will have the same pace therefore we will fall under the same influence of gravity. My density or mass in comparison the water with which I fall will not leave me with superior mass and therefore with superior gravity. The water mass will not have the waterfall more forceful or less forceful than what I do. The motion, considering all objects (the cup, the water and me) being unequal in density is not discriminated by gravity favouring the more massive to have those with more mass fall more rapidly on any basic grounds of having a mass advantage. That is what Galileo said and Galileo is accepted as truth.

There is no reason in heaven or Earth or in-between that will establish a coherency between the truth in Galileo where Galileo says that all objects fall equally to the ground and Newton bringing in his mass that is in place to have the job of falling done. Mass contributes nothing in the process of objects falling because if mass had a role to play, more massive object will fall more rapidly then less massive objects. That is one argument Newtonians try to avoid but that is proving the truth. Galileo very correctly determined that object descends to Earth at an equal pace and although admitting to this, science declares that Newton was never disproved even on one single occasion. In the light of this it is the truth that Galileo disproved Newton even before the birth of Newton because Newton was born at the time of Galileo's death. That makes any ignoring of Galileo by Newton not a matter of pure ignorance but it is then deliberate malice. That makes Newton's statements deliberate callousness of circumventing the truth. And the entire world not only accepted this ridiculous scamming but also is at the moment in agreement with labyrinth gestures of inaccuracy.

Galileo proved and nature confirms that gravity has not one single common thread with mass but mass totally depends on the worth of gravity. The process of falling confirms that mass does not establish or produce gravity since all objects hold gravity alike whereas mass is as individual as the fingerprints of people. Yet only Newton and Newtonian disillusions insist that it is mass, which is producing gravity because that gives those in physics an unfair advantage in realising what others find as senselessness. It is shown on television on many occasions where army battle tanks are dropped from aircraft and the tanks travel the same velocity towards the Earth as human soldiers do. The tanks do need bigger parachutes to restrain the momentum of the fall…yes that is true but the velocity of travel is the same as would a bicycle have when dropped from the same aircraft under equal terms.

I do explain this later on in **THE VERACITY OF GRAVITY.** There was a show on television where some young persons were dropped from an aircraft while they were sitting in the car. The persons got out of the car and into the car and pretended to push the car on the way while they all were descending to the ground. The persons falling, although much less massive that the car did not fall slower than the car did and neither did the car fall faster than the people did. Let your religious belief in Newton show how that confirms $F = G \dfrac{M_1 M_2}{r^2}$ while it totally vindicate all the threats Galileo got from his church.

Gravity might remove the radius but it does not employ mass to accomplish the destroying of any parting distance between the Earth and any falling object and that is been proven in no uncertain terms. Newton's mass has no validity in cosmic physics and falling to the Earth forms part of cosmic physics. All other physics form what is a part of man-made motion and life inspired motion. In that field of physics concerning the motion that life produce Newton was never incorrect and every aspect of his work is perfect. But it only involves the life inspired man produced labour of life where life is in control of the motion that comes about. That could be covered by the range of any form of motion from blowing gently on a feather to unleashing a rocket going to the Moon.

But that is a part of life where life is confined to the Earth and life as such is alien to the cosmos. In spite of my correctness, my work has been rejected because of this statement and only on the grounds of me accusing Newton of being incorrect where through the rejection of academics and to such a degree of rejection without merit that I have now become punch drunk. Should you decide to do it one more time, then be my guest, but before you do please take a challenge that I pass in your direction; make sure about you reasons and the accuracy of your reasons when you decide to again reject my work this time. This is no threat because I am far too small to threaten any institution. Rather see it as a challenge of your spirit of fairness.

A numerical sequence announced by J.E. Bode in 1772, which matches the distances from the Sun of the six planets then known. It is also known as the Titus-Bode law, as it was first pointed out by the German mathematician Johann Daniel Titius (1729-96) in 1766. It is formed from the sequence 0,3,6,12,24,48,96, and 192 by adding 4 to each number. The planets were seen to fit this sequence quite well – as did Uranus, discovered in 1781. However, Neptune and Pluto do not conform to the 'law'. Bode's Law stimulated the search for a planet orbiting between Mars and Jupiter that led to the discovery of the first asteroids. It is often said that the law has no theoretical basis, but it does show how orbital resonance can lead to commensurability. The importance that becomes known is the sequence the Ties – Bode law saw in the number arrangement of 3; 6; 12; 24; 48; 96 etc. The incorrect application of the Titus Bode law lies in subtracting the figure of 3 from 10 leaving 7. The other way of reasoning is to add four each time to the firs value of three starting with 3 and so on. The true significance of the Titus-Bode law is that it points directly to a circular growth of 7 stages. The 7 relating to 10 is a precise derogative of the Roche limit or the Roche limit is a precise derogative of the Titus Bode principle because he two systems interlink.

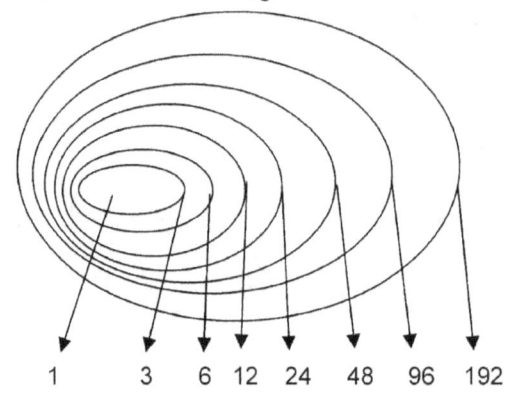

Planet	Mercury	Venus	Earth	Mars	Ceres	Jupiter	Saturn	Uranus
Bode's Law distance	4	7	10	16	28	52	100	196
Actual distance	3.9	7.2	10	15.2	28	52	95	192

Bode's Law:

Science underwrites the Big Bang, which is a science of promoting the view where the cosmos is expending and becoming bigger. They officially declare their accepting that the distance between objects is increasing. Yet if they also propagate that Newton's mass contraction ids absolutely the only way of thinking. This line of thought is outrageous because that would imply that something is adding to the Universe while we know that is impossible. Nothing can add to the Universe because the Universe holds all there can be and so we have to go in search of what we are incorrectly interpreting about the cosmic "expansion". Science is missing something vital and later on I will explain what they miss.

Please explain just how this line became the chosen spot for material that was scattered to gather!

From all the positions the dust could select, it formed at a precise distance from the Sun at a precise point from the Sun, How did the dust achieve that at random?

Let us use $F = G \dfrac{M_1 M_2}{r^2}$ and see what is true about the Newtonian vision of how planets (ands stars) form. It is said the solar system was one big sphere of dust and debris that was the result of what was in the past. I have written many books where I call this into question and therefore I don't wish to ponder. If it was a lot of dust and the centre produced the Sun, how did the planets evolve just five hundred million years later? Become so large where it is while Pluto and Mercury became so small where they are. Most puzzling is how the planets formed at the precise locations where they did. The Titius Bode law is a definite and is no presumption.

The Newton formula $F = G \dfrac{M_1 M_2}{r^2}$ is unable to explain the principle discovered by Titius and later by Bode and it is not coincidental. From this one can arrive at the origins of the solar system. These allocated positions completely ignore mass and that is why mass in the cosmos has no value. Every time science speaks of how the solar system formed they take it as the God given truth that the lot of dust was collected at one point that formed a planet where each has its space in relation to the others having space. Please don't just mention it as correct, but prove by using any of Newton's formula or then any other formula that this took place. Prove that the dust can collect that way and that the dust can select on specific point where the smallest with the least mass became so dense they are solids while the large ones with much more mass only got to become gas. Prove that this happened and explain how it happened and then after that feel offended when I accuse science of deliberate fraud.

As humans we might adopt mass as a tool to help with calculations concerning human perspective about human activities by Jupiter has no mass because Jupiter is floating in cosmic liquid and the space Jupiter claims is to the cosmos just more space whether it is occupied or otherwise that makes no difference to the Universe. It is a small pump spinning that reduces space in the allocated space of a much larger pump that is condensing so much more space. How did the chosen circles become the allocated chosen positions in which the dust gathered to eventually form planets?

Please explain just how did material start to gather from the inside as well as from the outside of the specific line. What prompted material to go in a certain direction and others go in another direction in such an orderly fashion as to form a planet in all of one allocation, considering the vast space the debris was scattered in.

Considering all, would it not be more prudent to call $\frac{M_1 \times M_2}{r^2} G = F$ a magic poison of witchcraft.

What prompted the dust to choose a specific line where the one lot would move to the outside and the others to move to the inside while the next lot decided on a very new line where that lot would then form a very new and different planet? Nine structured planets in all formed so nine lines were chosen from a cloud of dust. It is told that the centre Sun formed five hundred billion years ago and the planets formed four thousand five hundred billion years ago. With the Universe only thirteen billion years old, when did the gathering start? No one ever has ever come up to say how long the process of dust gathering took while it should be so easy to determine...just apply Newton's so correct formula $F = G \frac{M_1 M_2}{r^2}$ and from using the formula correctly, then determine how the debris did this lot. If ever there was a cocked up fairy tale with no substance other that trying to convey absolute mythology in the name of science, then this is as good as the others can get because there is a lot such in-discrepancies I point out in other books.

Te only valid formula is that of Kepler indicating space-time as space $a^3 = T^2 k$ time.

The value of the space-time revaluation is only applied to the occupied space-time belonging to the matter and not to space-time itself, but this leads to the concentration of unoccupied space-time, which in fact determines time in space.

You might think: **Very well, but what has all this has to do with the cosmos?**

Here is but a very few changes that comes about. What is the Universe?
The Universe consists of two factors, which is space and time. These two factors are inseparable, undividable and one single unit. The gain to one factor is the loss to the other factor. There is occupied space-time and unoccupied space-time, but there is NEVER NOTHING. What is presumed to be empty space is unoccupied space-time! That is the value of the Universe, which I named cosmic liquid, and is the current value of time evenly distributed through space.

The fact of "Nothing" can only exist in a person's understanding and perspective, but not in the Universe. Three factors rule the universe:

The Four Cosmic Pillars; The Result Thereof.

1. Densified space-time (matter)
2. Occupied space-time (atom of elements claiming space-time)
3. Unoccupied space-time (The value of time in space)

There are only two energy forms in the Universe. The first is heat and the second is life. No other form of energy exists in the four dimensions that the universe exists of. No force is to be found in the Universe, there is only balancing values.

There can be no such a thing as empty space. The Universe is time contained in space, which makes it space-time. Space has only one value, and this is to contain time and time provides space with a definite value. Space forms the history or the past tense of time and time results in space forming.

Let us for one minute leave Newton's surmising about Kepler's failure out of the picture and concern us with what Kepler found long before Newton thought about what Kepler found.

Kepler said that the space a^3 is equal to the motion T^2 of the space a^3 distant from a specific centre k. That then is $a^3 = T^2 k$.

Reading this mathematically encrypted coded formula of the cosmos given to Kepler and keeping it removed from Newton it reads as the space a^3 is equal to = the motion T^2 of the space a^3 in ratio to a centre k. If we bring in the full equation it will be $k^0 = a^3 \div (T^2 k)$ which means half of space is slid and half of space is moving. However also true is that everything through movement defines a value in relation to one point holding singularity k^0. What this proves is that gravity is the motion of space provided by time being the liquid. Please allow me to explain. In the formula $a^3 = T^2 k$ the space forms as the space is in motion.

Newton suggested that $\frac{dJ}{dt} = 0$ where he stopped time to have the motion of the circle demolish the work that the circle does. That means he got time standing still or being T^1 and the motion $T = 0$. Let us ponder on that thought for a while, while remain with the formula Kepler suggested it will seem that according to Newton $a^3 = T^2 k$ and in that T^2 then becomes **1**. Should that be the case then we have space going flat because $a^3 = T^2 k$ where $a^3 = T^2 \times k =$ forming a square instead of a cube, and the Universe we have is a three dimensional cube in every aspect there is.

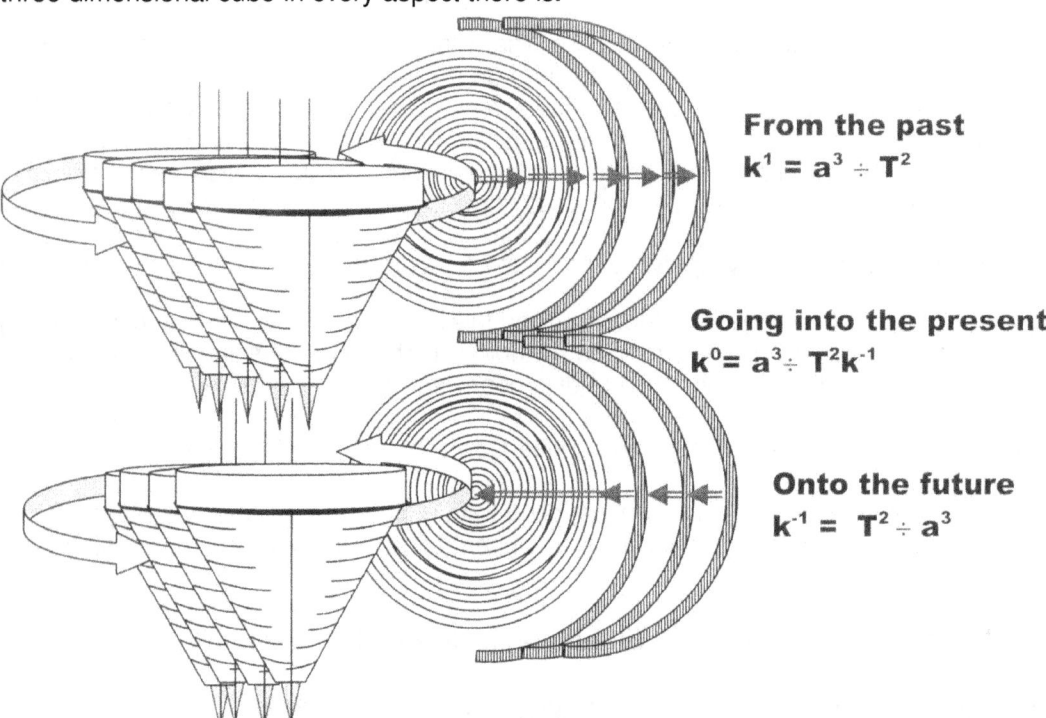

From the past
$k^1 = a^3 \div T^2$

Going into the present
$k^0 = a^3 \div T^2 k^{-1}$

Onto the future
$k^{-1} = T^2 \div a^3$

Time is taking any object by movement from the past, bringing it through the present and relocating the object by moving it all the time into a new position it will have in the future. The movement coming from the past is detrimental in locating a new position in the future and therefore time and gravity is the very same principle. Everything in the Universe is within the Universe because it moves and nothing that is not moving is in the Universe. Everything moves by the cycle of time and finds a new relocated place to move to and from as time moves on. Everything is driven by heat and heat controls the entire Universe. Science call heat energy because science has never thought further than their noses I are long. Heat forms space when expanding while space forms heat when contracting while space and heat is the very same aspect in reverse to each other.

When movement comes about the ratio between space and heat goes into an imbalance where the liquid that is standing as ($k^{-1} = T^2/ a^3$) totally dominates the solid holding space by the factor of $k = a^3 / T^2$. It still is the same thing being in reverse to each other.

Let us for one minute leave Newton's surmising about Kepler's failure out of the picture and concern us with what Kepler found long before Newton thought about what Kepler found.

Kepler said that the space a^3 is equal to the motion T^2 of the space a^3 distant or relevant from a specific centre **k**. That then is $a^3 = T^2 k$. Reading this mathematically encrypted coded formula of the cosmos given to Kepler and keeping it removed from Newton it reads as the space a^3 is equal to = the motion T^2 of the space a^3 in ratio to a centre **k**. What this proves is that gravity is the motion of space provided by time being the liquid. Please allow me to explain. In the formula $a^3 = T^2 k$ says the space a^3 forms from where **k** the space is in motion T^2. In $\frac{dJ}{dt} = 0$ Newton suggested that $\frac{dJ}{dt}$ is movement while he then said that being movement $\frac{dJ}{dt}$ is where such movement $\frac{dJ}{dt} = 0$ forms zero and then that is where he stopped time to have the motion of the circle demolish the work that the circle does. That means he got time standing still or being T^1 and the motion became $T = 0$. Let us ponder on that thought for a while, while reaming with the formula Kepler suggested. It will seem that according to Newton $a^3 = T^2 k$ and in that T^2 then becomes **1**. Should that be the case, then we have space going flat because $a^3 = T^2 k$ where $a^3 = T \times k =$ is forming a square instead of a cube, and we all know that the Universe we have is a three dimensional system in every aspect there is.

Symbol used to indicate the relevancy **k** is the formula Newton introduced as $G (m + m_p)$. This is just a longer and probably a more detailed manner of indicating **k** and to serve Newton's purpose of better defining of **k**, but it symbolises precisely to the point what **k** stands for nonetheless. I wish to draw your attention to the matter of Johannes Kepler's findings that Mainstream science considers as resolved and closed for many a century while it is not. My investigating Kepler helped me to resolve other unresolved matters but finding the success I had was only possible by using Kepler's work without Newton's fiddling.

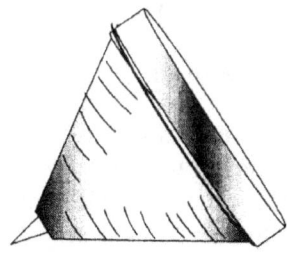

I too am well aware that at first glance you will immediately arrive at the opinion that the theme of my work has to be considerably below the standard of an intellectual Master such as you must be, due to the position you hold, and because of that, the normal research work you do. The work you now are dealing with is at a school level and argues about the most basic fundamental aspects of physics. Nevertheless, I hope that this writing may spark interest even at such a low academic level and grade in scientific sophistication and development because I am about to prove that I discovered:

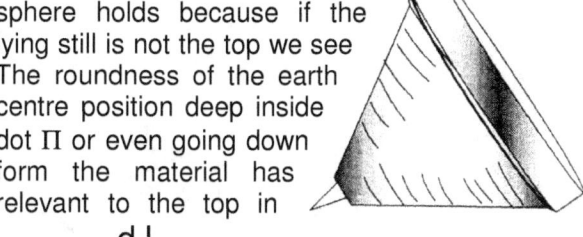

The top lying still holds the same singularity principle that the sphere holds because if the shape the top has. The top lying still is not the top we see but is the Earth extended. The roundness of the earth protects singularity at a centre position deep inside the Earth. However the top is a dot Π or even going down to a spot $Π^0$ and is only by the form the material has which puts singularity being relevant to the top in place. According to Newton it takes no effort $\frac{dJ}{dt} = 0$ to get the top from where the top was motionless to where the top is spinning. I say this on the work that Newton suggested comes about from the effort it takes the top to circle.

Even a child can tell that from a position of lying down to a position of the top spinning erect does not take nothing as an effort, because it is an effort to get the top spinning erect. Spinning the top puts the top in a new and independent Universe as the top then finds courage to fight the gravity of the Earth up to the last "breath" is fought.

With everything in either the form of a cube or a sphere all forms take shape in either one of the two with the second forming a strong relation. The fact of this is that it is eternity and infinity being divided by space in form and being space in form forms two potential possibilities. Space is merely a time delay of heat parting time in eternity from time in infinity. On the outside of space there is this part we all realize that has no end and was thought of as space but which is time in eternity. Science thinks of this, as outer space but is in fact time. Within everything spinning there is a line evoked that is infinity evoked and space divides as the space spins and by spinning, it parts these two forms of time. To reach infinity one has to reduce the line we call the radius to a point where such a line can no longer reduce. That will not remove the line because what ever was the circle remains the circle because although the line then is so small it falls outside the parameters of our Universe, the line still connect to Π, since the circle Π never released the line from duty. There it does not end because going one further takes form into total singularity where the lot is inside Π^0 and Π^0 is outside our Universe. At the point we find everything inside and forming part of singularity but that point is a spot that becomes the dot Π by motion establishing the line, which at that level is a dot. There can be no chance ever of removing the radius as Newton suggested in $\frac{dJ}{dt} = 0$ because removing the radius will not clarify the space borders. It is the circling of material that defines Π^0 from Π and clears the space circling at a pace as material because the potential of such a radius is lingering in parting Π from the value of $\frac{dJ}{dt} = 1^0$ which is what space in time is. The factor that **k** represents (**k= a³/T²**) defines Π from both sides of the value of 1^0 and that can never remove from the Universe because there then is no longer any part forming our Universe, because what forms our Universe is **(k= a³/T²)**. Yet it is there in all its potential to be used nevertheless. The circle can never remove because the fact of Π forming Π^0 which is 1^1. If only 1^0 continue, then material $\Pi\Pi^2$ is also no longer forming part of our Universe. By installing $\Pi\Pi^2$ it is as a potential to be a form and become an independent Universe **(k⁰ = a³/T²k)** whenever motion **T²** calls upon the circle **a³** in the duty to become a form by motion. The line in the centre of all spinning things that becomes the radius to form the circle that would be Π might seem to us being imaginary, but it is in infinity and with all mathematical proceeding at our disposal, it is a mathematically calculatedly factor remaining in infinity and stored as a potential to be used by what ever in which form it is required.

I am of a very different opinion about Newton's point of view where he declared that forming a circle moving $\frac{dJ}{dt} = 0$, doing such removes Kepler's relevancy factor. This places a value of empty space in which a top would spin and Newton missed the difference there is between a top spinning and a top laying on its side on the Earth. There can be no such a thing as empty space. The fact that space is valid removes an empty connection because space can be anything there is in space. The universe is time contained in space, which makes it space-time. Space has only one value, and this is to contain time and time provides space with a definite value.

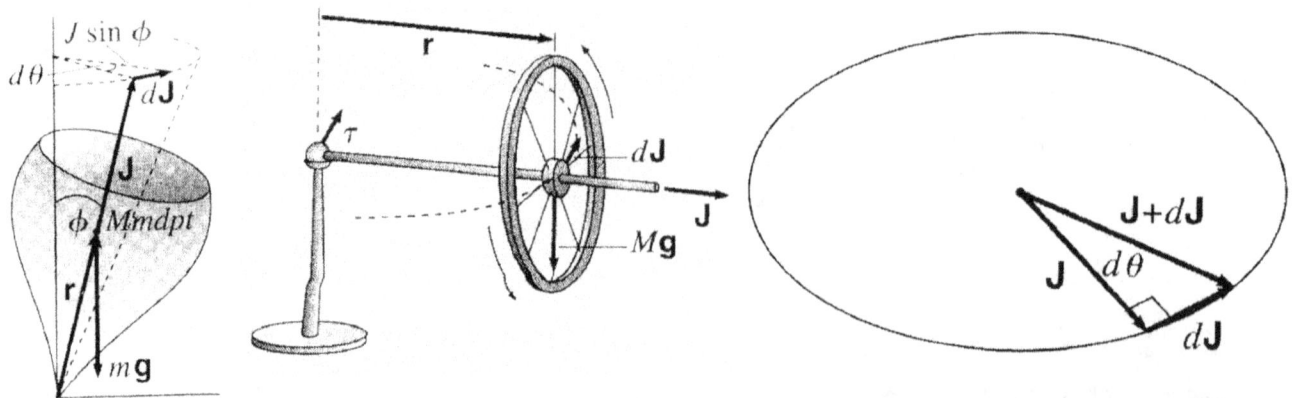

All of the calculations Newton made are very correct except the eventual and final conclusion Newton came to. Newton never understood the mathematical concept of time playing a part in physics. Time can never stand still because time is forever moving by establishing space.

Being the mathematical genius as Newton is so often portrayed as, Newton had very little insight into mathematical possibilities, because when he suggested that $\frac{dJ}{dt} = 0$ he made one huge mathematical blunder. Newton or no other person may place any two objects in a direct relation and have an outcome of Zero. Much surprising is that not one mathematical genius that came after Newton drew the correct conclusion that forming $\frac{dJ}{dt} = 0$ is mathematically not acceptable. Newton saw that dividing something into something else could bring about zero and that is impossible. In concluding that $\frac{dJ}{dt} = 0$, Newton found a way to replace Kepler's symbolic relevancy value of **k** with using the symbols G $(m + m_p)$. In doing that Newton painted a picture that has no real meaning except where Newton tried and succeeded to put mass into an argument that has no true validity in cosmic principles. This is just a longer and probably a more detailed manner of indicating **k** and better defining of **k** but it symbolises precisely to the point what **k** stands for nonetheless. I wish to draw your attention to the matter of Johannes Kepler's findings that Mainstream science considers as resolved and closed for many a century while it is not. My investigating Kepler helped me to resolve other unresolved matters but it was only possible by using Kepler's work.

I too am well aware that at first glance you will immediately arrive at the opinion that the theme of the book has to be considerably below the standard of an intellectual Master such as you must have, due to the position you hold, and because of that, the normal research work you do. I realise it is dealing with a subject school children learn but in that comes the issue that goes unnoticed Nevertheless, I hope that this writing may spark interest even at such a low academic level and grade in scientific sophistication and development because I am about to prove that I discovered:

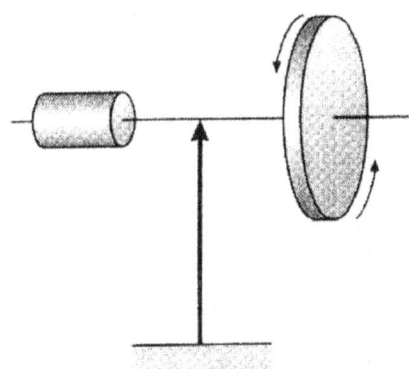

Newton did not think the situation through when he contemplated about gravity. Newton should have thought about factors keeping the gyroscope upright while the gyroscope is spinning. The gyroscope will fall on its side when not spinning and in that position the "Earth's mass" could play a part since the gyroscope fell on its side. However, as soon as the gyroscope started spinning, the balance shifted in favour of a position wherein the gyroscope stood upright. What then came about had the ability of keeping the gyroscope upright. This is rotational movement and I explained how rotational by the square of the double seven forms Π and Π is forming the curvature of space-time and in that bending of space-time we call the atmosphere that keeps the gyroscope square with the Earth and through that the gyroscope stays upright. That is evoking singularity that establishes gravity in relation to the Earth evoking gravity through also spinning.

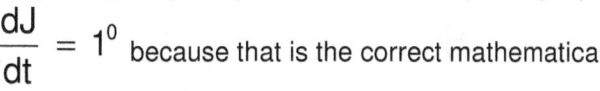

Newton found mathematically that the movement of the top by spin removed the value of the radius $\frac{dJ}{dt} = 0$ where quite the opposite applies. The spin of the top $T^2 = a^3 \div k$ positions the relevancy that **k** as a factor produces by initiating singularity k^0 on both sides of the relevancy $k^0 = a^3 \div T^2 k$ as well as placing singularity in relation to the spinning top $\frac{dJ}{dt} = 1^0$ because that is the correct mathematical principle coming from the equation.

The spin of the top does not eliminate the relevance of **k** but institutionalise the measure of **k**.
Trying to find a measured value for **k** is showing no understanding about what **k** is. The value of **k** is finding the space that **k** indicates in terms of what moves. The indicator **k** identifies the space a^3 that the top claims in terms of singularity k^0 that the movement T^2 isolates from the rest of singularity $\frac{dJ}{dt} = 1^0$. The value of **k** is dictated by T^2 as the movement isolate the space a^3. The measure of **k** is the relevance **k** claiming on behalf of the space **k**, which uses the relevance of **k** to limit the space spinning in accordance with T^2.

The Four Cosmic Pillars; The Result Thereof.

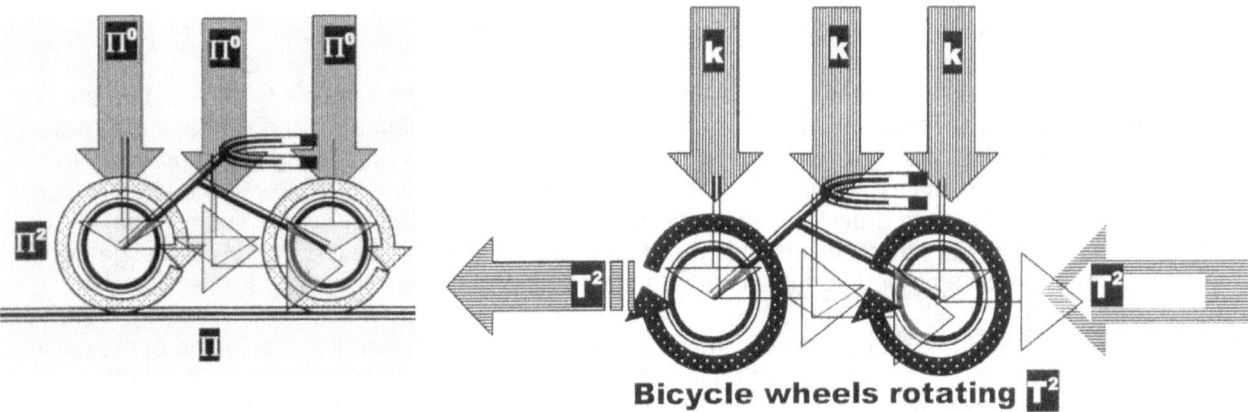

Let us have a look at the bicycle. It is said that the bicycle works on a balance and by science mentioning that the rider of the bicycle is applying a balance, in that the entire problem is solved. That is so typical of Newtonian simplicity about a very complex issue. It is the same as putting gravity down to mass pulling by some small particle called the graviton without ever showing any ability to look more intensely to find a solution for a very complex problem.

Bicycle wheels rotating T^2

Creating a graviton or creating dark matter to look for solves all the unsolved issues. Saying the riding of a bicycle is due to balance is the same as putting everything in the Universe down to mass taking charge of particles. As the wheels spin (T^2) the relevance of **k** leaves the bicycle firmly attached to the ground and in doing that it confirms the space in location (a^3) in terms of singularity k^0. Newtonians would call this having mass or whatever. Then having the bicycle moving forward in terms of individual cycling gives a relevance of (**k**) to the peddling power and the movement (T^2) then is about having momentum in relation to the Earth spinning.

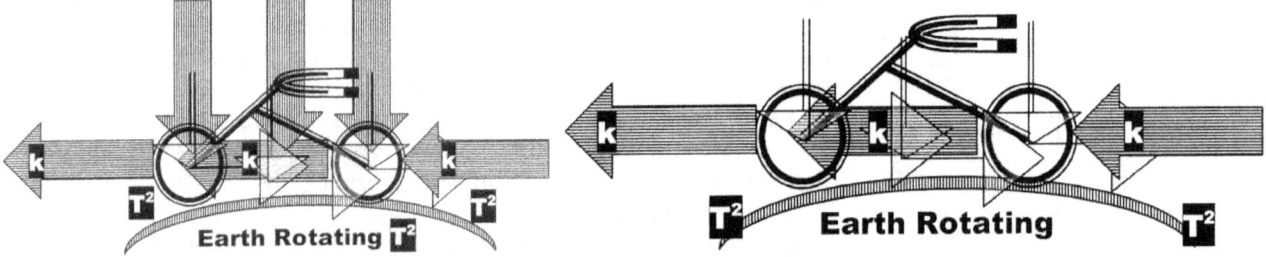

Earth Rotating T^2

Both times **k** indicates the same space, which is the bicycle, but every time the relevance of movement attaches some significance in relation to another situation where other factors bring measured values. It is not the measured value that **k** carries but the significance of **k** is in the value **k** brings the rest of space (a^3) moving (T^2)

Gravity is the principle applying by which movement is following three directional changes using singularity, which means it is going through three directional alterations at once. This is because singularity is $1^0 = 1^0 = 1^0$ and all three are then applying one and the same singularity rerouting.

Understanding the fundamental principles guarding gravity is rooted in this very idea and one must understand that going straight will always come to any object going in a circle while redirecting movement to where movement is not, which is in space instead of in time.

The value of space (a^3) (of the bicycle) in terms of movement (T^2) (peddling) in relation of the relevancy (**k**) (direction of movement) connecting to singularity (k^0) is what the relevance of changing of positional allocation is all about and that is the mathematical relevancy that **k** brings about. That is what Kepler's formula is all about. It is the space (a^3) moving (T^2**k**)

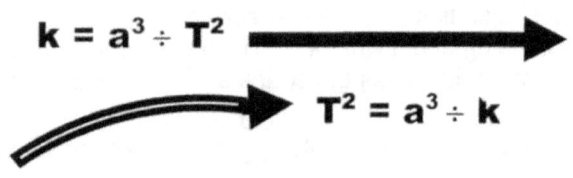

and relocating the space brings about space –time.

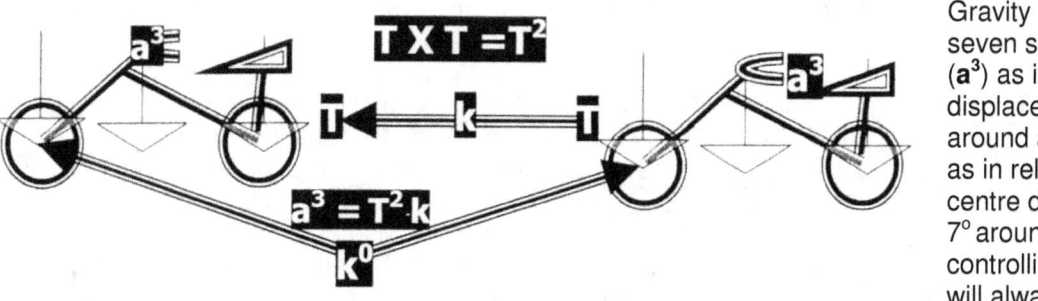

Gravity is the double seven shift of a body (a^3) as it moves (T^2) by displacement of 7° around an axis as well as in relation (k) by centre displacement of 7° around an axis of a controlling body. There will always be a **governing body** as well as a **controlling body**. The wheels of the bicycle is the governing body while the Earth spinning about its axis is the controlling body or the bicycle spinning around the Earth is the governing body while and the Earth going about the Sun is the controlling body while the Earth spinning around its axis is the governing gravity while the Sun having the Earth spinning around the Sun is the controlling body. There will forever be **k** forming 7° displacement while T^2 is forming the other 7° displacement. This is what Kepler's formula confirms in the formulated statement that $a^3 = T^2 k$

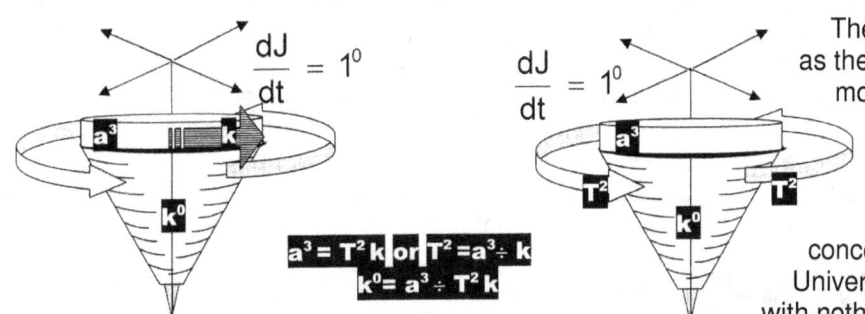

The relevancy factor given by Kepler as the symbol **k** receives validation and motivation as being that which place space between that which can never reduce and that which can never increase. The expanding of the Universe is a concept void of sensibility because the Universe holds everything, is everything with nothing able to add or to remove from the entirety of the Universe. It is the relevancy of **k** favouring a^3 as material claims more space because the shift progresses in favour of the density that material claims. The entire idea of the Hubble shift is the product of the relevancy of **k** becoming wider as time T^2 increases with movement by time increasing.

The top starts to spin because the spinning motion T^2 gives relevancy **k** to the space a^3 claims more as the space a^3 then forms with singularity k^0 extending to the limit that Π allows. The only process that can evoke the establishing of a^3 in terms of singularity **k** as well as singularity on the outside of **k** forming cosmic liquid is the movement of T^2. That keeps the top spinning upright and keeps the bicycle peddling upright.

Newton got it wrong about mass because the only way mass could be awarded to a body is if an object forms part of the solid material that spins and is pushed onto the surface of the Earth by the contracting cosmic liquid. Therefore Galileo is correct in assessing that all things fall equal because it is the space being compresses or being compacted and in that compacting things move towards the Earth until such a position as where the Earth stops or disallows material to move further down while the cosmic liquid that is heat contracts to a state it is on the surface of the Sun. It is not mass pushing down heating the Earth up because then please answer the question as to how can mass heat anything up by pushing down onto something else. The main thing that should be recognised is that when Newton determined $\frac{dJ}{dt} = 1°$ this did not annihilate the relevancy **k** but the truth is that it defined the importance of **k** forming the relevancy. **Sir Isaac Newton's** says that $a^3 = T^2$. I have to believe **Sir Isaac Newton's** when it is said that three dimensions are equal to two dimensions or in mathematical terms that $a^3 = T^2$ on no more grounds than that **Sir Isaac Newton** said so and without having any other proof to back the statement. Remember, Kepler never said $a^3 = T^2$, that is the part coming from the fantasy of **Sir Isaac Newton**. Kepler said $a^3 = kT^2$ which places three dimensions on one side holding three dimensions equal on the other side of the equation. There is a^3 on the one side of = and then there is kT^2, which is $k^1 \times T^2$ which is $k \times T^{2(1+2=3)}$ and that makes $a^3 = kT^2$ having three dimensions on the one side being equal to three dimensions on the other side. There is no way in heaven or hell that one can have the third power being equal to the second power or have a cube that is equal to a square, even if you are **Sir Isaac Newton**. There is no one on Earth that will tell me that $10^3 = 10^2$. There is a case that $10^3 = 10^2 \times 10$ or that $2^3 = 2^2 \times 2$ but never can it be that $2^3 = 2^2$. Not even when **Sir Isaac Newton** is doing the saying so. If one says that in the event where $a^3 = kT^2$ one may assume that $a^3 = a \times a^2$ or $k^3 = k \times k^2$ or even that using $T^3 = T \times T^2$ will also bring equality but never can $a^3 = T^2$...and then there are academics who try to convince me that $a^3 = T^2$ because **Sir Isaac Newton** was of the opinion that $a^3 = T^2$ and furthermore they expect me to also believe that it is true that **Sir Isaac Newton** has never been wrong on any suggestion and because no one could ever find **Sir Isaac**

Newton to be wrong, I have to accept that $a^3 = T^2$ and take it as the absolute truth without questioning this abnormality!

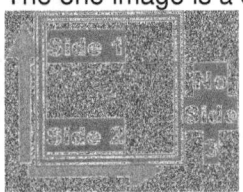
The one image is a cube with three sides. The other totally different image is a square having two sides Sir Isaac Newton said the two are equal while they can never be equal since they are one dimension apart. Sir Isaac Newton convinced so many generations of idiots considered as being the wise amongst the wise and fooled those to the point where these stooges are willing to believe they are wise enough to believe that a cube is equal to a square and only on the ground Sir Isaac Newton said so.

Sir Isaac Newton proposed and moreover convinced the world of science, and this includes every one and all members that should be the most intellectual bunch living on Earth in human form, that they and the entire world should accept that the inexplicable $a^3 = T^2$ is correct and that the biggest trick in fraud can be played on a bunch of fools all willing to be stupid enough to pretend they are clever enough to see that $a^3 = T^2$ and they are so stupid that they pretend to be so clever that they will accept that $a^3 = T^2$ which when translated in words means that two dimensions are equal to three dimensions. This boils down to someone being able to walk into a mirror become the image in the mirror that is two dimensional and walk out of the mirror becoming his three dimensional self again and still being exactly what he was and totally unscathed. This is the same as stating that a person's reflection coming back from the mirror is the same as the person filling reality while standing and looking at his image in the mirror. In this group hosting the most advanced minds man can produce there are a big enough bunch of zombies pretending to be mentally superior while being big enough idiots that are foolish enough not to think and not to ask questions but be small minded to the point that they will accept that a cube is equal to a square $a^3 = T^2$ just simply going on the say so of Sir Isaac Newton's Let's bring what Sir Isaac Newton interpreted as to what Kepler said in relation to Kepler's formula and see how much Sir Isaac Newton defrauded the work of Johannes Kepler and what in fact Johannes Kepler did say.

If it is true that $a^3 = T^2 k$ and we dismiss Newton's obscenity while going back to basic mathematical principles we find that: $a^3 = T^2 k$

That means the space will move in a straight line while it circles. That can only indicate a controlled expanding because $a^3 = T^2 k$ (also is); that the line between the centre k^0 and $k = a^3 / T^2$ is increasing at a rate of k because of a^3 / T^2 which indicates that expanding is happening at this very moment. The line k is not zero as Newtonian madness would suggest or the line is not 1 ($a^3 = T^2$) but the line is gaining by one dimension every rotation that is completed.

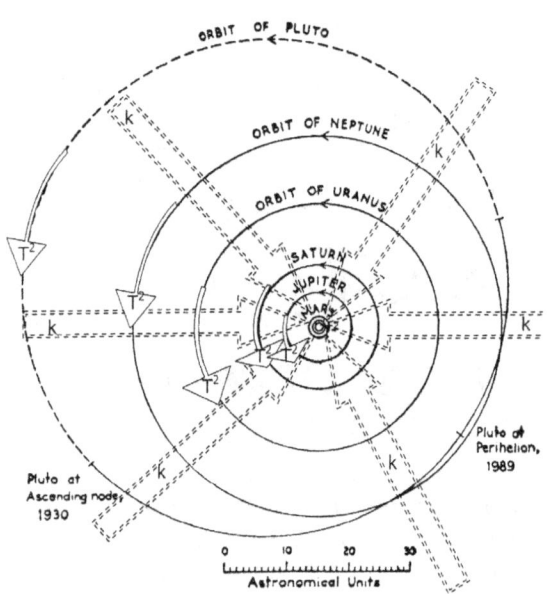

$T^2 = a^3 / k$ which means the time it takes to move is in ratio with the distance the space will move.

$k = a^3 / T^2$ which means the distance the space would move depends on the time allowed for moving. However if that is true and it is a mathematical statement then it also must be true that:
$k^0 = a^3 / T^2 k$ there is a appropriated centre

$T^{-2} = k / a^3$ Time can increase by manipulation as well as time that can reduce by manipulation. One can travel by increasing or decreasing the valid time.

$k^{-1} = T^2 / a^3$ The distance is moving between the objects. If the objects are not coming closer it must be the substance holding the objects that is coming closer. When we have a table indicating a shift towards the centre $k^{-1} = T^2 / a^3$ that does not involve any of the planets coming towards the Sun it would then show the space holding the planets are moving inwards because it shows definitely that something about the radius is decreasing and that means something is shifting towards the centre.

The space a^3 is rotating T^2 and the only other counter action could be when the space holding the rotating space is reducing $T^2 / a^3 = 299$ by the margin space is expanding $a^3 = T^2 k$ and $k = a^3 / T^2$ that proves the

Big Bang is in progress. Kepler told so much about so many by using so few syllabifications that a mathematician such as Newton was unable to comprehend the full implication

PLANET	SEMIMAJOR AXIS $A(10^{10}m)$	PERIOD T (y)	T^2/a^3 $(10^{-34} y^2/m^3)$
Mercury	5.79	0.241	2.99
Venus	10.8	0.615	3.00
Earth	15.0	1.00	2.96
Mars	22.8	1.88	2.98
Jupiter	77.8	11.9	3.01
Saturn	143	29.5	2.98
Uranus	287	84.0	2.98
Neptune	450	165	2.99
Pluto	590	248	2.99

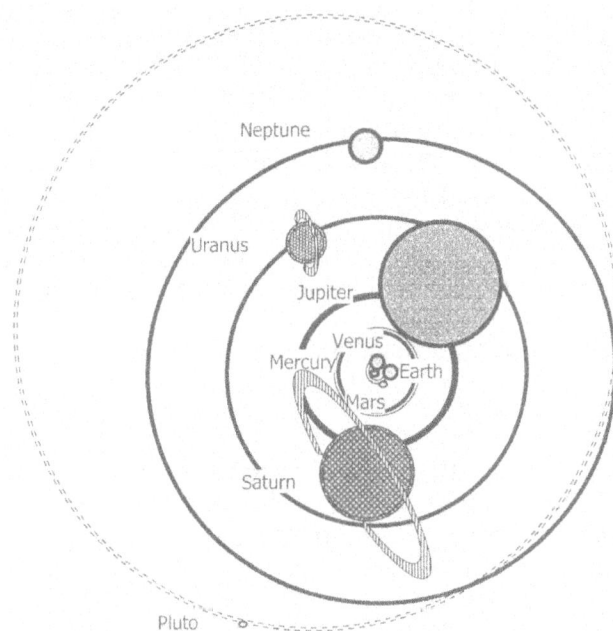

From Kepler's space-time $a^3 = T^2k$ formula we find that the relevancy of all planets $k = T^2/a^3$ in relation to the Sun is alike. This is only possible when the planets are floating in buoyancy because when in buoyancy all objects are equal in relation to the water holding them. There is no big or small but just those having specific density in relation. If there was no buoyancy mass or size would form some sort of resistance that would allow more and less restriction in some or other form to be present. In the sea and to the sea in that case there is no big fish or small fish but there is only fish. To us humans we think in perception of distance but in the cosmos space in the form of distance is the measure of time developed. The same goes for temperature and mass. Mass is good and mass is a product of the Human mind to put perception when it is needed but mass is dysfunctional in relation to the cosmos. All the planets are more or less 299 in ratio from the Sun, which makes the time affecting all the planets $T^2 = a^3 / k$ which is then $T^2 = a^3 / 299$ and in relation to space $a^3 = 2\ 99\ X\ T^2$. What this does is it puts all the space in ratio at an equal distance to the centre of the Sun and it puts the time in motion rotating at an even period around the Sun.

The Sun is at a level with all the planets parading past the Sun in the given ratio of 299. All the planets are precisely the same "distance" $299 = T^2/a^3$ from the Sun and has precisely the same "mass" $a^3 = 2\ 99\ X\ T^2$ in relation to the Sun and the rest also floating about the Sun while they all travel at the same velocity $T^2 = a^3 / 299$ around the Sun. The lot is in a bowl of liquid and it is the liquid that keeps a regard to the Sun where the Sun forms the solid as the regard to the liquid.

Johannes Kepler said $a^3 = T^2k$ which is the space a^3 in orbit T^2 is equal to the distance T^2 travelled in terms of a centre k that is in place.

Kepler said that the space a^3 is equal to the motion T^2 of the space a^3 distant from a specific centre k. That then is $a^3 = T^2 k$.

Reading this mathematically encrypted coded formula of the cosmos given to Kepler and keeping it removed from Newton it reads as the space a^3 is equal to = the motion T^2 of the space a^3 in ratio to a centre k.

What this proves is that gravity is the motion of space provided by time being the liquid.

Please allow me to explain. In the formula $a^3 = T^2 k$ the space forms as the space is in motion. Newton suggested that $\frac{dJ}{dt} = 0$ where he stopped time to have the motion of the circle demolish the work that the circle does.

That means he got time standing still or being T^1 and the motion $T = 0$. Let us ponder on that thought for a while, while remaining with the formula Kepler suggested it will seem that according to Newton $a^3 = T^2 k$ and in that T^2 then becomes **1**.

Newton changed the symbol of **k** by using the mathematical equated symbols G $(m + m_p)$. This is just a longer and probably a more detailed manner of indicating **k** and better defining of **k** but it symbolises precisely to the point what **k** stands for nonetheless. I wish to draw your attention to the matter of Johannes Kepler's findings that Mainstream science considers as resolved and closed for many a century while it is not. My investigating Kepler helped me to resolve other unresolved matters but it was only possible by using Kepler's work. This changed the aspect of gravity in cosmology fundamentally and as I am about to show most and totally incorrectly.

Let us for one minute leave Newton's surmising about Kepler's failure out of the picture and concern us with what Kepler found long before Newton thought about what Kepler found.

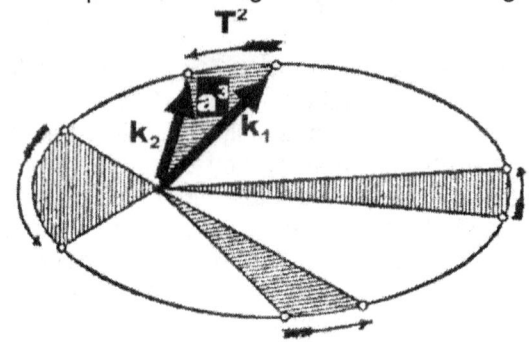

Kepler said that the space a^3 is equal to the motion T^2 of the space a^3 distant from a specific centre k. That then is $a^3 = T^2 k$. Reading this mathematically encrypted coded formula of the cosmos given to Kepler and keeping it removed from Newton it reads as the space a^3 is equal to = the motion T^2 of the space a^3 in ratio to a centre **k**. What this proves is that gravity is the motion of space provided by time being the liquid.

Please allow me to explain. In the formula $a^3 = T^2 k$ the space forms as the space is in motion. Newton suggested that $\frac{dJ}{dt} = 0$ where he stopped time to have the motion of the circle demolish the work that the circle does.

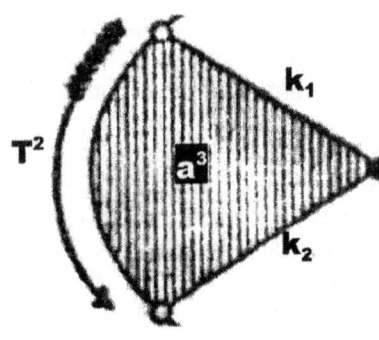

That means he got time standing still or being T^1 and the motion $T = 0$. Let us ponder on that thought for a while, while remaining with the formula Kepler suggested it will seem that according to Newton $a^3 = T^2 k$ and in that T^2 then becomes **1**.

When Kepler said $a^3 = T^2 k$ it mathematically reads that the space a^3 is equal = to the movement T^2 of the space in relation to a radius **k** connecting to a specific centre.

This indicates the value of space-time putting space a^3 equal = to the time $T^2 k$ moving.

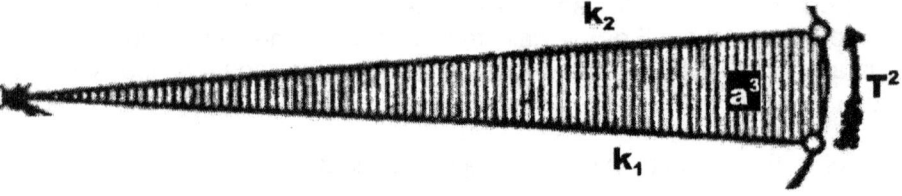

Two lines connect the value of **T** coming from one point and ending at another where this using of **T** is forming the line T^2 and that symbolises the movement between a point that k_1 indicates and a point k_2 indicates and that gives a value T^2.

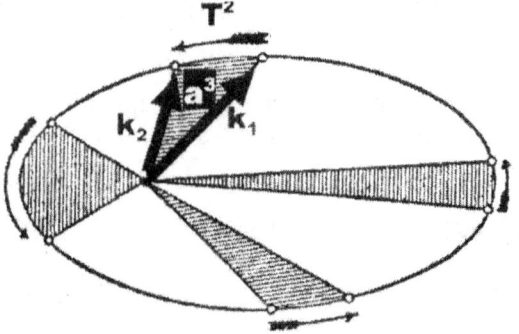

Nobody on Earth, not even Sir Isaac Newton has the authority to remove that factor **k** on any grounds there can ever be thought of from an equation Kepler produced as forming space-time or mathematically said $a^3 = T^2 k$.

When any person and that even includes Sir Isaac Newton goes about and takes onto him the authority to make changes to a mathematical discipline in terms of the authority such a person claims to have and as that person wishes, then the result is catastrophic. Certain things can't be changed by man and one of the things is the law on mathematics.

I just can't understand how the complex Newtonian mind works! You just can't wish away the figures used in Kepler's tables and dismiss the figures to nothing just because Newton said so! By dismissing that which is printed and has a pertinent value other that zero…then doing so one must realise that the man going by the name of Sir Isaac Newton was dabbling in nonsense and was going about committing one huge error. I can't understand the concepts of Sir Isaac Newton and I am proud not to understand Sir Isaac Newton for it does me proud.

PLANET	PERIOD (Years) (T)	MOVEMENT (T^2)	DISTANCE k	SPACE (a^3)	RATIO
Mercury	0.241	0.058	0.39	0.059	0.983
Venus	0.615	0.378	0.728	0.381	0.992
Earth	1.000	1.000	1.000	1.000	1.000
Mars	1.881	3.54	1.524	3.54	1.000
Jupiter	11.86	140.66	5.20	140.6	1.000
Saturn	29.46	867.9	9.54	868.25	0.999
Uranus	84.008	7069	19.19	7067	1.000
Neptune	164.8	27159	30.07	27189	0.999
Pluto	248.4	61703	39.46	61443	1.004

In the above table that Kepler configured as $a^3 = T^2 k$ we have three distinct factors combining to form a specific value that indicates space-time $a^3 = T^2 k$ and moreover shows that the Universe structurally is composed of in terms of space –time $a^3 = T^2 k$ and every factor as much as a^3 and T^2 as well as k has a part and a role in forming the eventual value of space - time $a^3 = T^2 k$. What did Sir Isaac Newton say happened to all the values under the column reserved for distance or then the symbol k? How did Sir Isaac Newton explain the values just disappearing? Why did no one ever think of questioning this unless every one participated in covering up fraud? The values are there and no person may ever discard the values in terms of then only applying the others unchanged.

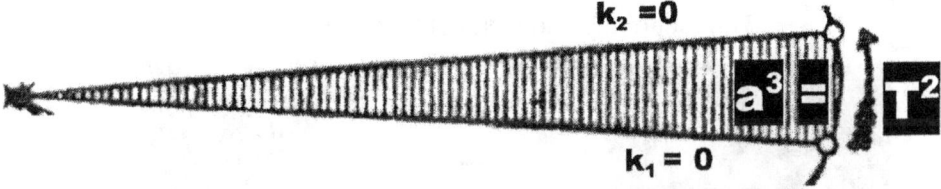

Those that wish to resort to using the excuse that Sir Isaac Newton implicated a ratio was in place, well that would not stand up either as a reason because Sir Isaac Newton used $a^3 = T^2$ which says the space a^3 is equal = not in ratio with but equal = to T^2. If one puts an equation it will also involve a ratio but the ratio then can show the symbol used to indicate equality = as being a ratio. Sir Isaac Newton put the space a^3 in terms of equality = to the distance travelled T^2 by disregarding the distance from a centre such as the illustration above will show and the illustration shows complete madness on the part of Newton because the distances k is in place and holds a value other than zero. This rubbish excuse Newtonians contemplate to cover the fraud their Master committed of only indicating a ratio is nonsense because Sir Isaac Newton clearly says $a^3 = T^2$ and that puts every value Kepler had in the column devoted to a^3 as being precisely the same (equal to =) the value dedicated to T^2 being in the column devoted to T^2. He never meant it to be seen as a ratio as Newtonians wish to prove. There is a ratio, yes, but that ratio hinges on k and removing k removes all the validity such a ratio presents. If $a^3 = T^2$ the cosmos would have corrected this statement by removing k altogether and then the cosmos would not have had to leave the removing of k up to Newton to do it. Surprising as it may be to science, but Newton is not God.

When Sir Isaac Newton says $a^3 = T^2$ that does not prove that $a^3 = T^2$. It only proves Sir Isaac Newton was the worlds biggest and best silver tongue devil and cheated an entire Earth load of scientists for almost four hundred years. He fooled the supposedly wisest humans we all think there can be to pretend to be wise so that they can hide their stupidity while they only focus on their stupidity by not questioning the validity of $a^3 = T^2$. Can you bring me one other con artist and fraudster that can manage such fraud as to fool four centuries of scientists in believing horse dung is fig jam? Those pretending to be wise showed how big fools they are because those fools tried to pretend they think $a^3 = T^2$ is correct just because Sir Isaac Newton said $a^3 = T^2$ and only those as equally wise as Sir Isaac Newton can be able to see that $a^3 = T^2$. It takes some doing to fool so many people for so long and leave all those fooled feeling good about themselves in that they are fooled while believing they are the wisest there could ever be born from a woman. Sir Isaac Newton was the biggest con artist ever to live and never again will the world experience an equal to Sir Isaac Newton. It is no small wonder that science is infested with atheism because science upholds disdainful lies based on mediocre understanding about truth applying as a reality and crooked science! Newton is all lies and shambles and reading this book will prove that.

The Four Cosmic Pillars; The Result Thereof.   Page 206   APPLYING AS THE SOUND BARRIER

If science cannot prove God's existence, it is not God that does not exist, but it is science failing and therefore it is then that specific view about science that should be re-examined since it is the view on science that is proving as being incorrect. This fact is what the so very brilliant and intellectually mindful Newtonian atheist should remember when they fail in their science altogether. That their science fails altogether and that failing it does in all its splendour, is a fact I am delighted to prove! The fact is Newton's views were never tested and that the Newtonian views on science were never challenged before and because of that Newton principles never withstood diligent scrutiny before. When **Sir Isaac Newton** is investigated even in the flimsiest of manners, well accepted facts seem to become very suspect, to say the least. This becomes evident when concluding all the facts this book presents.

Now, in this book, for the first time, Newton is tested and such testing is the proof you gain by reading that which I uncover. What I bring into the open are unseen facts, which I present you with as I take you on a tour through an avenue of facts I introduce in this work. The lack there is in sensibility concerning **Sir Isaac Newton's** principles this book proves. The theories of **Sir Isaac Newton** require proof, which was never given while God never needs proof and that is what science constantly seeks.

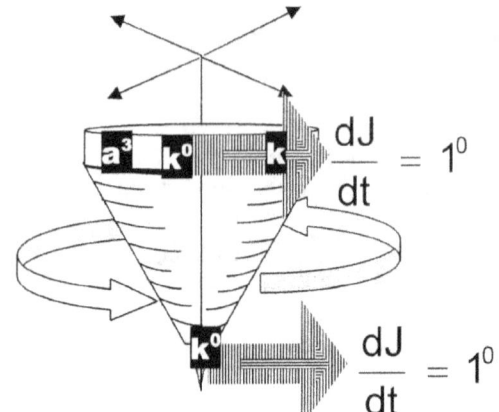

The solution to the problem that Newton found was not eradicating the factor **k** but it was defining the location of the factor **k**. His mathematical conclusion was not dismissing **k** but to the contrary, it was confirming the position that **k** brought to science.

The **k** factor that Kepler placed in his formulation of $a^3 = T^2 k$, placed the atom or material in the confinement of singularity Π^0 inside the centre and on the spot where infinity meets eternity at the end of the relevance brought on by **k**.

The fact of the line being present disqualifies the Newtonian conclusion of the representing **k** as line shows being present, but all is the fact that shows the of singularity on sides of the divide being in the centre and divided by **k** from the one forming outside material.

factor zero. The singularity most of the line presence both namely

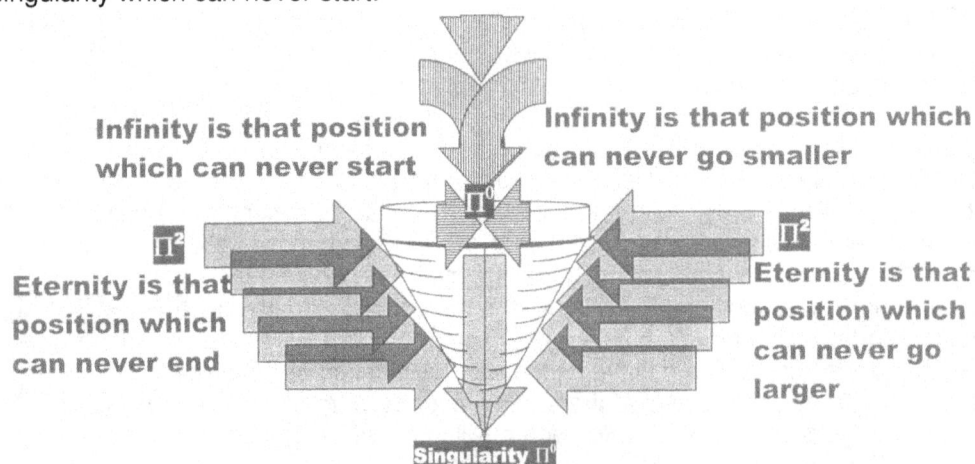

The factor **k** used in $a^3 = T^2 k$ positions the Universe in between singularity that can never end and singularity which can never start.

The first thing about gravity one must remember is that all things are spinning. There is no particle in the entire cosmos that is not spinning. The entirety of the cosmos is about everything spinning in relation to one specific point not spinning because that point not spinning at that point is excluded from space and is therefore not part of the Universe holding space.

In the **centre** of al **things spinning** a **line forms** coming from a **dot** that has **no space**. This is **a dot** that first forms and **that initiates space** that later forms. The **line running from top to bottom** of **the spinning top** comes from **a dot** that extends but just as in the case of **the original** dot **the line has no start and has no beginning**. When **reducing the radius** of a circle it would eventually **end at the point where the line is not holding space, placing the line there, while the line is not there**, because **crossing the line would land space in the very opposing quadrant**. As soon as **entering the line one has gone through the line**. The line **has no inside** that can **go even smaller** yet we know that the line must have some ability to be able to go smaller because we read that in the mathematical set up of Π.

The dot as well as the line that forms can't have a value of zero since the line is there and is present for all to witness. If **the line was absent** only then could **it be valued as zero**. Realising the value of this line holds the entire principle of gravity in perspective. The **line is there** so it **has a place** but it **dot** holds **no space**. The line is **formed from the** and since **there is no space it has to be singularity**. Since **singularity could have one value** and **that is having 1** as a value this **dot and line forms singularity**. In the line there is no space because **as soon as anything enters the line it moves right through to the other side** and the line by **itself is there without ever forming space to confirm its presence**.

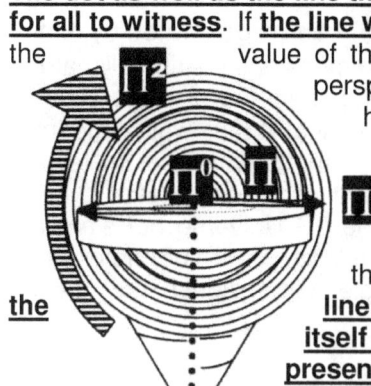

What evokes the line is the movement of a confined space that is filled with material spinning around a centre that holds no space. The line comes from a point that can never start but for material evoked by movement and in space it could never have a beginning except for the movement of material placing spot without space in place to never could have space. The spot as well as go smaller.

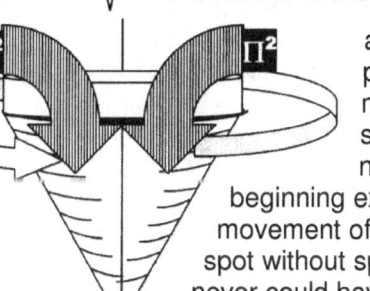

the form a line which the lien could never

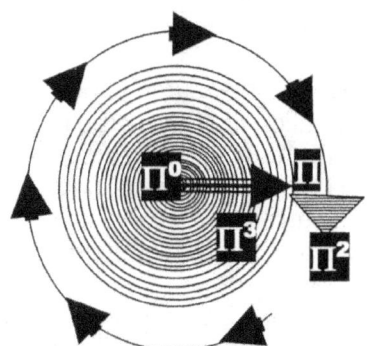

The line forming a circle confirms Π as forming. The movement of space measured as Π moving from one point to another point and that movement takes Π into a square value as $Π^2$. However without the movement the line does not form and therefore without moving the line forms no gravity and therefore gravity is all about confirming the line coming from the dot in the centre.

the product of space forming is then

Singularity can't move because singularity in infinity forms infinity 1^0 by the concept that it is the only immovable point in the entire Universe while it is the entire Universe that moves. There forming time is having one centre spot 1^0 that stands still in relation to the entire Universe shifting one position in relation to the one point never moving. Yet we know it is the centre singularity shifting with material although the centre being 1^0 can't shift. However singularity $Π^0$ always remains absolutely connected to Π through forming the connection that Newton saw as $\frac{dJ}{dt} = 1^0$ or what Kepler saw as $k^0 = a3$

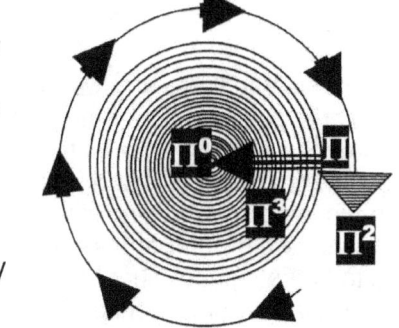

÷(T^2k). The relation between $Π^0$ and Π never changes because it remains connected although we know it is material shifting and spinning. From the perspective where infinity forms 1^0 it then has to be eternity at 1^0

that moves since Π never moves and therefore all movement is on the side eternity holds. That is why the Moon is part of the Earth although we know it is not. It is because the Moon always shows the same spot towards the Earth in relation to singularity 1^0 within the Earth.

This is the dot that has no start. It is 1^1, the part that released from the spot 1^0 when motion parted singularity. It came apart when motion unleashed the dot 1^1 that has no start from the spot 1^0 that has no end. It is the Universe born from motion that was driven by heat. It is still there because once anything is part of the Universe and forms a principle within the Universe it has n where to go but to remain within the Universe. Walk outside at any time and you are a witness to the result.

When you walk outside and look at the vastness of the blue sky or at night at the black night sky, you are physically part of singularity in the part of 1^0, the part that moved away from 1^1. You are within the part 1^0 that has no end because it has only one side, which is the inside. It is 1^0 going nowhere. It is the part that I named the spot that had the dot 1^1 moved away from

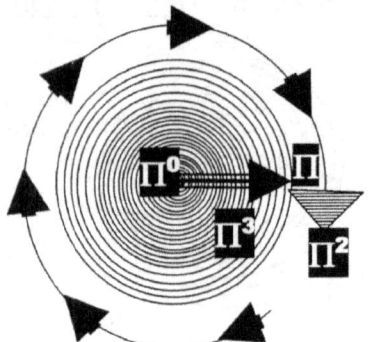

In the spinning top we find that singularity $Π^0$ can be generated by motion. But singularity $Π^0$ has no motion within the dimensions we find allocated to the Universe in which we live. Since the singularity found in the centre of the spinning top is in truth just a mathematical point, which means in mathematical terms the point with no sides cannot even be calculated as a factor since the measure thereof goes beyond what mathematics ever can calculate. Mathematics has a use within the 3D Universe but singularity that keeps the spinning top attached to singularity governing the gravity of the Earth, that singularity is truly single dimensional and beyond mathematical measure. It is singularity $Π^0$

By finding where it all began is equal to finding where the line began we have to trace the line in order to trace the development of the entire Universe. As seen the first development went beyond where mathematics may take us. The Universe did not become more but only focussed better on detail. What is was present because nothing can be new to the Universe that started out to be what it presently is.
An object can rotate in outer space as long as it can maintain a speed that will keep the object rotating in that orbit. No line can go straight because part of all lines in the cosmos is the curve. There is always some relation between the factor of how much **k** influences or how much T^2 influences and the combined unit determines a^3.

That reason too has its footing in the Titius Bode law because 7/10 is half of 10/7 and by going seven the factor also has to go 10. The compliment of this is gravity at $Π^2$

The high spin or slow time flow was ideal to seal the atom and produce extensive numbers of solids joined by motion of that which relative allow motion and that which was allowed motion. It was one in relevance to one. It was Π coming from 3 as much as 3 enticing Π by generating Π as motion.

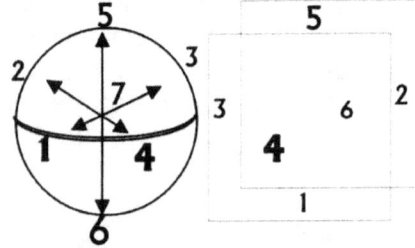

The distortion of Π in relation to 3 delivered a measured quantity in points of 1.0471. The distortion of 3 in relation to Π delivered a measured quantity in points of 0.9549. The total spread across both sides of the developing Universe came to 2.002. The including of 0.002 must be put to the interfering of singularity (1 - 0.998 = 0.002). The reasonable interpretation to make is that the Universe grew both sides of singularity.

In the Universe there are two forms of material. There is connected singularity spinning faster than the speed of light which is housed in an atomic cocoon we cal material where the material that through movement around a centre forms solids and the there are the unattached points holding singularity that holds no form and there for can structurally compromise shape to accommodate whatever form has to fit that space. That I call cosmic liquid and I have a feeling that is what science is in search of when looking for the "dark matter". The one forms a cube and is loosely connected by six sides connecting whatever for is required to form while there is material that structurally forms a sphere and therefore abides by the laws applying to a sphere.

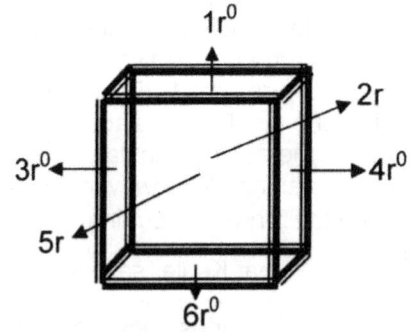

The Universe as big as it is constitutes of two forms derived from one fabric. The fabric is singularity and the forms are occupied space-time or material covering singularity in infinity and unoccupied space-time, which is cosmic liquid. There is liquid that will adopt any form and there are solids that remain rigid in form because the spin of material confirms the density of the material.

The singularity forming eternity 1^1 is without form and can reform to accommodate any form, which is appropriate. That ability qualifies singularity in eternity as a liquid and therefore I named it as the cosmic liquid. The cosmic liquid 1^1 stands in relation to cosmic solids 1^0, which is what materials are and materials, are strongly connected to singularity in infinity 1^0.

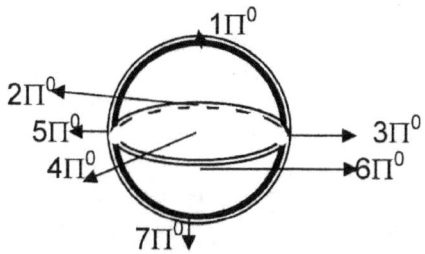

The sphere is directly associated with material which is rigid in form while the cube will house any form since the cosmic liquid is completely adaptable. The cube connects six sides at angles of various degrees and the form comes about from six sides loosely in contact. The sphere on the hand is controlled by six points on each side with two always directly opposing the other and the six points are connected by a precisely placed centre spot holding singularity and confirms the rigid ness of the sphere. The centre spot confirms all six points and this allows the seven spots in the sphere to form a unit of immense strength

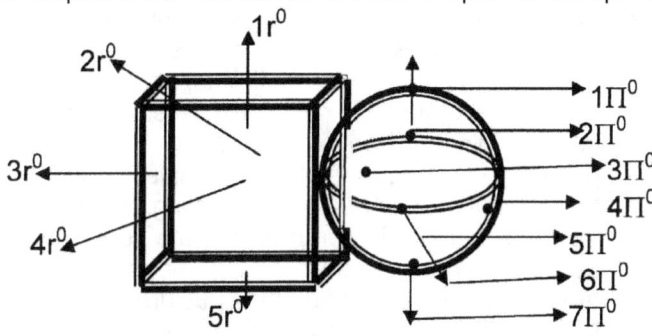

In the normally applying of gravity, we find contracting lines running vertically as the lines connect with the Earth centre. There are numerable lines running from the outer regions to the centre in diverting by 7^0. Motion provides extending of the 7^0 establishing the centre connecting points to the Earth to which it connects. When it is only the Earth providing the motion there is only one spot of space allocated to the bicycle in the time frame applying at the time. The Earth takes on a specific size by which it duplicates the space it holds in relation to what it renders the bicycle also sharing the space, which the Earth has. By that standard the bicycle also holds an exact relation of space and volumetric size related to its position within the Earth. **This is where the motion by duplicating the structure moving changes the dimension in equation. Motion is the duplicating of existing space from time in the past through time in the present towards in the future to time.**

Al movement in the Universe are about material (not mass) forming a sphere that turns or spins inside a cube where the cube has no distinct form or specific bonding other than having sides connected by angles. The movement of the material which is in a sphere turns around inside a cube and because the cube being solid connects with a specific centre that holds the centre equal in strength to all other six sides, unit is structurally sound and because of the strength of the sphere, the sphere will remove any point belonging to the loosely connected cube to structurally commit to the sphere. In that way the sphere will reduce the cube whenever contact is made through the movement of the spinning sphere. In short, the sphere is always stronger than the cube and will always remove any part of the cube and adopt it to become structurally bonded with the sphere. This has nothing to do with mass but is a simple result of the one type of form being more formidable than the other. This is a major part of gravity and main reason why objects are drawn towards an object such as the Earth or the Sun.

Every aspect of the formula that Kepler introduced applies more to the atom than what it does apply to any other sector forming the Universe. In fact it applies so well, it brings to mind that the atom is the Universe and what we think of, as a Universe is just the result of what the atoms form as a legacy. That which is smaller than the atom, the atom houses and what is larger than the atom, the atoms forms. It is the atom moving in a star that forms a group that forms the unit that gives singularity in infinity as well as in eternity the applying relevance which results in the applying gravity and in that becomes the star. The top is just mimicking the behaviour the atom applies.

Looking at the top from a vantage point taken from the top, it seems that the body structure of the top is solid and the air surrounding the top is liquid. The solid body is kept upright by the liquid. The top as a structure composes of solid particles that light cannot penetrate and that material cannot pass through. In that sense it seems to fit all the conditions we set for solidness. The top spins and it spins through the air that allows the top to spin seeing that the top has much more density than the air has. The air can't be "nothing" or even contain "nothing" since the air can produce sound and form wind as the top spins. Air is 600 times less dense that water and to "nothing" density can't be contributed because "nothing" is nothing. By having the same characteristics but much less density the substance forming air then must be less dense that water is but performing as much the same type of ingredient.

What can move is liquid and stands related to what cannot move being singularity. Since everything in the entire Universe is composed of singularity as used in everything forming the Universe in its most basic form, therefore singularity is everything that is immovable but also since everything is singularity, everything can move.

The main issue giving everything in the cosmos applying clarity is to recognise that **space is heat expanded** and **heat is space that is contracted. Heat and space is the very same thing but is the flip side of each other.** Everything in which the top spins and which is outside the space the top claims is liquid with the top forming a solid or so it seems to us. Well yes, in a way and not that much either. The top is a solid because the material forming the top, which is atoms, is in fact liquid singularity that is forming the top but within the atom the singularity is spinning so fast it contracted into a solid substance. The top is a pump that pumps heat or liquid from the outside inwards just like a turbine engine. The contracting of singularity from the outside is compressing the liquid on the outside to become solid within the atom. Every atom that is rotating inside the structure of the top is keeping the centre erect. The centre is totally motionless because all the atoms forming the top are moving around the motionless centre or infinity and the moving of the top circle is extending the centre-singularity of the top to the edge where the top meets eternity. The extending of singularity is holding the air as a liquid and being the liquid, the flow of the liquid keeps the top erect and spinning. The spin produces a cold in relation to the hot that the liquid is. Freezing to a solid-state represents cold and melting to a liquid represents hotness, notwithstanding human perception.

The part that remains unchanged and rigid never altering

Therefore space must be that part that always changes and moves

What is moving is liquid and what is not moving is solid. That is in terms of singularity. This seems to contradict what we see but we have to remember that a line of singularity moves from infinity and extends up to the end of the material at Π. The line consists of singularity where every atom hold a centre singularity which is centred in the very centre of every atom around which the atom is spinning and this line formed by every atom's centre singularity is an unbroken line from the centre to the end of material. Since the line holds singularity in place, and all singularity has a value of 1^0, therefore all allocated points holding singularity in the chain of atoms from the centre to the end where mass forms are valued as the same being 1^0. The material in proton numbers might be different, but the centre singularity is equal everywhere. This line holding singularity is never broken inside any material composing atoms and is constant throughout and is therefore rated by the centre singularity as unbroken and never changing. It is forever standing still. In contrast to that the singularity forming eternity is forever changing, taken from the perspective infinity forms, by the movement of Π remaining constant while outer space forms by going $Π^2$ and it is that which to us never moves are the moving factor to singularity where that which is that to us the part always moving.

From the centre singularity forming infinity or that which can never reduce more, everything has a reference in relation to another point where all other points have to move while any centre point stands still. That which is capable of relocating is forming a liquid in relation to that which is securing the position of rotation. Everything in the cosmos can move and yet not one particle in the cosmos can move. The cosmos stands divided between the eternal moving of eternity (which is what can never end) and the immovability of infinity (which is what can never star since it is eternally present).

The Four Cosmic Pillars; The Result Thereof. Page 211 **APPLYING AS THE SOUND BARRIER**

Everything around the top is liquid with the centre being a solid. However the solidness and liquid has cosmic standards and just as it is in the case of hot and cold, big and small, fast and slow, those are human applied standards where our standards and cosmic standards do not share any measurements. The cosmos has no hot and cold, big and small, fast and slow. So too, does cosmic notions about liquid and solids have a totally different meaning in cosmic terms.

There is a pumping interaction of space-time flowing towards singularity through every point that confirms singularity. Everything in the top that forms the material is also liquid. By providing motion the matter in the top doing so serves as the liquid factor that extends the space that singularity provide. The structure is composed of atoms. In the atom there is a governing generated singularity around which all material rotate. In the case of the atom all the rotating material forms the heat while the generated centre, which is incapable of rotating, and then forms the solid factor. This solidness extends through out all space that becomes material. Every aspect that is without motion stands in a relation of 1^0 and that which is relatively moving or changing location or finds a new position that holds 1^1. Everything that is standing still is 1^0 and everything that is moving is 1^1.

Gravity or motion is a constant relation that solids have with heat where heat forms the liquid and solids form space. Remember that **heat is space concentrated ands space is heat expanded**! Forming space (a^3) there is the rotation (T^2), but part of the rotation (T^2) is the lateral (k) progressing by rotation (T^2) to confirm the generated centre (k^0). That is Kepler $k^0 = a3 \div (T^2 k)$. The generation of material confined in space is formed in the rotation but the flow towards the stationary centre is the lateral and just as electricity produce a flow of time in relation to space collapsing, space-time by measure of gravity is using the same system to do the very same.

In cosmic terms there is no substance difference between 1^0 and 1^1 but for the spin that brings about such differentiation and it is a relation where one moves as the liquid partner then thereby the other is the solid factor. The spin parts the two factors, which are in truth the same composition forming singularity, but differentiating between the entirety of the cosmos and the infinity is movement. Both are not as much equal as they are precisely the same. Infinity cannot move and eternity cannot stop moving. By parting to form different substances, infinity had no movement while eternity was introduced, formed by motion as part of the cycle. This came in place at a point where one part stops moving in relation to all of that forming the other side that cannot move but does start moving. The factor that shows motion forms the liquid while at that moment the factor that does not show motion forms the solid. The measure of 1^0 is transformed to 1^1 and which ever forms 1^1 is passing the extending of space on to 1^0. Time forming eternity spins because everything about eternity spins and this happens in order to secure the centre singularity forming infinity. But also time moves and in that there is the linear factor mentioned by Kepler as **k** that always is part of cosmic motion. The centre is referred to by heat but heat also secures the centre by reconfirming the centre in the lateral as $k^0 = a^3 T^2 k$. But in both cases singularity is reinstating singularity by confirming the fact of singularity as it is referring one another. In the manner that 1^0 confirms a position in singularity 1^0 is supporting 1^1 by generating 1^1. By generating 1^{00} it is repositioning and reallocating a position by confirming 1^1.

Notwithstanding all the names given to the different concepts a body in motion becomes more massive than a body standing still. Science may call it momentum or whatever, but the faster a body moves the more compact such a body becomes and the heavier the body gets. A body moving is duplicating its posture or structure by the speed with which it moves. It is becoming more of what it is when it moves and by moving the body gets more massive in relation to the speed which by it moves. That is what Kepler's formula $a^3 = T^2 k$ proves. When either the movement expressed as **k** or T^2 increases, so will a^3 become more. By movement the mass factor rises to new values.

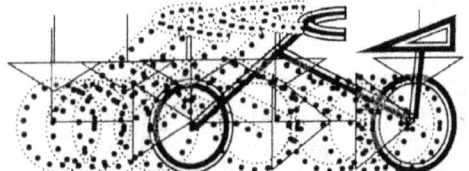

By not moving the bicycle is flat and by mass forms a part of the Earth through the mass it has. With only the Earth serving movement the Earth holds the **governing singularity** vested within the Earth's centre while the Sun holds the **controlling singularity.** The instant the bicycle starts to move independently it increases its share of space it holds within the Earth because by movement it releases from the singularity the Earth holds the bicycle with and the bicycle becomes independent serving its own singularity $k^0 = a3 \div (T^2 k)$. When the bicycle starts to move the atoms forming the bicycle form the governing singularity while the Earth then becomes the controlling singularity. This is very important because in this the Coanda effect comes to its right place as gravity. By duplicating the space allocated to the bicycle at a higher premium than the Earth does with the motion the Earth provides, the space the bicycle charge increases in ratio to that which the Earth charge because the bicycle maintain the duplication that the Earth grant but then still add to that space by enforcing more space provided by the addition motion that the bicycle adds. The motion of the bicycle not only extends the vertical connecting lines and not only changes the direction of the vertical connecting lines, but does both. The value added and the change in direction contributed is what brings about flying and moreover is the cause of the sound barrier. This I explain in detail later on.

From the allocated position we hold in terms of the Earth, the bicycle seems to be stationary when it is not moving with the aid of life. When we stand still in terms of the Earth' it seems to us that we are standing still and in terms of the Earth then we are not moving. However that is a human conception like mass is and is far adrift from a cosmic reality. We move as fast as the Earth spins. We move as fast as the Earth rotates around the Sun. We move as fast as the Sun rotates around the Milky Way. We move as fast as the Milky Way is rotating around another common centre because there shall be such a common centre that is allocated to order another common centre and this role diversification goes on running up to form eternity.

As soon as the motion of the bicycle enters the equation, the bicycle gets cosmic status because the bicycle generates time in the manner of parting singularity with time. The bicycle achieves cosmic independence and Universal recognition as an independent cosmic Universe unconditional to the fact that it is a cosmic alien entity providing the movement.

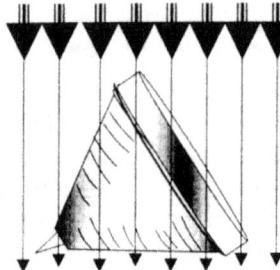

As the bicycle is standing still, it is still in motion but holds a position to the time the Earth provide where it fills a certain volume of space in that given time in motion where the Earth rotates. However, by gaining independent movement, the bicycle gains cosmic independence. It is Kepler's $a^3 = T^2 k$ where T^2 releases a^3 by the margin of **k**.

Let us now forget the fact that life is responsible for the motion of the bicycle and pretend it is all a cosmic affair. Let's remove the movement "life" adds to the equation and replace such movement on through cosmic principles introducing cosmic movement or time. Let's see what conditions apply when strictly only the cosmos plays a role and leave out the artificial contribution to movement that life brings on.

Since the bicycle moves faster than what it did when it was within the Earth motion, it now fills more space than what it did before the individual motion commenced. In having individual motion on top of the motion the Earth supply, the bicycle is filling more space than it did before when it was stationary.

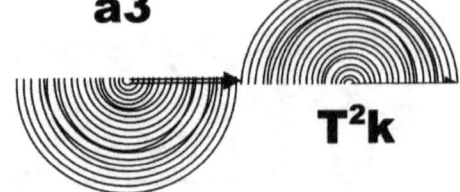

As Kepler stated, there are two positions to the Universe unit. There is always space confirmed by the motion thereof in relation to the motion thereof. One side of the cosmos would be in the fixing of the cosmos while the other half of the cosmos is about replacing what the cosmos removes in order to rearrange in the flow of time. That is what Kepler introduced when Kepler introduced the formula $a^3 = T^2k$.

The motion fills space the bicycle fills but also the bicycle fills space the Earth provides. It is filling more space within the Earth, (which is space of the Earth) than it did when only the Earth provided such space. The bicycle fills more space in providing own movement and therefore is bigger in its role as filling space that it had when the Earth had it filling space. When moving independent the bicycle fill more space than what it did before it started to move independently. When moving independent, it has more of the gravity going around than it had before when it was only dependant on the Earth to allow gravity going around. The bicycle grew bigger by the same margin that the Earth grew smaller in ratio to each other.

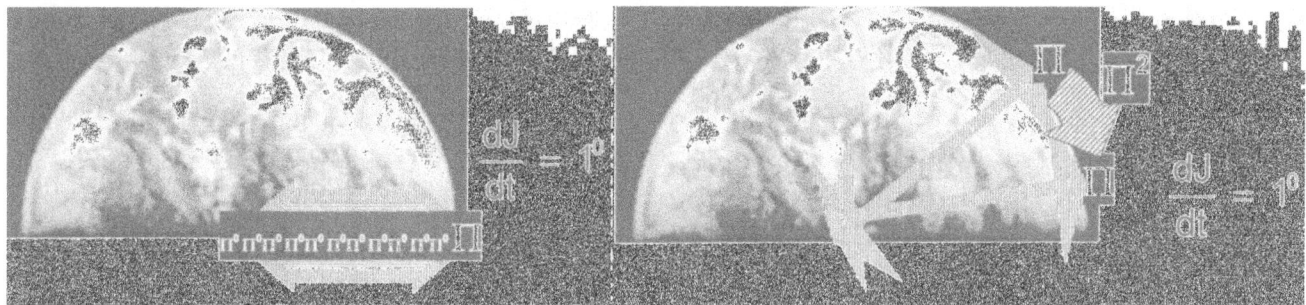

The bicycle or any object holding mass and is therefore without motion forms part of the space, which the within the liquid space but is on the surface which confirms it as forming a part of the space the Earth's material holds. The bicycle then forms part of the other side of the Universe that is part of the solid while being connected to the Earth. In that we find mass forming a value because when the object connects to the Earth, it extends the chain singularity forms from the centre to wherever the object rests. Then the atoms all link in singularity from infinity within the Earth to eternity where the object holding mass becomes the last link that links infinity to eternity. The link runs as a strain holding Π^0 Π^0 Π^0 Π^0 Π^0 and the last one linking becomes Π. Then the object is holding mass by connecting Π^0 to Π. However also by connecting Π^0 to Π, the object finds two principles in Π linking to singularity as Π^0 where the linking to Π extend into forming mass and forming the Π factor in the relation is why the object then finds mass. However, there is another part of having mass and that is because the object then forms part of the Earth by extending singularity Π^0 to Π, it then forms Π that connects to singularity in eternity and by connecting Π and relocating Π, that then becomes the principle we call gravity.

In the circle there are also two lines where each line holds one point to singularity. From these line crossing we find a point splitting the line in the two opposing directions, so in that way it is the same as half a square, but the third line indicating direction brings about a difference that distinguishes the circle from the square. The circle direction indicator is always Π placing the pointers at $r \times r = r^2$

By receiving the movement, singularity received a value outside eternity as Π^0 received edges. Concerning singularity that permits no space, the edges form at a point where space starts by introducing Π^0 as factor becoming Π. Granted the fact that where the edges are, it is so small there still is no r to present a circle since r is represented by singularity as Π^0.

Having edges where Π^0 duplicate to present the edges singularity lost the value of Π^0 to the value of Π^1 with the same value singularity had being Π^1 to the one side and Π^1 to the other side, the cosmos received the eternal value of the first dimension outside eternity. It was the square of Π^1 being Π^{1+1}. That was the first dimension outside singularity Π^0 where singularity has a value of Π^1 in the form of $\Pi^{1+1=2}$. The first claim to space had a value of Π^2. This applied to both sides of the claim to space outside singularity, and the double proton became the dominant factor on matter.

As singularity burst out into matter forming space as much as occupying space inside singularity, the protons started flying around, spinning around singularity, as each individual proton occupies matter in space

All spinning objects throughout the Universe are spheres. In the sphere (a circle surrounding a circle) we find the same six points away from singularity that is in the case of the sphere a centre point connecting what ever point to that specific point of singularity where the centre point has a value of Π and all the points hold a combining value in and out of singularity (meaning singularity commits both ways) a value of Π^2. All the connecting points confirm the point of singularity with the value of the infinite Π to the value of Π^2.

From this we can establish the fact that working with a sphere instead of a cube as well as working with a circle instead of a sphere will accomplish the same end result but have much less complications. Any line running from the point of singularity is indicating a value of Π as Π^0 confirms Π^1 as this claims space by rotation to the value of Π^2 on behalf of Π^0. This establishes the fact that any change over from a square to a circle will be changing the value from 3 to Π.

An object can occupy space-time at one point in the cosmos and only one point. At that specific point that object will claim space towards the individual singularity it holds by the value of 3 changing to Π. That brings about the Dimensional compensating point where matter change space-time from 3 to Π meaning it holds the square of space as a claim to singularity and not holding the space-time in a position of influence but under direct control of singularity by the value of Π^2. The sphere holds a value of $\Pi^2 \Pi$ to the cube being 3 away from singularity. In short, to form a cube requires three contiguous sides and to form a circle requires $\Pi\Pi^2$.

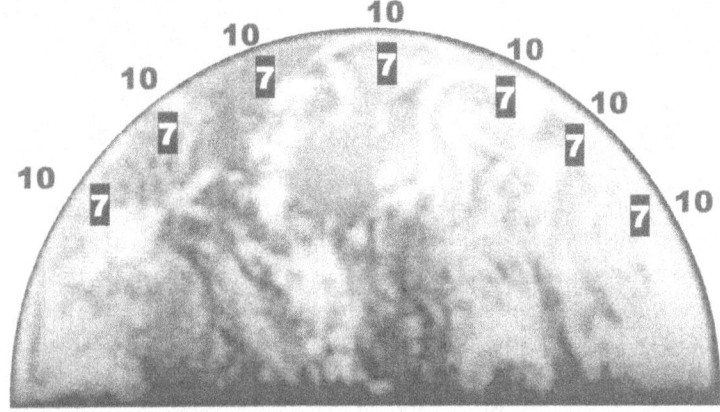

By applying singularity Π^0 in a position extending as it changes space-time from 3 to Π because of rotation, it puts the value of space-time in a square where Π connects to Π^2 as Π^0 extends to Π^1 and doubling in value to Π^2. Because any line is in fact acting as a double in avoiding zero it can never reach it commits in both direction flowing through singularity at Π^0 and past the space claiming point of Π^1 forming Π^2 it has to have a double Π^2 value and therefore the proton claims space outside singularity at a double square of Π^2 forming $\Pi^2 + \Pi^2$. Then the Earth housing the atoms forms a double square gravity value of Π, which is also a proton value of $\Pi^2 + \Pi^2$.

As I am about to show mathematically how 7 relating to 10 by the same action where 10 relates to 7, this relation forms Π^2 this happens in a double spin. It is an atom that forms the star. The atom spins around its axis. That is one Π^2. Also the atom spins around the star's livers another Π^2.

The lines converging from singularity holds a square to one another and that implicates the oldest mathematical principle that I know of, the law of Pythagoras.

Again we find the presence of a triangle holding a square. This holds space away from matter and therefore we are calculating the square of space depicting singularity and time (always in a square) away from the immediate claim on space by matter.

At the end of the space relevancy 3 where matter occupies space (21,9 / 7) is a border Π. That border is the exact point where space reforms to a square of time placing all matter (occupied heat) and heat (unoccupied matter) to a value of the square of time. That specific point is in relation to the square of the diminishing shield around the Earth. However it takes matter (Π) from the 3 dimensional positions to the square (Π^2) in relevancy to time in singularity. With time holding space in singularity the 4 sides of Π truly relates to form of the total square value of Π.

The "gravity" factor of Π^2 becomes one and only holds the square to the Π position as it holds space to singularity at a square (time dissolving space at a square) and the time value (Π) remains dimensionless in singularity at (1). It then is $(1)^2$ where the one becomes the space position ($\Pi/2$) representing time (1) at that point. This makes the position that time normally has Π^2 but directly links to the controlling singularity which we then give a value as Π^3 which then relates to the singularity position of space diminished from the three dimensional to times single dimension in the square. That makes the Roche limit hold the position of $(\Pi/2)^2$ when the neutron position of time (Π^2) links directly to singularity (1). This may only represent figures, something to accept through intellect but lying far outside the reality surrounding our everyday understanding.

Matter in relation (part of) with the total dimension of space.

$$\left(\frac{10}{7} \div \frac{7}{10}\right) = 2.04$$

$$\frac{1.4285}{0.7} = 2.04 \quad \text{Taking from both orbiting influences}$$

SPACE DIVIDED INTO TIME

$$\left(\frac{7}{10}\right) \div \left(\frac{10}{7}\right) = 0.49$$

$$\frac{0.7}{1.4285} = 0.49 \quad \text{Taking from both orbiting influences}$$

SPACE MULTIPLIED WITH TIME

$$\frac{7}{10} \div \frac{7}{10} = 1 \quad \text{and} \quad \frac{10}{7} \times \frac{7}{10} = 1 \quad \text{Therefore not influencing change}$$

THE PROCESS PARTED USING THE ROCHE PRINCIPLE

$$\frac{\frac{10}{7}}{\frac{7}{10}} \qquad \left(\frac{\Pi}{2}\right)^2 \text{ The Roche influence on Titius Bode}$$

$$2.04 \times \left(\frac{\Pi}{2}\right)^2 = 5.033$$

$$\left(\frac{\Pi}{2}\right)^2 \quad \frac{\frac{10}{7}}{7} \qquad 2.04 \times \left(\frac{\Pi}{2}\right)^2 = 5.033$$

$$5.033 + 5.033 = 10.066 \quad \text{from both objects}$$

SPACE DIVIDED INTO TIME

$$\frac{7}{10} \qquad \left(\frac{7}{10}\right) \div \left(\frac{10}{7}\right) = 0.49$$

$$\frac{\frac{10}{7}}{}$$

$$\left(\frac{10}{7} \div \frac{7}{10}\right) = .49 \quad \left(\frac{10}{7} \div \frac{7}{10}\right) = .49$$

$$.49 + .49 = .98$$

$$.98 \times 10.066 = 0.8 = \Pi^2$$

TIME SPACE $= \Pi^2 = 9.8696$

TIME SPACE $= \Pi^2 = 9.8696 =$ Space and time in a dimensional implication

I have shown where the proton value fits forming $\Pi^2 + \Pi^2$. I have also shown the movement of the neutron value of $\Pi^2\Pi$. Science confirms the value of the electron as 3. When taking these values into one unit to show the relevancy applying to the atom we find the atom forming a displacement value or a contracting value of heat as $(\Pi^2 + \Pi^2)(\Pi^2\Pi)(3) = 1836$ which is the "mass" difference there is between the "mass" of the electron and the "mass" of the proton. With the neutron forming liquid or movement, call it what you like, the neutron has no "mass" indicator. This is what relevancy an atom applies in outer space or when the atom is within a small gravity area such as what the Earth produces.

Let's return to the bicycle and see how all of these factors explain the cosmic relevancies applying.

As the Earth rotates around the Earth's axis, the Earth confines space that is in outer space to become layers. As the layers come closer to the Earth surface, the layers become denser in progression as the layers come closer to the surface of the Earth. This condensing of space depends entirely on the rotation of the Earth that is condensing or compressing the space in layers of air above and inside the Earth. This is what produces gravity. When an object has more density than what the air around it has, the material will not move into the Earth as a liquid but will be restricted in movement by the Earth surface. This happens when solid meets solid and the meeting of the solid restrains the movement of the body that was moving down with the space being compressed. It is said, "it is the gravity that drags you down". This "gravity pulling you down" is "what is giving you mass" and it is not the "mass that is giving you gravity". The entire action is dependent on movement because time depends on movement and gravity is time. When a body is in mass it is motionless but for the motion the Earth produces and therefore the body only moving by virtue of the Earth and having mass while doing so is classifying such body to the category of the solids. If the bicycle is lying motionless on the ground the bicycle is having mass because it is classified by the Earth as being part of what forms the solids. What moves form the liquid part that can move and what is stationary is solid. When anything repositions its location in terms of the centre singularity, that moving object becomes a liquid being in relation to the solid. With the bicycle being motionless, the bicycle is a solid.

The moment the bicycle moves, the bicycle switches sides and switches allegiances. The bicycle then becomes part of the liquid that the space confirms not as space but as the liquid that extends the space. The bicycle then is no longer part of $\Pi^0\Pi$ but then positions forming part of the liquids$\Pi\Pi^2$. It then forms the gravity adding space instead of the gravity confirming space. It becomes $k^{-1} = T^2 / a^3$ instead of $k = a^3 / T^2$. This is a part of the Coanda effect and I will return to this entire explanation later on in the book. What applied before does not apply any longer because the bicycle is then on the other side of the Universe. Yet, it is much more complicated than that because when the bicycle was part of the Earth extending solid space which is gravity in relation to contracting being $k = a^3 / T^2$. As soon as it moved it became $T^2 = a^3 / k$ which is relative to the old position $T^{-2} = k / a^3$. **This too has no principles that it can share in the mass applying gravity idea. It is about being stationary in relation to moving.**

Taken from the Earth's perspective, there are lines (**k**) running at 90^0 in line with the rotation of the Earth T^2. Then the Earth draws lines in the direction in which is spins and these lines all have a singularity value of 1. It is these lines that the top and the bicycle manipulate in order to "balance" or move up straight. The Earth gives the bicycle one line in singularity which to confirm its position as far as the space the Earth grant the bicycle to manage. That gives the bicycle a specific space to hold in relation to what the Earth has and in relation to what the Earth offers the bicycle. That is gravity. That however have no principles it shares with mass pulling mass. It is being on the one side of the Universe which space a^3 holds motion T^2 in relevance k or as Kepler said $a^3 = T^2k$. In moving, the alliances switches relevance from $k = a^3 / T^2$ to $k^{-1} = T^2 / a^3$ where the bicycle forms an legions with motion by duplicating the space it holds in relation to that which the Earth grants $k = a^3 / T^2$ and therefore becomes part of motion where motion forms part of time $T^2 = a^3 / k$.

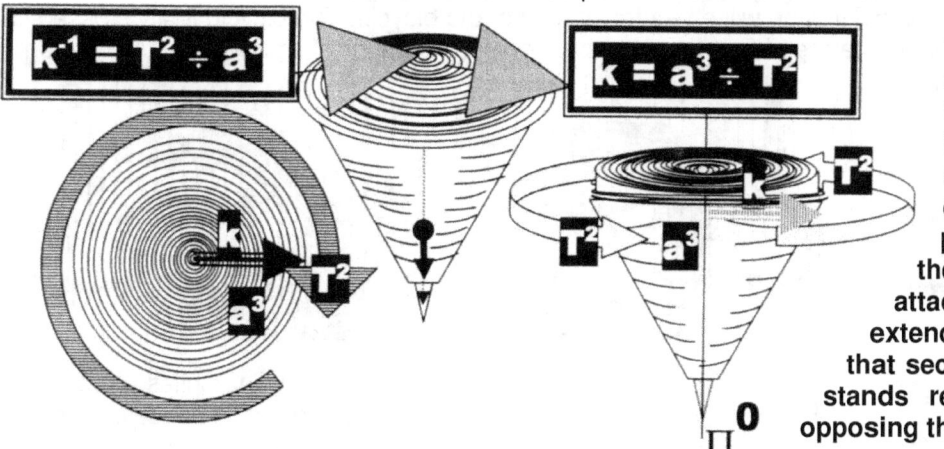

There are always two sides of the same Universe forming one Universe. There is the space extending to confirm the liquid by producing a solid and there is the liquid attaching to the space to extend the space by motion that secures the space. The one stands related to the other by opposing the other.

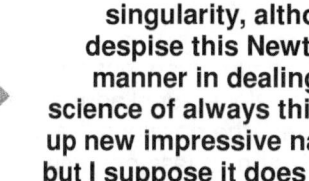

I have decided I have to name the two factors forming singularity, although I despise this Newtonian manner in dealing with science of always thinking up new impressive names, but I suppose it does make explaining a little less complicated. **Governing singularity** which I prefer to refer to as 1^0, I would call the **governing singularity formed by T^2** and the circle forming the cyclic spin or directional movement k that singularity I would call the **controlling singularity** again to which I like to refer too as 1^1.

When the bicycle is part of the space that the liquid or the motion confirms, it has to move positions and relocating it in terms of the governing singularity by implicating it as space in the motion part, which forms the extending of space. In that case the motion there are two forms applying singularity, which is k as well as T^2 where both produce the contracting lines in singularity that runs a reducing and reclining formation into the centre but the two are crossing by 90°. When the bicycle is apart of the space, the lines attribute to its space reposition and the space the bicycle then holds is reallocated by the movement the bicycle then holds.

There are always two opposing time lines forming one united space. The one is the line k and the other is the half circle T^2 where from those perspectives there then form the triangle a^3. This is a part of singularity, which forms by the value of a straight line, a half circle and a triangle all having an equal value in being 180°. By moving, the bicycle then forms the 90^0 cross-references to the allocated position it had before. It is $k = a^3 / T^2$ or it is $T^2 = a^3 / k$.

The moving of the bicycle involves duplicating the space the material forming the bicycle holds and from this results the changing in the position the bicycle has in relation to the space the Earth allocates in terms of movement and the position the Earth allocates to the bicycle. The faster the bicycle goes is actually the number of times such repositioning of the space the bicycle holds are in response to what the Earth allocates and what the Earth takes in a specific period of motion repositioning the entire Earth in terms of every aspect forming the entire Universe. When faced with the question of how the bicycle manages to stay upright it always comes down to charging the achievement to a balance…but a balance of what? What goes into balance to achieve the upright position? The bicycle repeats its position in relation to the position the Earth grants and when the repositioning of the bicycle is faster than the re-allocating of the Earth, the duplicating of the bicycle from one position to the next will sustain that the bicycle can cross the vertical lines faster than the vertical motion will effect the stance of the bicycle. The bicycle firstly crossed its allegiance by no longer forming a partnership with space but becomes a factor of liquid presenting motion.

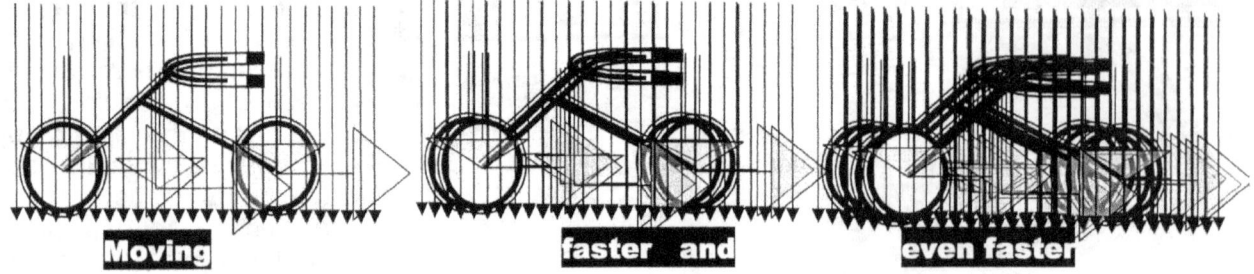

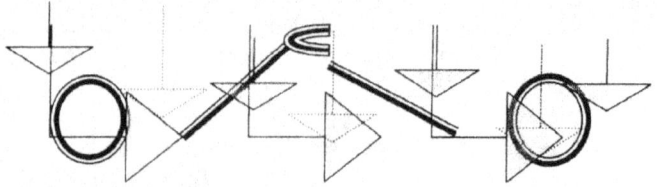

With the motion the bicycle applies, the bicycle is duplicating its position or its space it claims in ratio to more space than what it had while only being part of the Earth and moving only with the Earth. The motion holds more space because it holds less space per time unit and there are more units of space per time unit used to secure every position it takes. Since the bicycle is propelled at a faster pace than what it was when the Earth alone supplied the space and forced the time by establishing the duplication tempo in the time contracted, the positions the bicycle claims has to become more per time unit in terms of the Earth moving while also being less per time frame. By supplying more motion the bicycle grew larger in ratio to the space, which it had when the Earth was the sole space-time or movement provider. Then after accelerating the bicycle then has more space in relation to what the true status is in terms of the Earth moving. The bicycle holds a larger part of the Earth by which the Earth has to reduce the space it offers the bicycle. The Earth had to shrink in order to provide the bicycle with more space or the bicycle had to grow in terms of the size of the Earth.

In the normal relation that the bicycle has with the Earth when the bicycle is motionless and having a "mass attack" (or in the manner we think about the status of the bicycle in terms of only its position within the scope that the Earth provides) being motionless and then considering the position changing when the bicycle started moving, we find that the bicycles space increased rapidly as the Earth space decreased rapidly. With the new motion the bicycle finds much more duplicated space and that disturbs the ratio of space shared by the bicycle within the confinement of the Earth. When the bicycle is moving the Earth then presumes the bicycle to be much bigger than what it was when the bicycle had no motion other than the Earth moving. By moving, the bicycle physically got larger and this is a fact not only by relevance but also by actual annexing and capturing of space in any given period of equal ness. The Earth provides a certain value but as the bicycle moves faster the bicycle annexes more of the Earth space and that improves the size of the bicycle in a volumetric and physical measurement. It is a^3 that grows bigger therefore the Earth a^3 has to compensate by reducing that much actual space. In this changing of relevance comes a problem.

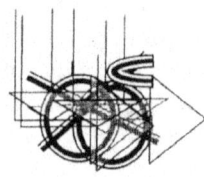

The bicycle does not even contribute a morsel of space when compared to what the Earth delivers. The Earth that actually became smaller resents it becoming smaller by the demand of such an outrageous exploiter such as what the bicycle became. To the Earth, the bicycle is motion and the motion is liquid and therefore taking space contravenes the being solid part. Being liquid is also being heat but being heat means becoming hotter when becoming more. To the Earth the moving bicycle is liquid motion and being more makes it being hotter and being hotter is therefore is more volatile. Therefore the Earth refuse to become smaller and the bicycle being space claimed cannot become hotter without destroying its independent molecule unit.

The Earth takes the position that was before the motion of the independent object came about previously and held by the sun. By establishing the directional motion in accordance with the k^0, which the Earth then provides instead of as previously provided by the Sun, relevancies replace previous ones. In this the bicycle has to lift from the surface of the Earth and become part of the liquid in the air. The bicycle will become airborne and start to "fly" like we see racing cars and aeroplanes do or tend to do.

In normally applying gravity, we find contracting lines running vertically as the lines connect with the Earth centre, which is the gravity we confuse with mass. It is a state of contraction and is the result of space being confirmed in relation to the motion of liquid time. Motion provides extending of the 7^0 establishing the centre connecting points to the Earth to which it connects.

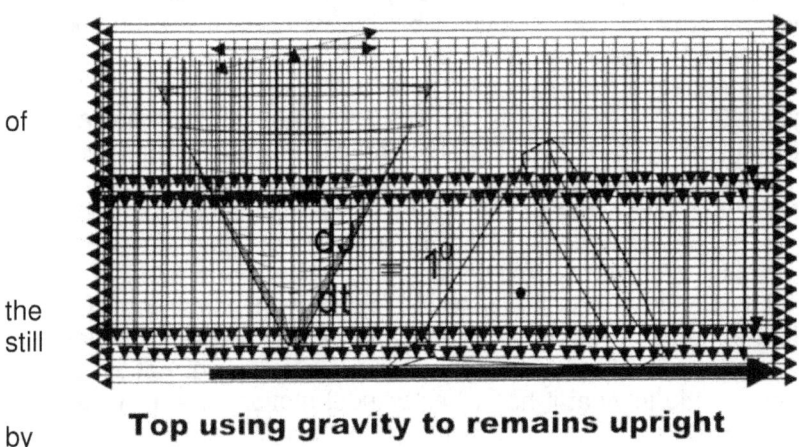

Top using gravity to remains upright

of

the
still

by

The lines to which I am referring is the lines that the top uses to spin and which the bicycle uses to and because gravity it uses a double 7° inclination that makes the lines in singularity cross by 90°. Those lines crossing each other that forms a singularity grail comes about forming what the result is of the Coanda effect and all virtues not understood by science butt attributed to the Coanda effect. By spinning the lines are commanded into forming singularity. This comes about putting the relevancy factor **k** represents in place.

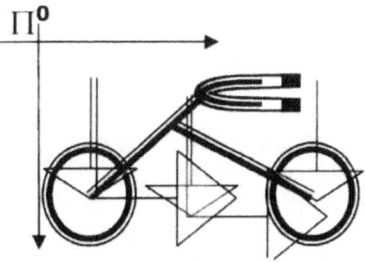

The motion of the peddling provides the bicycle forward thrust that connects with the downward thrust that is already confined to share space with the Earth and that is due to the Earth being in spinning motion. The bicycle can only have independent motion if and when the bicycle has the correct number of atoms filling the space by movement in sequence with time and connecting the downward movement with the peddling movement of the bicycle holds the connotation of singularity in infinity and singularity in eternity and it is the establishing of that relevance that will grant the bicycle independent motion. If it were a cosmic body without life supplying energy, the bicycle would require the concentration of a lot of heat within the centre to initiate the first movement. With a star starting to move the relevance between infinity and eternity must differentiate to such an extent that movement will result from such a large discrepancy. That places the **k** relevancy factor at a premium in order to charge the spin T^2 as to get the space a^3 released from capture and that is how stars are born. Without the unit being able to concentrate the correct amount of heat differentiation on the inside compared to on the outside that will enable the space to generate motion. Having life supply the movement to the bicycle does not count in cosmic terms because such motion under cosmos standards does not exist. In fact under cosmos standards the entirety of the bicycle unit does not exist. Only when the unit that forms the bicycle structure and holds the number of atoms, which will produce the amount of motion through which the required heat will be subtracted from time in order to generate the gravity or motion needed, will the bicycle under cosmic standards gain motion.

$$T^2 = \frac{a^3}{k} \qquad F \neq G\frac{M_1 M_2}{r^2} \qquad \frac{dJ}{dt} \neq 0$$

$$k = \frac{a^3}{T^2}$$

$$k^0 = \frac{a^3}{kT^2}$$

$$\frac{dJ}{dt} = 1^0$$

Earth surface and direction of rotation

Before that the cosmos does regard the bicycle (or the top or anything on Earths that holds "mass") as the Earth and only the atoms standing independent from the Earth holding independent movement under those conditions will the cosmos seem to regard the bicycle as not being part of the Earth. Ignoring the manipulation of life because as a factor life does not exist in the cosmos or exist to the cosmos rules of movement will the bicycle confirm movement by obtaining the required displacement to launch an independent duplication of all its atoms from one location to the next location in the required time synchronization and only then can the cosmos regard the bicycle in such a manner as forming independence through movement. In the real cosmos not influenced by life can the atoms forming the bicycle unit be able to condense from time enough heat to the centre as to give drive to the fledgling star in order to provoke independent motion. By appreciating the required differentiation (k) between infinity (k^0) and such differentiation release the fledgling star as a cosmic unity (a^3) and allow motion (T^2) to sustain the star's independence. Only if all of those factors being correctly in place do the cosmos put the emphasis on the unit as forming a star. But in order to be the carrier of such independence the bicycle will have to grow. On its own as it is there is no way in hell that the bicycle can get fired up and start going as a star does by cosmos style. If not for life it will go nowhere but be consumed by the Earth to become a part of the Earth in time. In the cosmos the atoms forming the unit provide the motion and only when the total effort of the combined unit atoms manage to move abruptly and with the required confidence can the gravity be generated where the gravity generated is the motion of the entire unit. That means by having motion the fact of having motion grants the bicycle duly respect from the Earth. The bicycle having motion of any status enlarges the bicycle in atomic space in ratio to what it has when without motion in relation to what the Earth has when only the Earth provides motion and when the Earth stands in size in relation to a static bicycle.

Science never separates the two thoughts such as what movement is a product of life from the idea of what is the result of the cosmos bringing about movement. When the bicycle is motionless, the bicycle is part of the Earth by gravity applied resulting in mass. The bicycle is cosmic. The Earth rotates while the bicycle gets a free ride. As soon as life steps in and brings about separate and artificial motion the bicycle moves as an individual cosmic entity would. Although moving along with the aid of life, meaning it moves artificially the bicycle still use the support of the motion that the Earth provides. With life aiding in the bicycle having more movement then as far as movement goes, it will inevitably do better than when it was just another part of the Earth having mass on the Earth as long as the motion that life provides is not in conflict with the motion the Earth provides. The bicycle becomes an object with the ability to transform the direction of the Earth's domineering motion by redirecting gravity and in that find the ability to change the direction of gravity. **When gauging what happens we must also admit that it is highly unlikely to find running rocks on Venus or moving craters on Mars. The motion that applies to the bicycle is an extension of the second force in the Universe, which is not part of the Universe and only affects a very small part of the Universe within the Universe, which is a tiny sphere we call life. Life giving motion is an alien product and gives an unrealistic adding to the Universe. The motion however has nothing to do with mass but is only extending what was to what will be through what is. It is refurnishing what will be with what was before it now is in the present.**

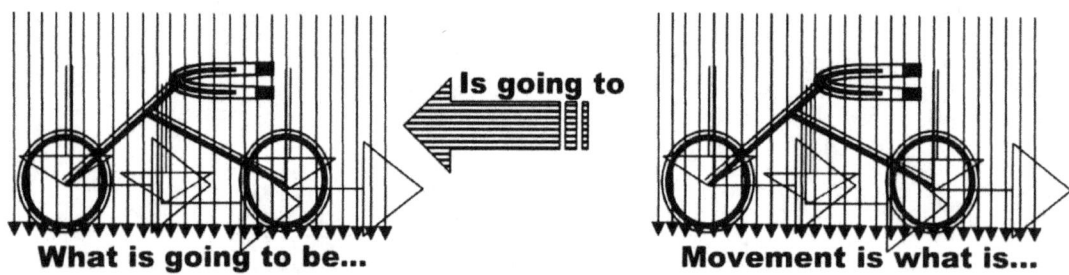

What is going to be...   Movement is what is...

When the motion of the bicycle accelerates, such points that are forming by the increase in motion then form connection points that extend to match to motion, putting a standard of duplication per time unit in terms of what now is extra movement to what previously was the norm. More space fills in the same period or the period reduces to match the filled space. The motion then contributes by increasing the space factor to keep the commitment with gravity valid. The bicycle breaks its form but because it is structurally bonded, it keeps its form in the cosmic fluid that holds no specific form.

Then finally I wish to kill the last myth about the "sound barrier".

The Four Cosmic Pillars; The Result Thereof. **APPLYING AS THE SOUND BARRIER**

Never is any object going to travel faster than sound.

The Concord did not use to arrive in America at a specific time and the sound that followed only caught up with the moving aircraft at a later arrival time and due to the sound travelling slower the slower travelling sound was able to land some hour and twenty one minutes later on another runway because the Concord beat the sound by some hour and something. The Concord did not once travel faster than sound because according to cosmic law such a task is not possible. The Concord formed its own sphere within the Earth's sphere and by doing that rejected or displaced the space in the Earth in which sound travels.

The Cosmic code applying to sound is as follows: **10 (Π)(Π^2)(4) =1240.25**

10 This is the space in which sound moves and is already compacted by gravity therefore is the result of gravity and not the process of gravity forming. The cosmos functions by gravity which functions by Π that form Π^2 by the movement resulting from rotation. The value of Π is either 3(7) + 0.991 which is reminisce of $3\Pi^2$ which I have already explained, or is produced as (10) + (10) + 1 holding singularity + 0.991 formed as singularity is growing in time moving on to become space. Since sound is part of the atmosphere and not something moving within the atmosphere, sound will hold 10 at value instead of 7.

(Π) This forms part of gravity forming in relation to solid where solid relates to the governing singularity placed in the centre of all things spinning, which forms infinity and holds a value of Π^0.

(Π^2) This forms part of gravity forming in relation to solid where the liquid forms movement relating to the governing singularity placed in the centre of all things spinning which forms infinity and holds a value of Π and with Π^0 because this always forms a direct linking line with Π and it then places therefore all movement onto space making space the liquid part that always changes and holds eternity to form Π^2.

(4) It is the mark of time or said more direct, it shows that the maximum of four lines that are used to convey sound.

=1240.25 km / h

One must take into account that the only constant there is in the entire Universe, is the reality that there is no such a thing as a constant anywhere in the Universe.

Nowhere on Earth is gravity at a precisely equal value because gravity is heat cooling or space heating and heat tends to fluctuate at any point by being colder or hotter than it was or what it will be. Gravity is also density and that too, fluctuates as circumstances change all the time.

When an object holds mass or in other words form part of the Earth by only moving with the spin of the Earth that follows in a circle around the Sun, there forms line arranged by the relevancy the Earth put in place. The position of the Earth moves in a fashion as to relocate the Earth (with whatever holds mass and forms part of the Earth) from one such a line forming singularity to another line forming singularity. This is gravity and that is movement according to cosmic discipline.

In calculating the Cosmic Code in reference to the "sound barrier" the object moving is doing so while being within the atmosphere and being attached to gravity by the spin of the Earth. As the object no longer connects with the governing singularity, the body in motion becomes the governing singularity holding a centre singularity in which all the atoms forming the unit unite to form such a governing singularity. The Earth becomes the controlling singularity placing the Sun as the directional singularity forming the direction of movement, just as the Milky Way forms the directional movement in the case of the Earth forming gravity. This gives the Earth influencing the movement a value of seven that replaces the link of Π.

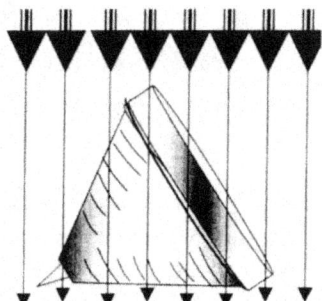

The Four Cosmic Pillars; The Result Thereof.

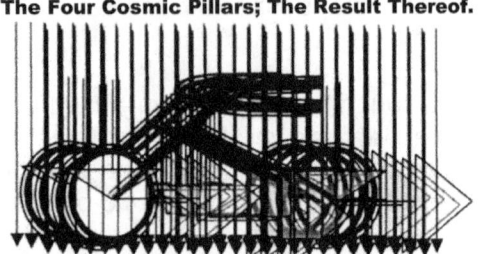

What is the effect of movement

When life brings to the Earth more movement than what the Earth produce, the duplication of the moving object ends to follow more lines at a certain rate than is there when the Earth move by itself. In other words there is more duplication of the body moving by movement that what the Earth normally will produce. The lines get used by more relation of the body moving and the duplication produces more mass because there is more movement in which the moving body is structurally duplicated. In this is the "<u>sound barrier</u>" principle vested which I will explain.

When any object is not moving the object form a part of the object, which holds the first object, captured. The cosmos disregard the existing of the first object in the event of it being stationary. However, as soon as motion applies to any stationary object, the cosmos grant the object having motion a position of existing by recognising it as an independent Universe that is entitled to all the privileges granted to a Universe.

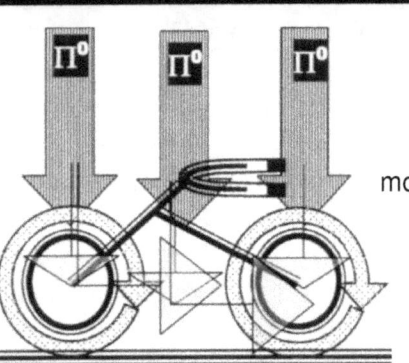

I have previously explained about the 7° rotational change in directional movement due to the curvature of the Earth. I also explained about singularity Π^0 extending to Π where the Earth curves and movement takes place. I have explained that the movement as such in cosmic terms are Π^2. The fastest speed at which a body can fall is $7\Pi\Pi^2$ which is 217.04 km / h. The speed derives from that relevant value connection that science made in terms of space forming weight in relation to time the Earth holds. Any object moving below $7\Pi\Pi^2$ or at that speed will remain on the Earth and will follow the curvature of the Earth. These lines are placed in ratio to the spin and as long as a moving object do not exceed $7\Pi\Pi^2$ = 217.km / h, no one would take notice. However, in recent times even motorcars are able to exceed the limit. By travelling at 217.04 km / h there is another` spacing which isΠ^0. Should the moving object move faster than $7\Pi\Pi^2 \times \Pi^0$, say to $1.2\Pi^0$ or $1.5\Pi^0$, the ability of the car follow the curvature of the spinning Earth becomes to problematic.

Earth Rotating

The moving object will become airborne and then follow the curve of the Earth. In doing that, the object will abort the Π as a value and replace the Π with the value of 3. In this the 3 repents total movement of time in having one: having Π^2 as the Earth rotates around its own axis, two: having Π^2 for the second time as the Earth rotates around the Sun and thirdly: a third Π^2 represents the movement that singularity governing (the centre of the Earth Π^0 never moves in relation to Π formed at the end of the Earth's curvature, then finds movement to the value of Π^2 by finding that everything in outer space is moving in relation to a permanently fixed Π because Π^0 permanently connects to Π. That brings the total applying values of Π^2 to three, hence $7(3\Pi^2)$.

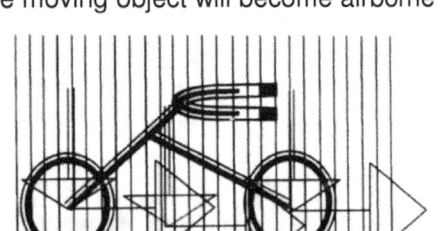

A ratio comes about between singularity in the centre and singularity formed as space in the time zone. The singularity within the centre Π^0 forms an alliance with the curvature of the Earth Π from which science derives the use of the value of mass. Also space in time which science call the atmosphere and which space calls outer space and has a value of Π^0 holds a link with Π by the means of Π^2 activating such a ratio.

In this ratio we have Kepler's formula indicating the balance there forms between movement T^2 and space in relevance a^3 / k, which puts the above mentioned ratio in mathematical equation. To keep the space in contact with the space the Earth holds in relation to the movement of the Earth requires a velocity in relevance of $7\Pi\Pi^2 \Pi^0$. After exceeding that sped the travelling object will start to leave the Earth or be retained to the Earth by forming wings that would cause a downdraft and push the car onto the Earth by excelling the Π factor. If not, then the ratio will become $73\Pi^2 \Pi^0 = 207.26$ km / h. The three comes about from time forming. Time can never stand still notwithstanding any Newtonian concept of time being able to stand still. When time stands still the dimensional implication would be that the Universe would fall into a single dimension as is the case in a Black Hole. Time takes one dimension from the past (1) into the present (1) and onto the future (1) and in doing that time moves in a space of three. Time is always as much part of the past as forming the present that moves into the future and in terns of that time forms a relevancy ratio of 3 instead of Π, which is what time in infinity forms by having material move. But explaining that is a very long story that is better left for another book.

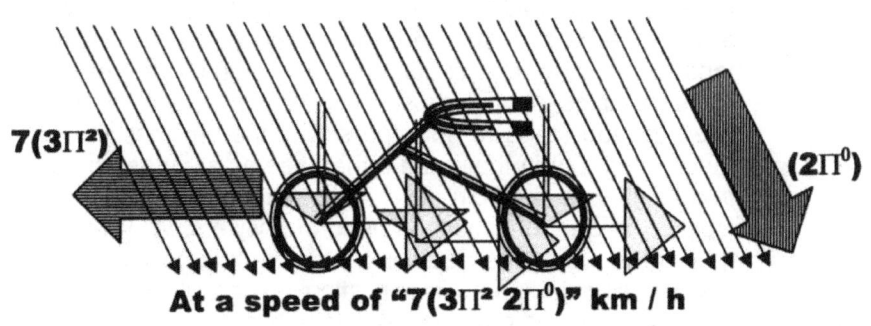

At a speed of "$7(3\Pi^2 2\Pi^0)$" km / h

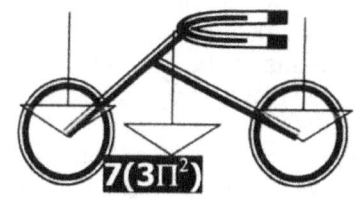

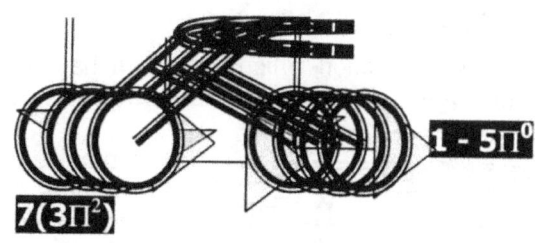

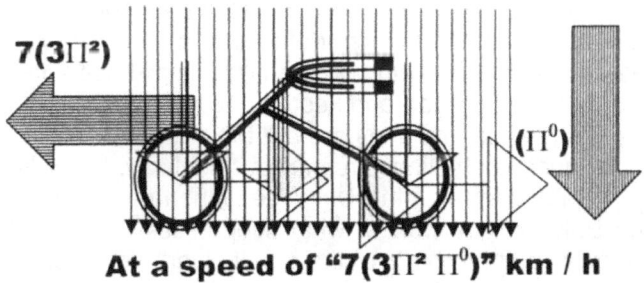

At a speed of "$7(3\Pi^2 \Pi^0)$" km / h

The movement will tolerate $73\Pi^2 \Pi^0 = 207.26$ km / h because this is as fast as the linear factor of falling objects move. As the movement is below or up to $7(3\Pi^2\Pi^0)$ the vertical lines forming Π^0 are used to duplicate the body of the moving object in terms of the movement the Earth sustains. As it exceeds the mark of $7(3\Pi^2\Pi^0)$ the movement can't follow the curvature of the Earth as the duplication then progresses in a more rapid flow than the Earth allows the placement of the vertical Π^0 singularity. At say $2\Pi^0$ the structure of the moving body takes up one line extra in relation to what the Earth permits and this would put the moving object in a space a^3 applying a different relevancy k because of the more rapid movement T^2. That will leave the moving object in more space than would the object normally use and by using more space it would have a bigger relevancy to cosmic liquid that it would have under conditions applying when it was travelling at a speed the Earth normally sustain. These lines do not hold space but merely represent space, which then could be hold by objects with space. In another book I show that what we think of as outer space is the history of time, which is space-time but that is not space. The movement places the moving object in more lines holding singularity, which places a bigger relevancy on the movement of the Earth.

That is the same reason why a top will try to lift when moving very quickly and why the rotor blade of a helicopter would allow the helicopter to lift from the ground. It is the rotational speed that overcomes the downward motion of outer space r the atmosphere or cosmic liquid, name it what you will, and when the rotation of the wing is higher than the gravitational cosmic thrust, the helicopter overcomes the burden of relevant movement that trusts the helicopter onto the ground. As the speed advances above and beyond what the Earth would permit, the Earth puts up more restrictions by providing fewer lines so that the heat levels in the craft will rise. The tendency of the Earth restricting the top by gravity and the top rebelling against confinement by the Earth gravity through spin is an everyday occurrence. Such behaviour by the spinning can be exhibited any time a top is vigorously spun. Every schoolboy has been a witness to this behaviour by a spinning top.

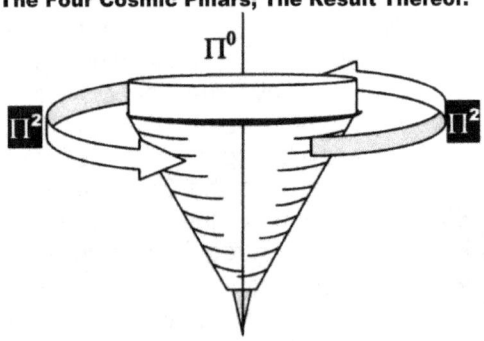

When the top starts by spinning to fast it is clear that the top is in a fight with something that is restricting its spin. This we can see by the re-aligning and the swaying of the top as it manoeuvres to try and circumvent or overcome the restriction. As the top spins, something starts to tarnish and erode the spin. When the top starts to spin too slowly the top tries the same manoeuvres, but in that case it then seems as if the top is in a struggle to keep the spin alive. These manoeuvres that the top displays, triggers questions in need of answers. Why would the top stand upright when spinning? It can only be that the spin activated singularity into manifesting the gravity of motion by evoking singularity within as to grant the top cosmic space and proclaim cosmic relevance.

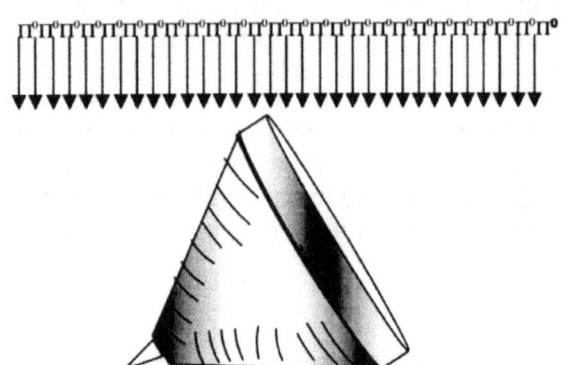

There is literally on Universe in difference between the top lying not spinning and the top standing erect by motion of spinning. The top in spin by individual motion is ostentatiously independent, spinning in a Universe independent of the rest and with so much vigour on occasion it finds the ability to match a sound produced by spin that is a product of the vigorous motion of independent spin. The spin, it would seem, finds a "life" by motion. While without spin the top is "dead" and part of another Universe that entombs it in the other Universe's gravity by restricting any and all forms of motion, except the motion which the Earth has and when the spin releases the top from the Earth's confinement it grants the top independent worth through that motion the top displays by doing so.

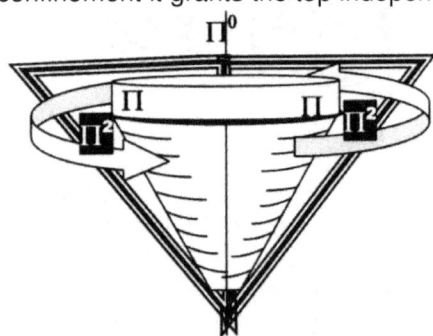

The other dominating object being the Earth has contraction lines so thin it is not in the Universe we have, still the lines run inward towards and through the body of the top. We can prove this because we can see that when the top comes into individual motion, the top sanctions one of those lines in the very centre of the top's spin in order to allow the top to establish an independent Universe where the top then finds control to dominate all the lines running to the centre of the Earth by commissioning those line to help stabilise the body of the top and keep the top erect. Even then when the top uses the line, the lines are still too small to be part of the Universe we have. These lines are singularity just as much as the lines are part of singularity. Every line to which I refer as forming singularity forms the body of the centre of the Universe. When the top is not spinning, the lines running towards the centre of the Universe are part of what suppresses the top into forming part of the Earth, but when spin brings the relevancy of the body of the top, it clearly releases the top from the Earth's restraining and having independence making the Earth then an alien Universe. But motion allows the top to command the lines and the top then uses the lines to stay erect.

The lines run through the top at 180^0, which forms 90^0 with the body of the top lying down. When lying down and having mass, the top seems "dead" as it surrendered its long-term position of independence and would eventually totally succumb to the Earth's gravity and in the end become part of the Earth's space-time forming art of the Earth's body by relinquishing even the structural independence it now seem to have. The top will become buried within the Earth. Finding motion brings a "life" into the singularity of the top, which gives the top a reason to fight the Earth by fighting for independence. Receiving the ability of motion invigorated the top into independence of a cosmic nature by the motion it received from the motion of all the independent atoms forming the independent unit. The points forming the sides of the sphere is structurally dominating the cube as the top then becomes a spinning sphere using the effort of all the atoms forming the top in a combined unit. The atoms combine as the top to still maintain a coherent structure with independent atoms becomes the motion that releases the top from the capture of the Earth.

Other aspects concerning gravity have to commit to the breaking of space. This might sound wrong but that is what happens when the "sound barrier" breaks. When expressed extremely crudely it is linguistically put as follows but is very bluntly stated. Yet, it still is the best way to explain the basics of the sound barrier. The bicycle is the compiling occupier of the independent space within the Earth's atmosphere forming concentrated space or time concentrated to be more exact that is holding the motion in duplication where the motion is continuing from a facet going to the next facet by duplication of what was through what is to what is going to be. While the bicycle is filling the space, in motion that is part of the space holding all aspects within by the atmosphere and the atmosphere is holding all that is in it together in the atmosphere of the Earth. That is time performing in ratio to space filled by material. Because the conflict the gravity experiences by having motion within motion, gravity first tries to break the movement of the object that is in independent motion. As the motion continuous, as motion extends the struggle of both leads to the situation where the Earth try then to contain the object within its space. As the object continue to move faster than the Earth can permit $7(3\Pi^2\,5\Pi^0)$ the lines of duplication breaks off. This it does by the breaking of linking lines, which is what forms the connecting devices that forms the sound waves in the adjoining space called the atmosphere. The atmosphere does the breaking of the relevant **k** on behalf of the object in motion since the moving space holding the object in motion as a unit shows much stronger bonding in structure unifying. We experience such breaking of space as the breaking of sound, which is showing motion or gravity differentiation.

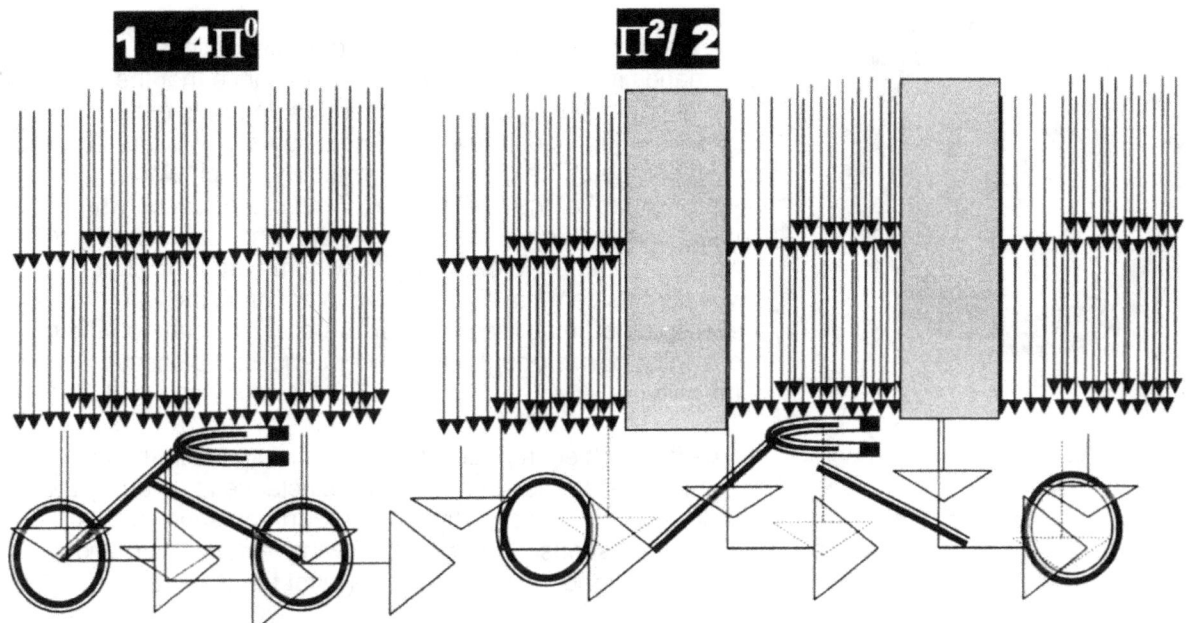

The lines simply break all links with the Earth and the Roche limit forms "gaps" to appear ($\Pi^2 \div 2$).
If we fit a wing to the moving body in order to make contact with the air, we simply enlarge the relevancy forming space to the ratio of the body. In short, we make the body seem bigger as we increase this ratio of $\Pi\Pi^2$ and by increasing that ration, we enlarge the Π factor. As I said before, the stature in relevance, of the flying object that moves beyond the speeding limit of $7(3\Pi^2 \times \Pi^2 \div 2)$ by doing so increases in size to the extent that it outgrows the Earth relevancy in accordance with the applying governing singularity which in relevance not in truth, becomes smaller.

The Earth becomes the controlling singularity applying the measure of movement onto the bicycle forming the governing singularity since the bicycle applies rotation. To the bicycle the movement of the bicycle replaces the role that the Earth previously performed while the Earth then becomes what the Sun previously was.

Let's move on and use a jet to describe what happens in the flying object atomic relevance applying when the object starts to move above and beyond using the Earth surface. In cosmic terms the flying object within the sphere of the Earth becomes so big that the object has the ability to sequestrate its own heat source that starts to drive whatever is moving forward.

This means that a star becomes so generated that it acquires so many atoms spinning that all the atoms spinning forms an ability to compress space within the moving object so that the object can start moving. This we know is not the case because life brought about the movement, but in the cosmos life as a factor is so alien that life as an energy source just doesn't exist.

The Four Cosmic Pillars; The Result Thereof. Page 226   APPLYING AS THE SOUND BARRIER

In normal cosmic terms there is no form of life that can muster enough energy to get something moving within the gravity domain of another object and therefore that means the object must be driven by what drives everything else in the Universe, atomic heat accumulation. Every atom is a little pump that pumps heat form from outer space into the star and it is the atomic movement that brings about the movement of a star or a planet. In cosmic terms that are the only way a body can bring about movement. We know that this process is not applying because it is the artificial form of energy called life that manipulates the movement. The linking between the atoms forming the flying object and the centre of the Earth changes in dynamics completely because the flying object then sequester its own centre taking charge of the governing of the atoms while the Earth then takes charge of that which acts as the controlling singularity of the atoms. By the size the Earth holds, the Earth will not allow such a renegade collection of atoms to grow bigger within the Earth's sanctioned time in space in accordance with the Earth generated governing singularity. In cosmic terms life has no meaning even to the Earth, which is a cosmic structure, which means the Earth complies to cosmic laws and not with Newton.

Since the generated singularity that the flying object commands it still has adhere to the dominance that the Earth's controlling singularity generates. In terms of the Earth, the Sun holds the controlling singularity while the Earth takes charge of the governing singularity, which places the Sun in control of the Earth's orbit. That is the reason why no object can ever leave the solar system, and that aspect gives the solar system the capability to become eternity. That is why comets come and go without ever hitting the Sun or ever flying off onto the next star system. This state of affairs changing brings about new consequences of a cosmic magnitude.

$7(3\Pi^2)$ $1\Pi^0$ $2\Pi^0$ $3\Pi^0$ $4\Pi^0$ $5\Pi^0$

The flying object reduces it proportional space in the face of the Earth showing such a strong reluctance to abide by the will of the smaller craft flying. The Earth crushes the craft flying in response to the craft growing in duplication and when the response is more than what the craft can withstand, the craft crushes by reducing space. Then liquefies its structure. That is what happened to Challenger when it entered the Earth and was by turning material into liquid that then became vaporised in time or as science call it "thin air". "breaking of the sound barrier" comes as a the aircraft displacing too much space through too much space or going too time in comparison with air or space being available at that level. In order to compensate for the insufficient space being available at such low altitude, the atoms forming the aircraft has to compromise in space by reducing in volumetric occupied space and have heat move to the atmospheric space. increases the quantity of heat surrounding aircraft drastically. In other books such as **Open Letter on Gravity** I explain this in much detail.

Position allocated to sound traveling according to the Doppler effect.

the craft liquefied gas and The result of (going fast) in

the This the **An**

$1\Pi^0$ $2\Pi^0$ $3\Pi^0$ $4\Pi^0$ ▭ $\Pi^2/2$ $5\Pi^0$

Position allocated where the aircraft is positioned in relation to k.

As I have shown, gravity comes about as the Earth is spinning and forming an axis that results in singularity manifesting and that forms a relation between space and time. forming singularity in infinity comes about Π resulting in Π^2 forming. The process of this manifestation is a relevance of a double triangle using the principle of Pythagoras, which is fifty going around the Earth's axis in fifty when going around the Sun's

By from

axis. When another object that is moving while it is being within the space the Earth contracts, it then also has to move faster than when the Earth spins alone and in that, the moving object has to have a source of energy that has it moving independently. Should this independent object then also move faster than the space the Earth contracts while being within the space the Earth contracts, the principles applying to gravity are bridged and the laws of gravity are compromised. While such an object is moving it brings in another Π^2 into the equation, which at first is represented by Π^0.

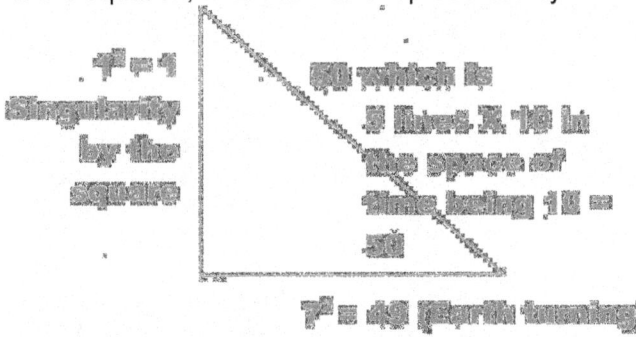

The independ3ntly moving object has to move at or less than $7(3\Pi^2)\Pi^0$. If the object moves at exceeding $4\Pi^0$, which is then $5\Pi^0 \times 10\Pi^0 = 50\Pi^0$, it fills more than all the space the Earth claims by spinning and in doing so, then create gravity. This is because of the Lagrangian point's position in accordance with singularity distributing space in relation to time (7 / 10). Therefore, by moving above 5 x 10 the object no longer duplicate the Earth singularity by its full worth and therefore it has to release singularity in 1 x 5 space divided by the time forming as 10 (the space that the Earth holds in relation to singularity applying to the moving object by time duplicating space) and therefore the object moves in another line not claimed by the Earth. Therefore, the sound can no longer follow singularity in which to travel. The object uses one more line that the object sustainers and the sound use the other four lines that the Earth fills and sustains.

The lines confirming singularity to which Newton incorrectly referred as being $\frac{dJ}{dt} = 0$, which when it is correctly sated then would be $\frac{dJ}{dt} = 1^0$ are in everyday use.

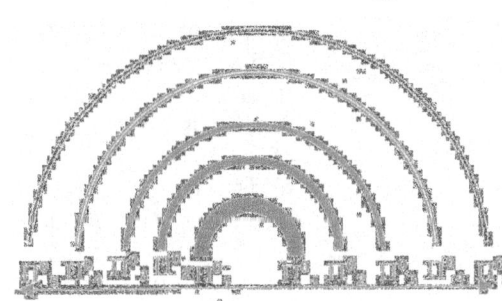

We are witnesses of these lines in everyday occurrences because it is these lines that carry light. I am not at this point explaining that part but they carry gravity in many forms other forms as well.
It is these lines that sound uses to carry momentum and to amplify sound. The sound nocks the one line that carries the distortion onto the next line and this process carries on until the knocking effect is so diluted that it is no longer noticeable. The ongoing process of distortion of the lines by movement we call sound. The use of lines or the employment of lines shifts and the amplifying reduces as we go into higher zones in the atmosphere.

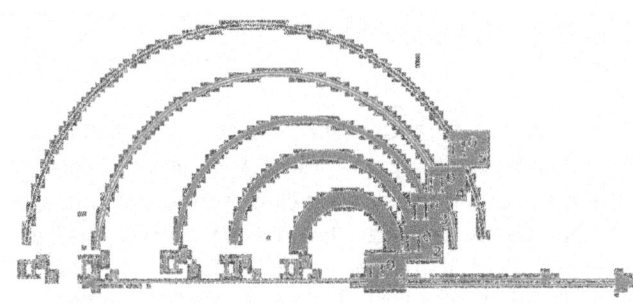

The use of lines or the employment of lines distorts the trend these lines follow when sound moves with these lines. The sound runs outwards evenly, losing energy as it travels from one line to the next line. However, with the object in motion it concentrates the lines that sound uses in the direction in which the movement takes place while the lines are in use. This occurs naturally but the effect becomes more distorted when such artificial or motorized movement takes place since the movement then starts to extend above and beyond the normal applying movement these lines normally could carry. The motorised movement has to move in one direction, which will compromise the speed the sound travels in, in both directions. It will accelerate the speed of sound in one direction while extending or retarding the sound in another direction. This has a name (as usual) without any Newtonian ever understanding what is going on in terms the process applying. It is called the Doppler effect. The movement of the object distorts the natural flow of sound in two directions (forward and backward travelling) and in doing so it forces the movement of sound in relation to the carrying of these lines to expand towards the back as the use of the lines expands in the opposite direction in which the movement takes place.

That is the reason why we in South Africa collide with game. The game at night is use to the normal travelling of sound and cannot distinguish between sound coming from a still standing object and a vehicle travelling at say 120 km / h. The game hears the sound and believes it to be very distant, while in fact the vehicle is very close.

As I said before, time holds 10 places in the fifty leaving space in which to travel five positions hence the Lagrangian five points. The five points form heat n the surface, but expands as the heat becomes space at higher altitudes. When travelling at ground level, the concentration of the lines does not permit such a relation between the liquid and the solid. The entire exercise has a foundation in the Coanda effect.

The breaking continuous line of singularity Π^0 by the forming of Π through moving as Π^2 results in gravity forming as a border indicator. It is not the speed, which would find a relation to the ground when travelling that holds the measure of value, but the relation to displacing space that forms the measure of value. At the bottom where the space is denser and the heat is more, one would find the displacement of space will result in the breaking of the sound barrier. However, at the very limit of space, where the Earth starts the compacting process of singularity, we have the displacement of space so expanded that such displacement requires $7(3\Pi^2)(\Pi^2 \div 2)(\Pi^2 \div 4)$, which in terms of the movement of the Earth would measure 2523 km / h but is in fact a reality value of only 207 km / h.

The sound carrying will never use five lines or even two lines.

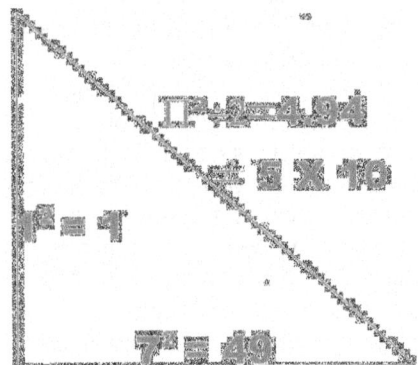

Gravity is Π^2. By travelling at a speed above gravity's limit (the Roche limit), then by doing so it splits gravity (Π^2) into two parts, hence $\Pi^2 \div 2 = 4.94$ and this it does by placing the moving object in another line in singularity, which is another line than the immediate line that the Earth moves in at such a low altitude. The sound barrier does not apply to speed as such, but it applies to lines forming singularity (1^0) displaced by movement when going through movement in space in time, thought to be space-time.

The real issue at hand when all is considered is that what is referred to as the speed of sound has very little to do with sound or with speed but it has everything to do with heat and the applying density in space while moving during time through space.

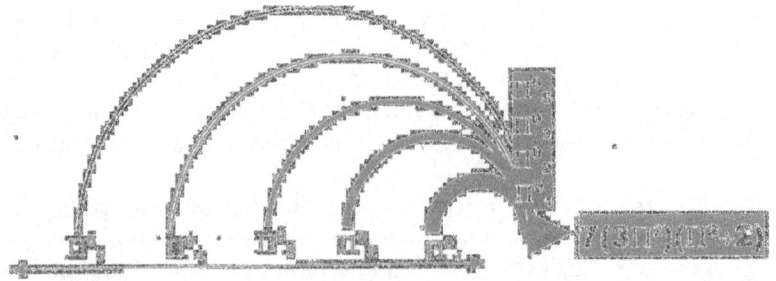

The Four Cosmic Pillars; The Result Thereof. Page 229 APPLYING AS THE SOUND BARRIER

Sound, just like everything else moves in a straight line just as much as it moves in a circle.

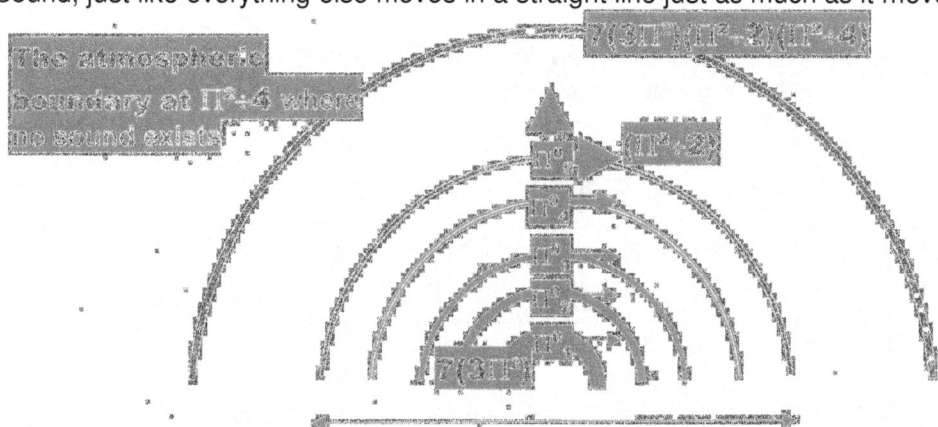

The lines connecting to the centre singularity within the Earth then connects to a centre singularity that forms in alignment with the arrangement that the aircraft's atoms form as they point a governing singularity that represent the accumulated singularity of the atoms of the aircraft. While travelling at $7(3\Pi^2)$ or below, the singularity connects in aligning with the Earth rotation and no compensation is required. As the speed exceeds $7(3\Pi^2)$ the alignments shift $(1\Pi^0)$ position meaning one line and this shift allows the shift to realign up to $(4\Pi^0)$ positions. This is the way the duplicating of the atoms forming the aircraft stretches in space holding material and the atoms form smaller particles spread over a wider area.

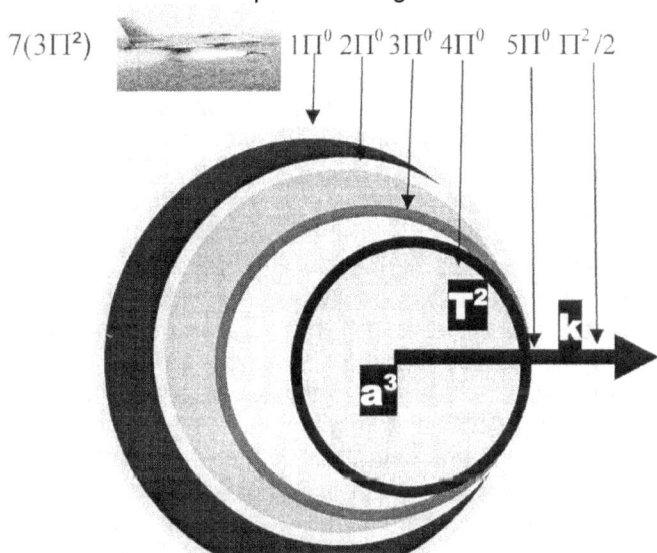

This process is the same as when a parachute opens behind any moving object. While the parachute opens, it defines more space in relation to atmospheric singularity or time and doing that restricts movement. The lines, which a parachute uses to restrict the movement and slow down the moving, the object is in the very same lines, which the aircraft realigns as to reposition the singularity it holds in contact. It is advisable to remember that although the singularity forming the atmosphere does not actually move, it is the reference in movement that brings about relocation of position and such relocation is what translates into movement transferred to atmospheric singularity.

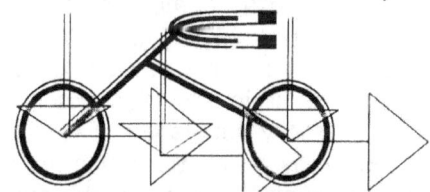

At a speed of "y" km / h the bicycle is standing upright by the movement it applies but the construction of the bicycle is still matching the form it has in relation to the movement of the Earth. Understanding this is very important when trying to understand cosmology and moreover the laws presiding in cosmology. The structure is in harmony with the Earth moving.

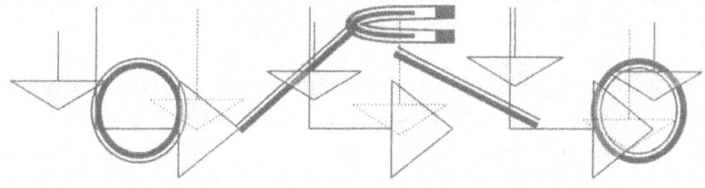

At a speed of "10y" km / h the ratio of material in space favours space going through more space by the faster movement in a comparable time frame. That means the bicycle takes less time to occupy a specific point of space but it uses more space in the same corresponding period in time. The bicycle uses one of these lines to travel between barely moving and $7(3\Pi^2)$. As it moves faster the bicycle begins t claim more space than what normally is used for the size the bicycle normally holds. The bicycle is getting bigger than what it was when having mass. That is why the "momentum" part becomes so much bigger when travelling.

The formula is $F = mv^2$, but in fact it should read $F^3 = mv^2$ which then is exactly what Kepler said when Kepler said gravity is $a^3 = kT^2$ and it is this part of using more space in the third dimension that improves the movement (F^3) of the bicycle. The occupying of more space in terms of the Earth allowing space (4Π) is the momentum part (v^2) in the normal physics formula while the mass replaces the Π. This is when thing go normal and everything seems to be under some sort of control with no wild horses starting to run wild. Then comes the part where everything goes array and the craft flying steps across the border limit formed by the Roche limit.

According to standards that the Earth applies the bicycle body moves up to $5\Pi^0$, which in cosmic terms are not allowed. Holding more than five lines to the one leading to the Earth will result in the Roche limit activating.

The sound barrier is not connected to speed but to a ratio between movements of material in confined space formed by compressed cosmic liquid or atmospheric space or air, which still is cosmic liquid. There is no true and precise applying $7(3\Pi^2)$ =207 km / h. It is a ratio between space or air and material moving through air. This ratio would be remarkably different in summer than it is in winter, and also it would be remarkably different at the poles compared to what it is at the equator. This cosmic liquid or air is a "sticky" "hindering" substance that forms a "drag" causing unbelievable friction when it is in a compressed state, which is when it is forming heat. When it is more expanded forming "colder air" or more space it becomes more tolerant to movement and allows movement with much less restricted properties applying. The distance travelled might remain the same but the density in air is much different and therefore the tolerance to movement is better suited for high velocity movement when going through "thin air" than it is going through "lower altitudes".

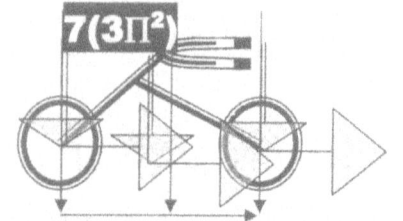

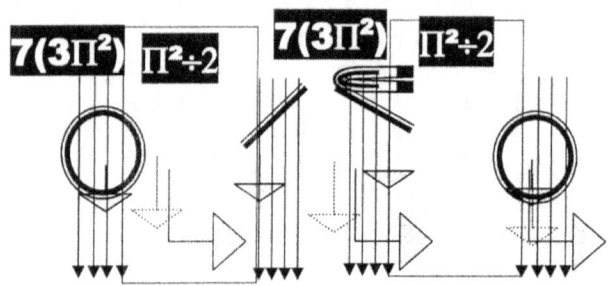

However, even in "outer space" this never goes away. Explaining why the reasons why this never goes away will take up too much space and must be left to another book on another day. It is well explained in the book **AN OPEN LETTER ON GRAVITY.** The difference in properties between dense and expanded air are considerably obvious when an aircraft fly higher. Only at higher altitudes does flying at high velocity become viable and this is because of the restrictive nature of the cosmic liquid that the material has to move through. No aircraft can fly at the sound barrier at a height of 30 000 km because there is no speed such as the "sound barrier" or Mach 1. Referring to this as a speed Mach 1 is referring to a ratio of liquid that has to be occupied in a specific time and when the air is thinner, this movement is less restricted as compared to when it is dense. Remember that space forms heat when compressed and heat forms space when expanded. Going through compressed air or heat becomes energy sacking because the Earth already compressed the air down below at ground level and with the aircraft going through the air in such dense conditions makes the movement fight what the Earth already established as a thick sticky substance. It is like swimming through syrup and swimming through water where water will allow movement with much more ease than the movement syrup would allow. However, going through air at high altitudes makes the compressing of air less arduous because the air is still thin and the compressing effort takes less strain than when the Earth already did the task as it has down below. There is a ratio forming between compressed air and movement, which allows air to form a ratio. The density of the air allows movement through which objects can move and in the ranks of the compressed or expanding air it is much more tolerant than going through the compressed air at low altitudes.

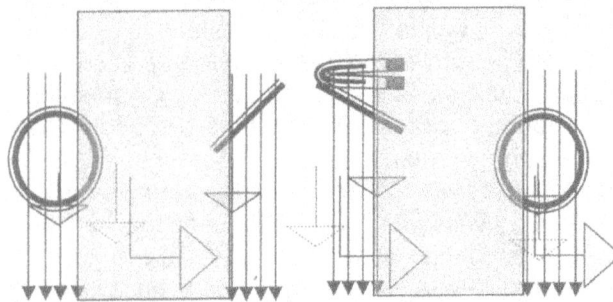

At low altitudes the movement may ask for more space that which is not available because of the dense air. In order to duplicate the positions it holds surpassing a thousand kilometres of space in one hour requires a specific area of space while the Earth provides only $7(3\Pi^2)$ and the rest the moving object must sequestrate from the cosmos by developing heat as it reduces atomic occupied space.

At high altitudes the expanded air leaves more lines available and the lines being less compressed leaves more space to use and therefore there is up there more available space to be asked for by the excessive duplicating flying object. The duplicating is less strenuous because the ratio of space available is forthcoming and has not have to be solicited from own initiative by the atoms moving.

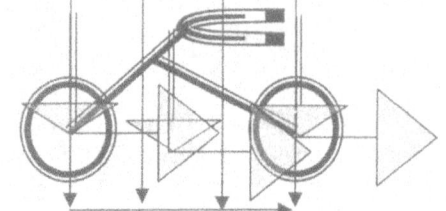

It is vital to remember that there is no longer contact with the ground to have as a ratio where the ground is a fixed form. The ground is a steady supply of precise space whereas air forming the ratio when flying is not. The air can expand and contract by heat changing the environment.

Down below the aircraft might be able to "break the sound barrier" because there is not enough lines to comply with the demand that the flying charges into use and at that stage one will arrive at conditions where at low altitudes one might find there is not enough space for the flying object "breaking the sound barrier" to go through as the space is compressed. There might be a lack of lines to capture and there might not be enough lines to occupy in relation to the movement that will hold such lines in movement and then the lines matching the Earth will break down its contiguous formation. At that compressed state we think of as altitude the lines forming are not sufficient to form a comparable formation as to form a link between the flying object holding the governing singularity and the Earth forming the controlling singularity. However, this situation changes as more space become available at higher altitudes.

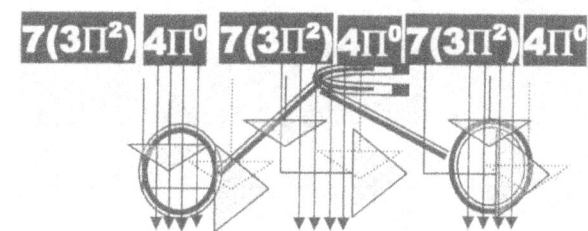

The there is the ultimate flying applying while the flying craft remains restricted to the Earth in as much as flying while being part of the atmosphere. The point is Π (3.1416) x 10^3 km up in the air. In order to maintain status of being airborne a speed of $7(3(\Pi^2)(\Pi^2/2)(\Pi^2/4)=2523.km/h$ has to be in place. At a relevancy of $7(3(\Pi^2)(\Pi^2/2)(\Pi^2/4)=2523.km/h$ just staying in the air is equal to $7(3(\Pi^2)$ down on the ground.

Flying at $7(3(\Pi^2)(\Pi^2/2)(\Pi^2/4)=2523.km/h$ is not being two and a half times the speed of sounds because up there sound has no speed.

The relevancy of displacing space-time requires 2500 km / h or dropping back to a level that applies to the ratio the aircraft then would travel at. To explain the relevancy of flying at $7(3(\Pi^2)(\Pi^2/2)(\Pi^2/4)=2523.km/h$ this is explained as follows:

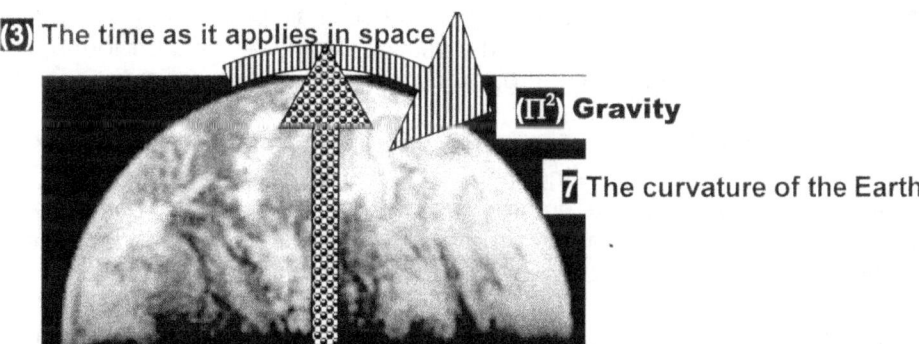

In this relevancy or ratio we find the very same factors that would apply to any craft being airborne.

7 The curvature of the Earth as the earth rotates to form gravity.

(3) The time as it applies in space where the past flows into the present and onto the future making time a contiguous flowing dimension.

(Π^2) Gravity. It is what comes about when seven and ten intermingle to form gravity as I showed a while back.

But this effort goes much beyond the limit set by the "sound barrier". It takes the Roche limit one step further than just setting it at an atmospheric level. It brings in a cosmic value that applies within the boundaries of the gravity field of the Earth.

This is the minimum speed that could be travelled when flying at the ultimate relevancy while still being in the Earth's atmosphere. When said as it is without me trying to get more technical than my abilities would allow then this is how it should read: up there above there are more little unseen stripes holding singularity than an aircraft travelling at 2550 km / h would allow. Again it depends on heat and not actual distance covered by going in a precise km. / h.

(Π²/4) The Roche limit as it applies reaching the boundary of outer space

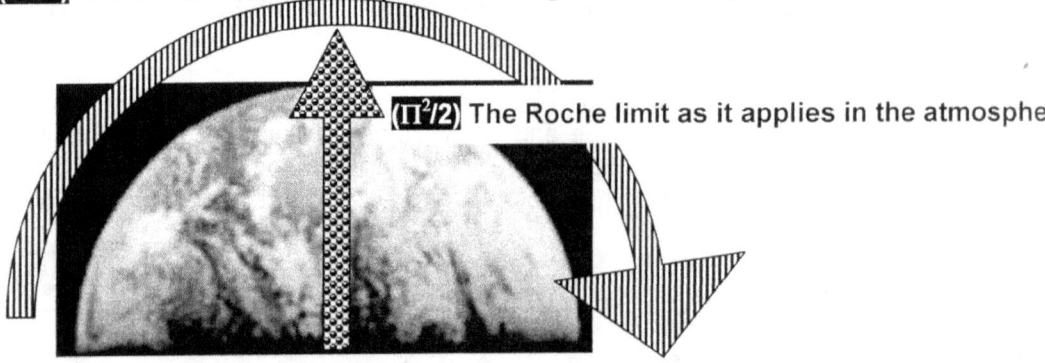

(Π²/2) The Roche limit as it applies in the atmosphere

(Π²/2) The Roche limit as it applies in the atmosphere

(Π²/4) The Roche limit as it applies reaching the boundary of outer space

=2523 The speed that must be achieved to remain airborne.

The maximum speed I could find that a boat can travel on water is $(7° (3Π²)(Π²/2)) ÷ 2 = 511$ KM / H or thereabouts. One should remember it all depends on water temperature and the temperature of air as well as a flow in the water or no movement in the water. The reason why $7° (3Π²)(Π²/2))$ ÷2 it would divide by two is because (and in this case I am guessing), is because water is half a solid and half a liquid and holds half the properties that a solid holds and half the properties that a liquid holds. When the solid gets too dense for water to have submerged while travelling at an excessive speed is because the solid becomes partly submersed while lifting out of the water as the speed increases. This happens because the boat or solid increases its volumetric space in relation with the water and therefore the material becomes more in relation to the water it is in contact with.

By duplicating and becoming more the body of the boat performing as the solid simply lifts from the water and in that way reform the density relevancy. When I first started investigating my research, I found that every time I came to a situation where in evaluating I was introduced to space-time or by investigation I came across the neutron's characteristics not showing mass or displacement restriction, I found both the characteristics of space-time and of the neutron to be identical to water. In both instances I got the idea I was dealing with water or fluid. This is when I was formulating the Cosmic Code as it applies in stars. The similarity between water and space-time as well as the neutron was truly uncanny and I could never understand this until I placed outer space in relation with a cosmic liquid and then realised what happens in the event of the Coanda effect, which is the role the neutron plays within the atom.

This is separating what is of the Earth from what is part of the Earth and this is a highly unnatural occurrence. It is breaking the boundary set by the governing singularity on the controlling singularity and the departing governing singularity has to overcome as much as over ride the position between the governing singularity and the controlling singularity. The one part confirms the object being part of the earth, the other part confirms the object overcoming the grip the governing singularity has by placing the object into the realms of the mutual singularity and the third part is separating the departing object by diverting it from the Earth.

The achievement in this is that life and not the cosmos secures sufficient liquid heat being released to a value where the liquid heat could transform the projectile into the upper atmosphere and even beyond.

I am the first to admit that the calculations introduce with the atomic factor relation on the Universe has little mathematical value other than indicating some relevancy and some predictability. But you cannot use it in precise calculations because the heat on the day of the day will apply different measuring standards to what science perceive as constants. A kilometre will be slightly longer or shorter an try to tell a pilot travelling at two thousand five hundred kilometres an hour a few meter nearer or further has no importance! For instance launching an object into the outer space will require an atomic relevancy of $4Π²(7(3(Π²) (Π²/2) (Π²) / 3600 = 11,2$ km per second. Should you wish to make it an applicable factor, one will have to multiply the second object's proton relation with 10 making the total formula stand at $4Π²(7(3Π²) (Π²/2)(10Π²)/3600 = 112$, which will indicate the cosmic factor value of independence, the same value that applies to Newton's $F = G (M_1+M_2)/r²$. By using $= 4Π²(7(3Π²) (Π²/2)(10Π²)/3600$ you cannot put a rocket into orbit that is true.

(4Π²)} Becoming an independent spinning object within the Roche limit of the Earth.
÷ Putting the independence in terms of finding release from the Earth confining atmosphere **{6² x 10²}** Representing the curve of the atmosphere (6²) in terms of space (10²) **=11.216** km / second. The speed one has to travel to escape into outer space and go beyond $\Pi^3 \times 10^3$

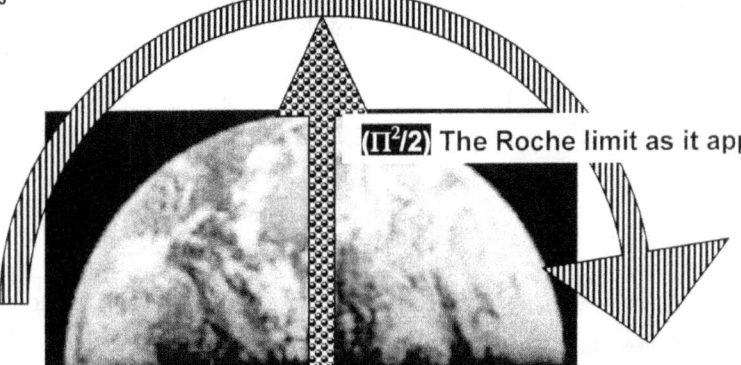

{7 The curvature of the Earth. {3 The time as it applies in space.(Π²) Gravity

{7 The curvature of the Earth. {3 The time as it applies in space.(Π²) Gravity

{7 The curvature of the Earth

{3 The time as it applies in space

(Π²) Gravity

(Π²/2) The Roche limit as it applies in the atmosphere

(4Π²)} Becoming an independent spinning object within the Roche limit of the Earth.

÷ Putting the independence in terms of finding release from the Earth confining atmosphere

{6² x 10²} Representing the curve of the atmosphere (6²) in terms of space (10²)

=11.216 km / second. The speed one has to travel to escape into outer space and go beyond $\Pi^3 \times 10^3$

Restricting us to the Earth is the laws keeping us in place and confined to the Earth, and therefore breaking this confinement would be our ability to bridge the factors keeping all of us in our place on Earth.

We first have to equalise and better the Earth's movement resulting in gravity, which is the seven degrees of displacing by rotational movement **{7 The curvature of the Earth**. Then we have to do something about the movement by Π^2 of the Earth spinning about its axis, spinning about the axis the earth establish as it spins around the Sun, and thirdly the axis the earth forms as it transfers all movement from the Earth holding infinity as the governing singularity to the cosmos holding the eternal singularity. The last Π^2, which I am referring to, is through change we see in all of outer space in reference to the centre spot we hold which is the centre spot we inherited from the centre of the Earth or with us moving then forming our individual centre spot in relation to everything out there moving. That then becomes the three forms of gravity establishing our confinement, which we serve as subjects of the Earth. That then explains the three values that Π^2 use to form time running from the past through the present and onto the future while placing us all as individuals in the centre of the Universe. **{3 The time as it applies in space.(Π²) Gravity.**
(Π²/2) The Roche limit as it applies in the atmosphere. To find release from the confinement the Earth captures us by we must at least break the sound barrier for starters by breaking the atmospheric Roche limit. .
(4Π²)} Becoming an independent spinning object within the Roche limit of the Earth. If we wish to circle the earth, we have to bridge the confinement the Π factor (not Π^0 which is the governing singularity) but that which makes us part of the Earth by forming a bond with the Earth in bringing all things within the atmosphere to connect as Π. The full compliment that keeps us in the rotation zone the Earth established is (Π²/4) The Roche limit as it applies reaching the boundary of outer space, which any flying object had to fly at to come to the limit of the atmospheric space. This must be bridged in order to neutralise the

confinement space and then on top of that we still have to produce by movement another

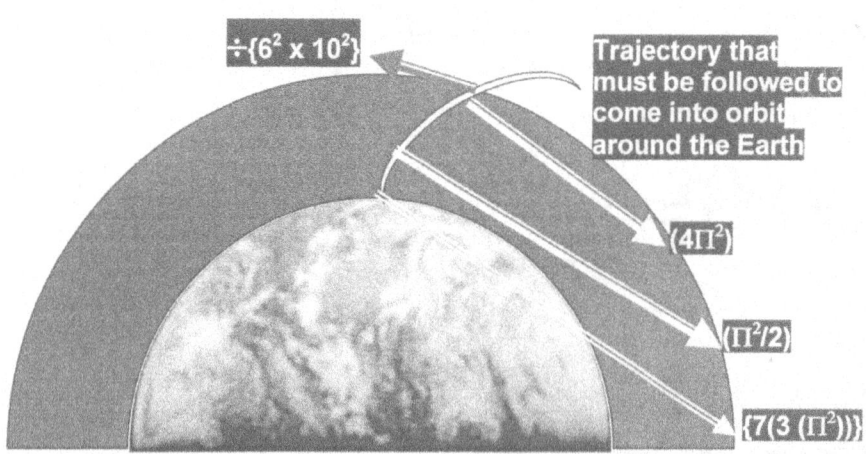

The last hurdle to cross is the form the Earth displays in terms of being a sphere while spinning in a cube. If anything wishes to be part of the cube, it has to break the bonds the sphere puts on it by circling. The relevancy material has in the Universe is six, but I do realise going this path requires a lot of explaining and I am trying my utmost to keep it short. Let's go another route that is simpler, but with less explaining and exposes more Newtonian styling of leaving everything unfinished.

We are spinning through 360° when spinning and while spinning we are spinning through space which I already showed has a value of 10. When multiplying 360° with 10 we have 3600 and 3600 is $6^2 \times 10^2$. This breaks down to become one putting the Earth in from being a governing singularity to become a controlling singularity and the object orbiting the Earth is completely detached from the space the Earth captures.

The required speed needed to escape the Earth's gravity is 11.216 km / sec. The underwriting of this, which forms part of the Cosmic Code, is captured by one formula. All object that is placed in space, which Newtonian science call outer space is rotating and as it does so all objects then are by rotating at 7°.

> In any **right triangle**, the area of the **square** whose side is the **hypotenuse** (the side opposite the right angle) is equal to the sum of the areas of the squares whose sides are the two legs (the two sides that meet at a right angle).

It is putting seven in the square of a Pythagoras triangle and the square singularity forms in the other side of the square sided triangle.

Then the sum total of both squares will be the value of the hypotenuse, which in this case then are fifty. Because singularity is equal everywhere, the square of seven duplicates and the total then becomes one hundred. The square root of one hundred is ten and space holds the value of ten while material moves by 7.

Matter move by 10 in relation (part of) the diverting of the straight line by 7 forming the total dimension of space. Everything that is moves at the sane time it is. Everything moves straight as it comes from the past and goes by 10 into the future after it formed singularity 1.991 and then diverted from the straight line by 7.

This will bring about space-time development that applied since when the Universe was fresh and new, hot and lively and full of spinning power. As the objects cannot destroy one another because the Roche limit turns the one lot into liquid, there is constant growth time introduces known the humans as the Hubble Constant affect that came into place the instant time came into place. Each structure had to build a Π value, to maintain and secure its relevant position until the end of the final eternity arrives. Said with using other words, each object will secure its place in future as a Black Hole or Proton Star. By producing a value of Π separating the Π^2 from three and thus ensuring itself an own atmosphere it will have to revert to space-time development that was in place at a time that even preceded the "Big Bang". I will produce once again the value of Π from its position of Π^2 (time) and matter. Through that the Titius Bode law comes into affect of 10/7 or 7/10, depending on whether space or matter holds a superior position to time. From that stance, all objects will relate to one another by the value of $\Pi^2\Pi$ and seen in a whole sale total 7/10 or 10/7.

One can also argue that matter holds the value of 7, always 7, in the position of space (10) and the combined space-time value will relate to time in singularity during that specific instant of time duration ending one eternity with the intervention of infinity (1). Placing all this array of words into a mathematical solution or formula $\Pi^3/\Pi^2=\Pi$. Space ends the eternity of time to the value of 1. (Time is the movement (spin rate) of heat in space). When a structure holds heat to time in space the relevancy becomes $4(\Pi^2+\Pi^2)$ therefore when the object retains time to space it will have a value of $4\Pi^2$. This puts all the sides IN TIME SHARING WITH THE OBJECT AT $4\Pi^2$ to the Π^2 of positive space-time displacing. The reference to $4\Pi^2$ means it rotates through the full circle of 4 points forming Π. All objects will relate to the $4\Pi^2$ in the same way. As I showed before, if space-time extends beyond the critical of $(\Pi/2)^2$, it will attach as the Titius Bode Principle of 7/10 or 10/7. When an object is in the atmosphere (the atmosphere being the fluid value of the Earth) it holds a position of $4\Pi^2+\Pi^2$. That means the object is completely secured by the positive space-time displacing of the Earth. When not secured it becomes $\Pi^2\Pi$ and I have shown that Π is the concentrated value of matter $(7)^2$ in time (Π^2). The atmosphere, we know through actual orbiting entry, holds an entry limit of 21,991° as a maximum and 7° as a minimum. That means to become part of the atmosphere density of the Earth, the object has to be space (21,991 or less) and prove to be matter (7) before the Earth will accept it. If holding a position of less than 7°, the Earth will discard it and if it is more than 21,991 the Earth will find the relevancy to be higher than the space it holds in a fluid time.

That places the object in a relation of $4\Pi^2 + \Pi^2$ (because it is part of the solid Earth) in a position acceptable matter holds (7) within the confinement of Π (21,991/7). That means the object is part of space (22,991) acting as matter (it holds an acceptable own proton structure) 7 relating to the Earth in the position the Earth allows of $3\Pi^2$. With the space position of the matter in the parameters of 21,991 it relates to the Titius Bode law as a factor of one. The object has the space value of 10 plus the space value of ten, in that instant of time (1) complying to the Earth's space (10) reduction (Π^2) formulating $\Pi^2/10 =, 99$. That makes the object complying with the full agreement as laid down by the Titius Bode law. The object is, no matter where it is, travelling at a rate of 7 $(3\Pi^2)$ in the space of the Earth (21,991). This will be agreeable to the parameters of the Titius Bode law as long as it remains within the space depleting "gravity" limits of less than Π^2. In accordance to the Lagrangian atom layout, anything less than 5Π is manageable and is in effect less than Π^2. When it exceeds $5\Pi^0$ it will start opposing the dimensional equilibrium space holds of 10Π, therefore it will (according to space) exceed the linear point of fluid, which is $10\Pi/2$ (space going in a straight line). By exceeding the straight-line value of $10\Pi/2 = 5\Pi$, it will then categorize itself in the position time holds to space, and that value is $10\Pi^3$. By exceeding $5\Pi^0$ it will start to defy space and this will automatically bring about an individual space-time relevancy of $\Pi^2 \Pi$, establishing its own proton position in the confinement of the Earth's space-time of $7(3\Pi^2)$. Thus while it holds a relevancy of Π and Π has the relevancy placing it at a value of Π^0 because the neutron space link to time is 21,991/7 in the atmosphere, the Earth will tolerate individual movement of up to $7 (3\Pi^2) \Pi^0$ to $7 (3\Pi^2) 5\Pi^0$. Beyond that point, problems start arising as the object is not complying to the Titius Bode law any longer.

Moving beyond $7(3\Pi^2) 5\Pi^0$ the Roche factor comes into effect and this is all in the space-time depleting zone established by the fluid atmosphere of the Earth. That means the Earth still holds its own 21,991/7 liquid base, but the object breaking the Titius Bode law bring in its own value of $\Pi/2$. This value of $\Pi/2$ is in the liquid parameters of the Earth where the Earth removes the linear factor value of the liquid from 1 (Π^0) to Π. That means there is a dual for supremacy of a liquid position is as much as $(\Pi \times \Pi/2)$ becomes $\Pi^2/2$. That places the sound barrier at $7(3\Pi^2) (\Pi^2/2)$. At the value of $7(3\Pi^2) 3(\Pi^2/2)$ more implications will come about, because the object will establish an own space within the space-time of the Earth and this will lead to the object whether joining the time of the Earth $4\Pi^2$ or securing itself an individual space and time. Obviously the heat supply will be insufficient to bring about the value the object needs to place it in outer space $(4\Pi^2(7(3\Pi^2) / (360° \times 10))$ so the aircraft will forcibly join the core of the Earth, crashing on the ground.

From the sound barrier and the effect the sound barrier has on matter, one can measure galactica and how that "growth of space" seems to become a reality as the Hubble Constant indicate.

When the object holds a position above Mach 1 and Mach 2, the temperature to the outside of the aircraft rises dramatically. It comes as a result NOT OF FRICTION WITH AIR but because the atom in size shrinks and casts some heat from the inside parameter to the direct outside. This is the same as that which we may find to be the value of the part that we can see from the outside in the core part of the galactica. The structure of the craft becomes a solid or a proton. The craft holds the value of $(\Pi^2+\Pi^2)$ in relation to the Earth because it maintains an own space value in the related time of the Earth. The craft applies a demand for additional heat, because to all cosmic purposes, it shows a higher resistance in sharing space and time. Place this in a human context and the aircraft has a higher mass because the atoms get more dense because the movement is greater than because the atoms became smaller than... Science holds the movement of the structure in the category of momentum, but momentum is only linear "gravity" and mass is circular "gravity". In order to claim the space-time occupation that the craft demands, will mean an excess of protons in order to bring about the space-time occupation required for that strong relevancy to heat. The cosmic law does not allow for artificial heat production and life has no normal place in the cosmos. To the cosmos it is all a relevancy of supply versus demand.

In order to stipulate where I refer to space –time my referring to space-time have to be by the use of a signal. Therefore I decided to use the $ as a sign by which I shall refer to space-time.

That is where the Mach principle finds a solid foundation. What separate time and space in joining singularity is the matter and matter is heat. However Mach placed all emphasis on mass and mass is only the resistance of matter in parting with own space to join another object's space. Part of the resistance of mass or parting with individual space is to apply the Roche limit that will remove the fluid value of the object from $\Pi^2\Pi$ to a value of $(\Pi/2)^2$. This will improve the density of the solid considerably because it removes the Titius Bode principle of matter in space as it eradicates 7 to 10. I have shown how 7 to 10 are the development of 21,991/7 and this of course is Π. By removing Π from the equation there no longer exists space in time but a value of half time to time $(\Pi/2)^2$. The influence that this brings along is a huge increase in time value of occupation of space-time that the object initiates. Where the object normally holds a relevancy of space to time in as much as $(\Pi^2+\Pi^2)$ $(\Pi^2\Pi)$ (3) = 1836, the atom adopts a new relevancy of $(\Pi^2+\Pi^2)$ $(\Pi/2)^2$ = 97,4. The normal mass of the proton will be (within the confinement of the Earth) 1,672648 x 10^{-27} and the electron will be 9,109534 x 10^{-31}. With the alteration of the neutron and electron relation the "electron" value of the total object becomes 18,85 times larger. That means the relation between the proton and the neutron has a normal relevancy of 1836, but is only 97,4. That means above and beyond all the electrons' space-time demand, the structure as a whole, has an immense demand of 18,85 times larger. All this happens within the atmosphere of the Earth.

The structure holds a normal value of 7 $(3\Pi^2)/10^2$ if one wishes to replace the Earth confinement of matter to the outside. Therefore it holds a relevancy of 2,07 that makes it a cosmic relevancy that helium holds. Helium is an element that surrounds itself with heat in a gas value. When the structure reaches an own velocity of 7 $(3\Pi^2)$ $(\Pi^2/2)$ $/10^2$ it then will have a value of 10,25 which holds a comparable mass (space-time integrating resistance) equal to that of Boron, at 10,811. At mach 3 the space-time demands will become $7(3\Pi^2)$ 3 $(\Pi^2/2)$ $/10^2$ = 30,778.

This places the relevancy where that particle can demand own space in own time due to the density it acquires from space. It has all the properties that a star in space requires, but it will still be confined to the Earth. In that way the object then must acquire a relevancy of the full Roche limit, without the protection of the Titius Bode law. The Newton cosmic relevancy will reduce to nothing $T=(\Pi^2) + (\Pi^2) + (\Pi^2) = 9$ and subsequently it will find itself in a huge struggle to defend its space-time or demise its position. With a relevancy of $3\Pi^2$ it holds the same value as liquid heat, which is light. The $3\Pi^2$ comes from the proton $(\Pi^2+\Pi^2)$ adding another neutron space value of Π^2 and this will then be (proton $(\Pi^2+\Pi^2)$ + neutron (Π^2)) = $\Pi^2+\Pi^2+\Pi^2 = 3\Pi^2$, the same as the photon at 29,6. Beyond 29,6 the structure will have to claim individual space-time value. That is the same value as that of cosmic structures still within the inner-core of the galactica. The structures in the galactica holds a value of 7/10 $(\Pi^2\Pi)+(\Pi\Pi^2) = 28,8$ and to space it holds a relevancy of $3\Pi^2$.

That actually means that the objects that are still within the galactica core, shining because of the outside value of the particles being $3\Pi^2$, are in fact burning silicon, or if you insist, it is glass. In the very centre, where professor Hawkins presume Black Holes to be, is the very opposite, holding structures that have carbon inner core and therefore holds a neutron value of $\Pi^2\Pi$, with no proton core development at this stage. They are still in eternity beyond the "Big Bang" and therefore beyond light. This value brings us right back to the aircraft that applies an own space-time relevancy of Mach 3, or $7(3\Pi^2)2(\Pi^2/2)$. This relevancy places the object holding a star status core of $4(\Pi^2+\Pi^2)= \Pi^2\Pi / 5$ that is the value of a star holding an element of carbon. That is still two eras away and they are not yet even in the present times in our field of vision. To the Earth holding an iron core $4(\Pi^2+\Pi^2)$ in a galactica 7/10) an object in a space time position of $\Pi^2\Pi/5$ is completely out of space, out of time, out of era. It will apply the Roche limit with such ferocity, as its own space-time occupation will allow. According to cosmic law that object has a relevancy of less than that of the Roche limit $(\Pi/2)^2$. The space-time relevancy that the Titius-Bode law allows is (7/10=0,7) and (10/7 = 1,428).

This is happening where time finds space; where time forms space and in the smallest that there is in the confinement we think of as the Universe. We should also see this when we see the Earth is turning or the Sun spins or the planets orbit. We should see where Π forms Π^2 that results in the forming the border of Π^3 when we look at things we are able to see, but when we look at what we are able to see we must know that this that wee see is just a reflection left over from the actions of the very minute things we are unable to see.

With the fact of the speeds I use in the sound barrier explanation, all the figures are only relevant below 500 m above ground level, and ground level being sea level. Any point above 5 x 10^2 becomes a changing factor to the relevancy of the Titius-Bode Application. In a previous part I pointed out about the inclination of the atmosphere and that is true, but that view apply the way we see physics through the eyes of Newton. That mentioning of the changes of relevancy is only an introduction, in the same way I introduced the relevancy of the Roche-law and the relevancy of the Titius-Bode law indicating the way Doppler interprets the Titius Bode law. The way the Titius Bode law forms gravity in conjunction with the Roche limit is as follows:

$$\left(\frac{10}{7} \div \frac{7}{10}\right) = 2.04$$

$$\frac{1.4285}{0.7} = 2.04 \quad \text{Taking from both orbiting influences}$$

SPACE DIVIDED INTO TIME

$$\left(\frac{7}{10}\right) \div \left(\frac{10}{7}\right) = 0.49$$

$$\frac{0.7}{1.4285} = 0.49 \quad \text{Taking from both orbiting influences}$$

SPACE MULTIPLIED WITH TIME

$$\frac{7}{10} \div \frac{7}{10} = 1 \quad \text{and} \quad \frac{10}{7} \times \frac{7}{10} = 1 \quad \text{Therefore not influencing change}$$

THE PROCESS PARTED USING THE ROCHE PRINCIPLE

$$\frac{10}{7}$$
$$\frac{7}{10}$$
$$\left(\frac{\Pi}{2}\right)^2 \quad \frac{10}{7}$$

$$\left(\frac{\Pi}{2}\right)^2 \quad \text{The Roche influence on Titius Bode}$$

$$2.04 \times \left(\frac{\Pi}{2}\right)^2 = 5.033$$

$$2.04 \times \left(\frac{\Pi}{2}\right)^2 = 5.033$$

$$5.033 + 5.033 = 10.066 \quad \text{from both objects}$$

SPACE DIVIDED INTO TIME

$$\frac{7}{10}$$

$$\left(\frac{7}{10}\right) \div \left(\frac{10}{7}\right) = 0.49$$

$$\frac{10}{7}$$

$$\left(\frac{10}{7} \div \frac{7}{10}\right) = .49 \quad \left(\frac{10}{7} \div \frac{7}{10}\right) = .49$$

$$.49 + .49 = .98$$
$$.98 \times 10.066 = 0.8 = \Pi^2$$

$$\text{TIME SPACE} = \Pi^2 = 9.8696$$

TIME SPACE = Π^2 = 9.8696 = Space and time in a dimensional implication

One should not try to focus on an image of such a spot or dot because there is no image where time forms singularity and singularity meets space. The line dividing the cosmos and that run through every particle, no matter how large or small is beyond our vision; it is where singularity becomes space, where 1^0 forms 1^1. Such a small line, being so small that it is not even noticeable in any way of looking at it and is so small it is not part of the cosmos we can see, but yet it is large enough to part the cosmos into sectors. It splits the biggest there is into particles and we are not even able to notice the precise location of such a split that is small that the line forming splits where time originates. In truth there is no top or bottom, front or back, left or right that we living in 3D can see. We shall have to use a general conception brought about by intelligence to visualise the point where $\Pi^2 = \Pi^3 \div \Pi$. Your intellect tells you about such a spot where singularity forms in infinity, but that is all indication anyone will ever find because that spot is on the other side of the Universe (quite literally). From the centre of the dot there is a top and a bottom spot. From those points there is connection with four quarters that lines up with three points. That produces six connecting points that are all aligning to the one centre. Because it serves big and small, hot and cold equal and alike, and it is the smallest cutting the biggest into equality, size is of no issue. Size is what man makes of it. In the Universe there is no size in hot and cold, large and small. ...And that dismisses all prominence of what we ever wish to give mass. For the smallest there is, singularity is serving the largest there is equally.

Our instincts, our logic and our calculating process all indicate that the sphere holds a centre point from where six evenly positioned point's position matter to form. That point where material form is $7/10(\Pi^6 \div 6) = 112$. Using the formula $F = G\dfrac{M_1 M_2}{r^2}$ it indicates to a force pulling objects closer, where each force is coming from each centre point the body in question has. If it did not come from a centre $1/r^2$ would not become $r^2=1$ in the formula. We have $r^2=1$ where every atom dismisses space as it is spinning and from the spinning the action finds refilling from outside coming towards as it flows inwards directly in a circle towards the centre (because there is always a double connection to gravity in motion) of the cyclical sphere. At one point in the precise centre, the dismissing of the total ability of all the components within the structure finds a peak and at such a peak the dismissing forms the biggest influence on all points at the outside. That places the border forming of the group selecting the unit by motion and such reducing becomes inherent part of the form of the sphere spinning to create cycles. In nature the only form provided outside singularity is the sphere. The contraction that causes the reducing of space must commit the two bodies towards a point in each case being spot on in the middle, not withstanding what direction the force is applying, the body will draw to the centre when being part of the unit. Only when the heat ratio promotes more duplicating than dismissing due to the spiral cyclic in relation, will the elements hold heat on the outside contracting the heat to the inside. Only in that way will that relation with heat counteract the dismissing and neutralise the overall effect of the collective dismissing.

If the Universe spins around a centre point holding singularity, and singularity confirms the centre of the Universe, then every particle that spins holds the centre of the Universe 1^0, making the number of universal centres immeasurable many, and every atom and sub atomic particle presented outside the atom in smaller bits, are all not pieces of the Universe but they are a Universe surrounded by many Universes. Every point holding 1^0 are equal to 1^1 notwithstanding the size singularity represents. If every atomic particle, no matter how small is holding the centre of the Universe, then the gravity is coming about from that point holding 1^0

because that is where the gravity that is applying in the Universe are applying contraction. There was a beginning that saw a radius between objects so small the size will never again repeat. That is where everything had to start. The diameter of the particles were also representing infinity but that should not be a contributing factor surely...because where it all started the main focus point is that particles were as cramped as it shall never again be repeated. This was when only eternity formed value in relation to infinity.

We find the circle coming about from a line going straight to form the Coanda effect even reminiscent of what we see in a galactica. If the galactica indicate the process as used to form galactica and keeping in mind the cosmos hardly ever change, then it is quite fair to say that the same process was used to form sub atomic particles, then atoms, then elements, then later stars and finally galactica and super giant stars that now are Black Holes. If the Coanda effect is so vividly part of what is the left over of the Creation process we then know how it started and why it started. We know that the mass differentiation of the atom between the electron and the proton is 1836, which in more correct definition is displacement and dismissing of space-time $(\Pi^2+\Pi^2)(\Pi^2 \times \Pi \times 3) = 1836$ it is an atomic relevancy that projected from the atom to the star directly without loss of any translation. If the formula $F = G \dfrac{M_1 M_2}{r^2}$ ever applied, which it does not, then it only applies in a very specific range, and at a very determinable point the formula does not affect objects in the air. After such a point one will find satellites able to orbit, be it part a definite pace that matches the rotation of the Earth. Still...below such a point orbiting objects will orbit too slowly and come crushing down to the Earth. The fact that the top tries to expand into space when spinning too fast confirms this fact. As the top hit the ground after being thrown in a spin it starts to move around in small circles while circling around the axis as it forms in a vigorous manner as if the top is suddenly too energetic to stand still and that is precisely what happens. This surging into the air by leaping from the ground shows it trying to find a new dynamic and this is a very important sign, which is of most importance. With all the excitement and no where to take it the extending of the drive line it reacts by going into the air as the relevancy runs down the singularity inwards towards the newly established governing singularity that keeps the whole job erect.

That is why the top is spinning in the first place. The more assertive the spin is in velocity the more reaction there is from the lines running towards the centre and extending the expanding outwards. In real terms the space of the top expands as the spin is in contact with more time in space during the same time in period and a bigger unit fills the space in which the top spins. In this the space in which the top spins has to expand as well in order to compromise for material relevancy growth to fit the newly acquired singularity governing the motion.

The support that the spinning top finds in the established governing singularity keeps the top spinning in an upright stance only supported by the controlling singularity that takes charge of the spinning space-time. It is the relation there is between the governing singularity and the controlling singularity that enables the spin the object has.

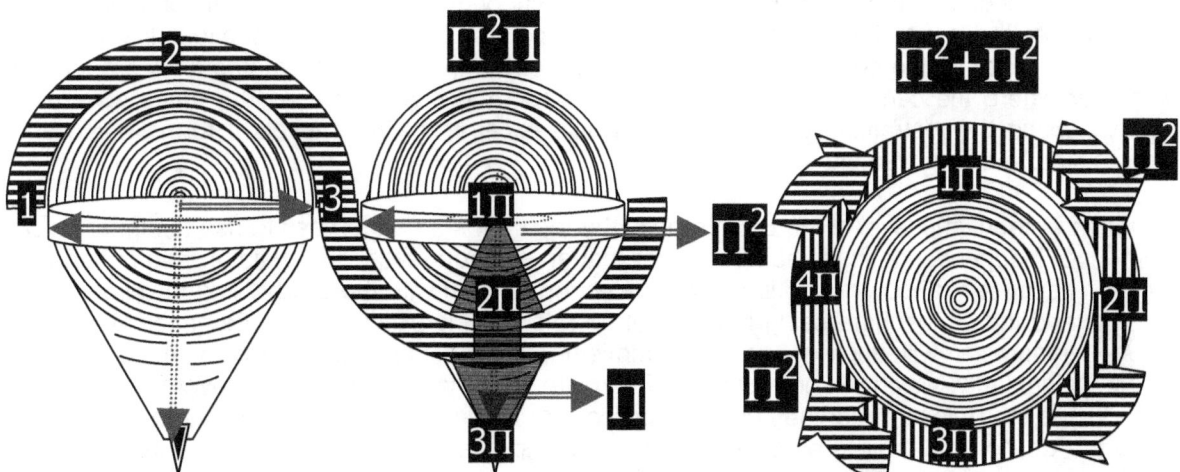

The heat that should supposedly under cosmic law drive the spinning top will come from the governing singularity accumulating the heat in concentration by the contraction or cooling ability the top singularity acquired. But in this case the spin is a result of life's ability to manipulate space-time and lead cosmic events. Instead of the tops atoms forming the spin driving the top, life instigated the spin of the top. The heat that would establish such a drive in motion in real cosmic terms would require a lot of nourishing and sustaining from a large number of maintaining atoms that produce a large flow of space-time.

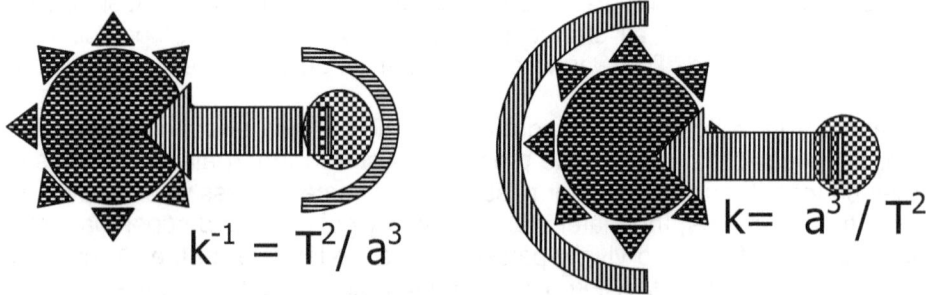

With sufficient energy the top gets into a fighting mood making the top very reluctant to give up this newly established freedom. The behavior now attributed to the top is normally the manner how a star develops in the galactica cocoon in the galactic center and is in principle how the fledgling star gains its birthright to leave the nest of the cradle of the galactica. It is by the atoms forming a sum total in movement that can support the generating of the required gravity to secure the movement that the heat would require as a drive that would unleash such a spinning star from its frozen galactic womb. In the case of the top the forming of the drive is done by life where the rotation can establish the center singularity that is governing the controlling singularity. In the top this drive comes about from life intervening but this same method implements the release of the new star from the blanket of heat within the center of the galactica that covered the star up to the time of release. This forms all part of the sound barrier that is in effect how stars liberate from the inside of galactica. Before the movement confirms a centerline the "fetus" star is frozen solid and pins by the movement the galactica establishes. Just as the spinning top would do, the spin free the fledgling star from the cocoon and by spinning past the Roche limit the star gains independence.

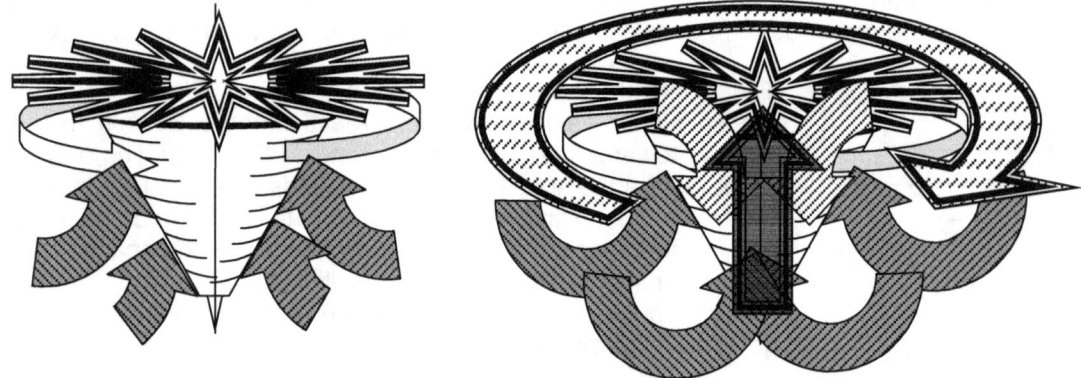

The example we can gather from the top shows how desperate a governing singularity can become and how such an exited singularity can put up a fight for freedom and independence. The top is in a fight for independence while the Earth is restraining the independence. The top is in the role of the fledgling star and the earth presumes the role the galactica would fill. This will also be the case when a star is born from the galactica center cocoon. The fight goes on until the Earth suppressed the last bit of motion that the top had and the top uses the last motion it had to defy the earth control, but this is only because life placed a limit value of movement into the spin of the top. In the case of the infant star the atoms from a spin that evoke the center where the center forms a line that will begin the star's cycle as the star's governing singularity. By fighting the restraining the young star would excel in the heat levels it forms within and thereby grow in stature and movement.

When the motion exceeds the level of the Earth gravity the top shows an eagerness to rise to higher levels of independence in the same manner that an electron reaches into higher rings of energy because the top with motion is in an electron relation with the Earth filling the proton role and the atmosphere being in the neutron role. This is exactly the same principle that the infant star will apply to generate movement within the atoms forming the controlling singularity, which will promote atomic growth, and that generating of movement will have the star release from the centre galactica. Let's quickly establish events as they translate singularity from a dot to a controlling entity that is commanding space-time through the establishing of a separate individual drive. The motion comes about which proves to be that which generates the gravity that drives the individuality in the top.

In the sphere centre is a spot that has to be there mathematically by measure of $(\Pi r^2) / (\Pi r^2) = \Pi^0 r^0 = 1$. In order to provoke the line into action motion is requires movement confirming space just as Kepler indicated where the space becomes equal to the motion and the motion is equal to the space $a^3 = T^2 k$.

The Four Cosmic Pillars; The Result Thereof. Page 241 APPLYING AS THE SOUND BARRIER

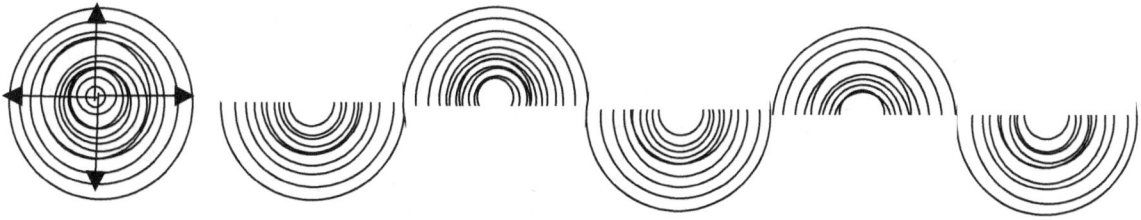

On the time line that forms the divide will establish a part formed in space and a part formed as movement. That is what we learn from Kepler's formula.

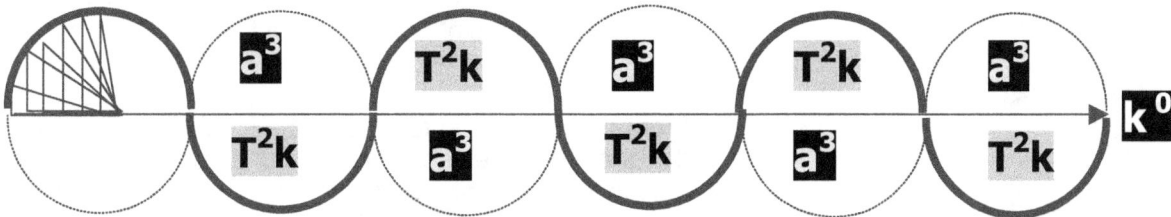

By the motion and the singularity the top evoke a graph from where the graph runs along the line of time.

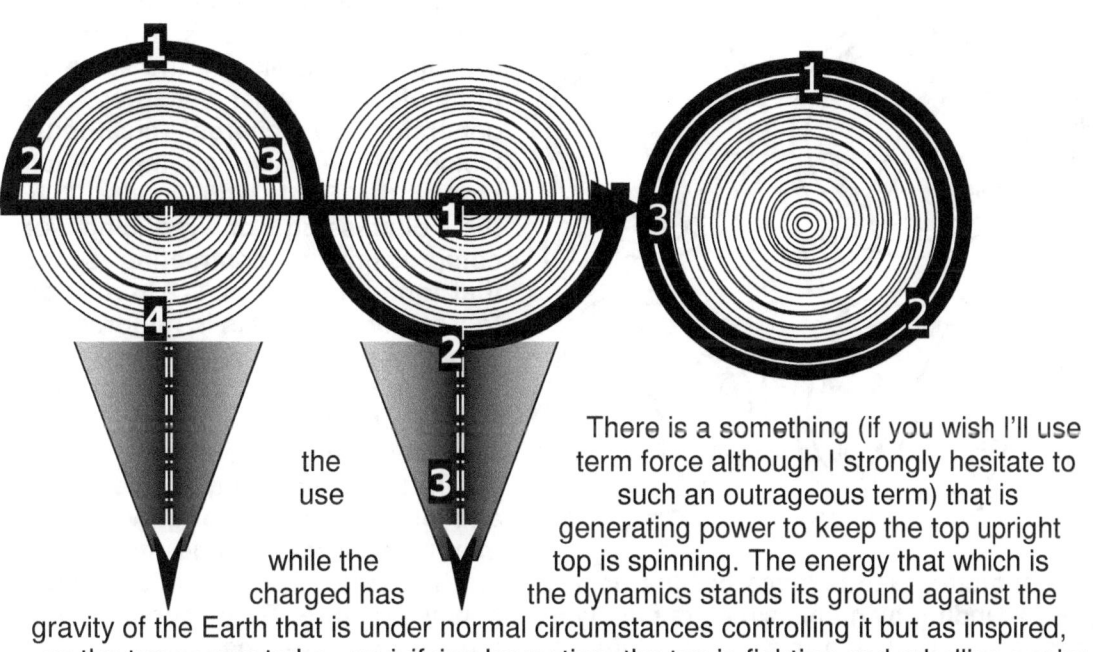

The balance is a control of motion that is established as a flow of space-time supports the ends (4) holding time while this generates the space (3) singularity containing and creating the space (3) in which the spinning takes place.

There is a something (if you wish I'll use term force although I strongly hesitate to such an outrageous term) that is generating power to keep the top upright the use while the top is spinning. The energy that which is charged has the dynamics stands its ground against the gravity of the Earth that is under normal circumstances controlling it but as inspired, as the top seems to be revivifying by motion, the top is fighting and rebelling against the Earth gravity.

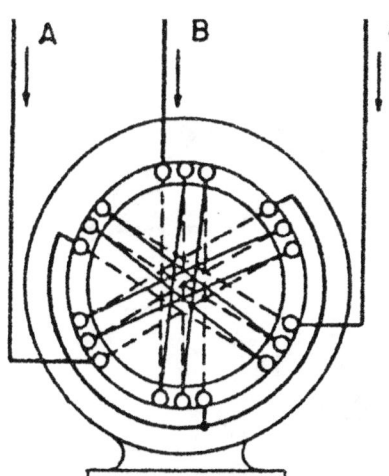

The top is self-driven, as an electric motor would be. The difference between it and an electric motor would be the origin of the source from where the energy comes which drives the spinning top. The top stands upright as individual as any self-propelled object can be. Although gravity is retaining the motion of the top, it is not contradicting the motion. It is not combating, but is merely suppressing or restricting the level of the motion. What we would think of as air restriction is no restriction because from the restriction comes support that keeps the top standing on a very thin needle edge. The top should tell us so much about nature if we would only listen and learn and not tell nature what we think nature should tell us. If Newtonian would only stop to dictate to nature and instead learn from nature when nature tries to tell us what we should know. The top is busy with the same process as charging electricity.

The Four Cosmic Pillars; The Result Thereof.

APPLYING AS THE SOUND BARRIER

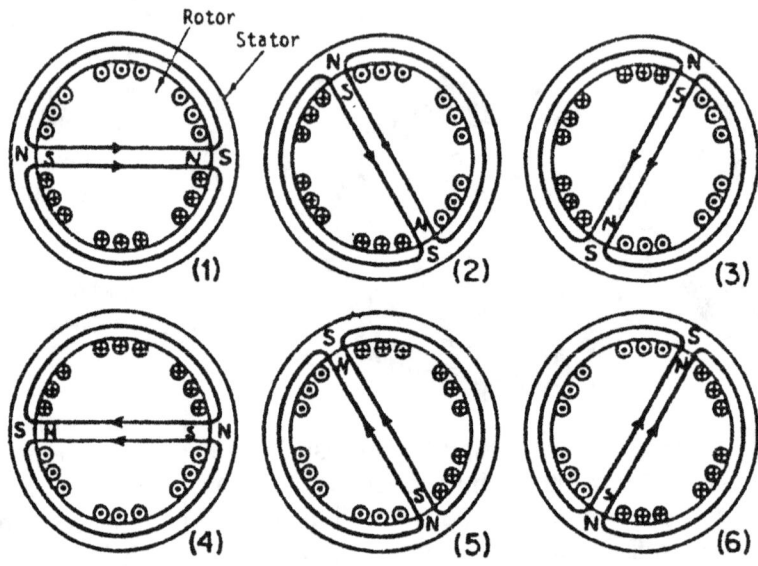

being the Coanda principal.

To generate electricity there has to be movement between iron and copper. In order to generate gravity there has to be iron moving in relation to copper while this is spinning in space. There has to be $k = a^3/T^2$ and there also has to be $T^2 = a^3 / k$. It is the rotation of iron in relation to space that forms the displacement of heat from space being 112 to iron being 55. If you wish to have the formula it is movement of iron that reduces space where space is $10/7(4(\Pi^2 +\Pi^2))$ and the spin of iron displaces or diminishes or reduces the space to a proton value of $7/10(4(\Pi^2 +\Pi^2))$. This puts electricity and gravity equal. I explain more about this in the Cosmic Code.

From the motion the top inspires by creating a situation that the top can establish a force or an energy, which is able to keep the top upright, is equal to the establishing of gravity and electricity. It is the very same principle

That which charges the top to stay upright uses the same principle as what makes the Earth spin and this principle which also is the same as principle as what charges the generator to charge electricity, which is the same as that which charge the Earth with gravity. It is the dismissing or reducing or compressing of contracting space that forms a factor of 112 in relation to a factor of 55. This is no force but it is merely a flow of space-time, which is contracted and condensed by motion and the duplication that the movement establishes brings about the conducting of electricity and also the forming of gravity. Man may name it gravity and then name it electricity or name it motion and balance but the names is just like hot and cold, which means that it is labelling and branding by name, which is merely a man, made culture. To the cosmos or in the Universe the process in principle is the same thing. It is what started the cosmos. It is what drives the cosmos. It is the engine giving motion as it is giving discipline in the cosmos. It is producing space to heat and heat to material.

It can be only singularity that is keeping the top upright. One must remember that the part doing the balancing, that is creating the space in which the top is able to spin, that is establishing the necessary time distance in which the top can spin, that is establishing the time difference that the top can use to apply the motion that extends the time is singularity. The line that is evoked is not in real terms part of the Universe because the line has no sides, can't move at all and is the only substance that is there fall all to see while being very much invisible.

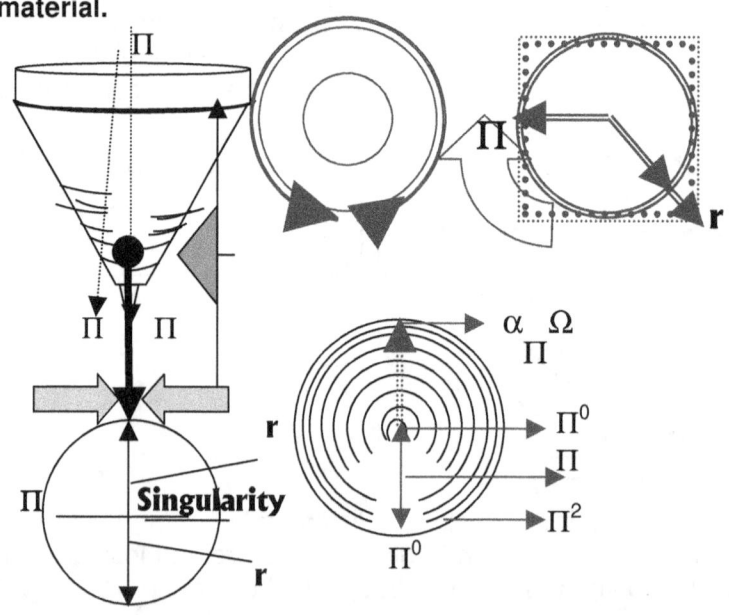

The line has the responsibility to establish everything that is in the Universe while the line by own measure is not a functioning part of the Universe and yet it sets the absolute control of what is going on in our Universe. Without the line being part of space in the Universe, it is what drives the Universe in time and in accordance with time. The line is invisible, in detectible, has no space, can be called into action any place and anywhere from where the line immediately establish control of what is in the Universe and creates matter in space in time in the Universe in infinite detail while the line is even less than infinite.

We have to recognise that gravity is a balancing of motion where rotation has to refer to $k = a^3 / T^2$ the linear motion of space $a^3 = T^2 k$ and therefore forming $T^2 = a^3 / k$. The proton must compromise in its motion to enable the electron to move, as the electron has to provide the proton space while the electron is moving. Therefore the spin involves space already established and the space contracting is space of other parts that overheated and in relevancy liquefied to fluid heat. Therefore the expanding becomes part of the growth as it becomes the contracted space. It is this effect that we gave the name the Coanda effect and is as much part of gravity as contracting liquid to solids is part of gravity. The spinning top again is as much proof that the Coanda effect is a product of gravity than flying is proof of the Coanda effect where it establishes antigravity through motion. As the top spins a centre is established with the motion of spin activating the centre not spinning. The centre comes about, as the centre remains motionless while the rest is spinning and the centre becomes an additional part being part of the motion.

That is how stars are born and that is how material became space. Time left its mark as time moved on and what time left behind is space forming material and also forming outer space. However it starts with time forming space in deliberately starting motion. It starts where space starts and where time releases space. This process is as apparent as the spin process of the top is. All the characteristics are still present.

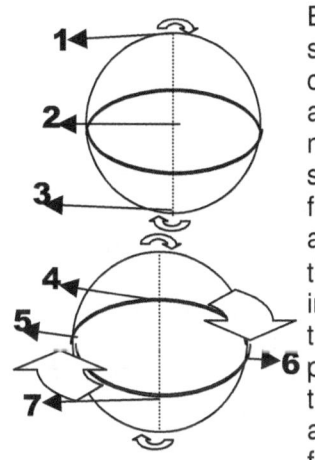

By not having motion the line forming in the centre also have no space as the space that forms extends from the line serving the three points forming the centre and growing to the outside. Where there is no motion, there is no space and where there is little motion there is little space. The only space the line may relate to can be a point that is on the border of the sphere that is crossing singularity and connecting the two edges on either side of the sphere that is forming the sphere. That means the line from one point holding singularity to another point holding singularity that line will cross the centre line which gives the line in singularity valid space-time to control. From this line a centre Π will initiate a circle to the value of four points moving in a circle. The four circling thus form the moving points will relate to the centre point evoking the two points that is motionless yet that also relates to the centre point as a result of the four points moving in a circle. Singularity forming the line does not have the ability of motion therefore singularity does not hold space. Yet singularity that forms the circle in movement may change the relevancy applying, but will never stop moving in a circle. Singularity is also eternally indifferent to motion and motion can excite singularity but singularity cannot be shifted by motion.

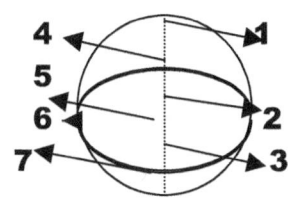

Every time motion takes place, the centre line holding 1, 2 and 3 stands still while evoked by the four points forming the circle. There is only generating going on and this generating takes place where space is reduced to the rim however there are four points that does shift by movement. The points shift from one location to the next location by generating space-time.

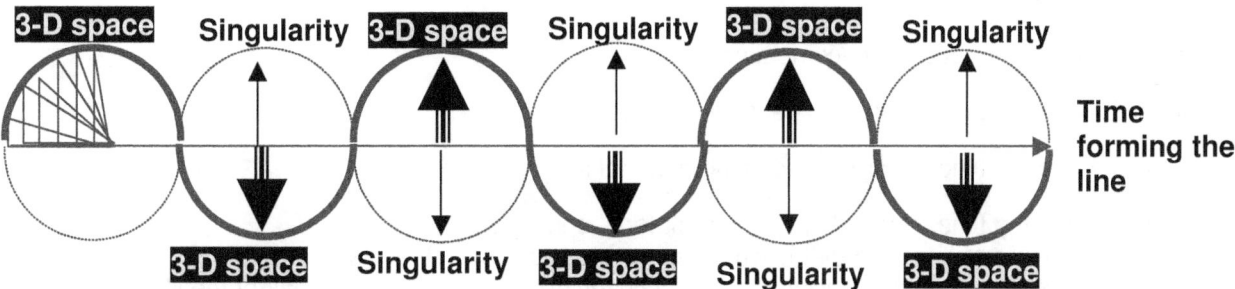

Space forms as a result of what time leaves behind as time soldiers on to bring repeat to what was before that then forms the history of time left as space.
In the roundness of the spin four points form as movement forming eternity that relates to immovableness of the three points forming infinity.

The Four Cosmic Pillars; The Result Thereof.

APPLYING AS THE SOUND BARRIER

This is how the smallest there is become part of the biggest there is. This is how the cosmos generate from singularity that which forms space.

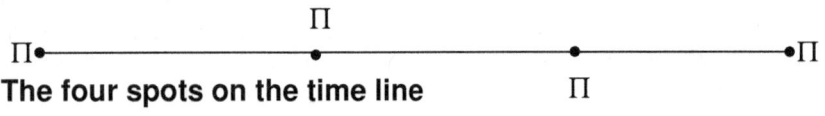
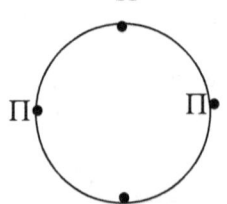

The four spots on the time line

The three dots spots on the space line

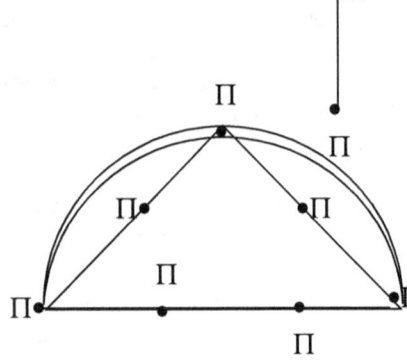

The time is by moving putting every aspect that is going to be in relation to every aspect that was before by being what time is in the present. The line is confirming the half circle forming what represents space, which was confirming the triangle where singularity then forms the very same of what is going to be. All that becomes form were 180^0 while representing singularity and then from that all that forms space were half of what singularity holds. Even the line is half with the growth being 0.991 on the one side and the other side of the Universe forming 1. Every aspect of the Universe is still part of the flat Universe that by going into relevance become 3-D and every line is a triangle as much as it is a half circle. There is no motion in infinity to form time and there is no space to set time in eternity.

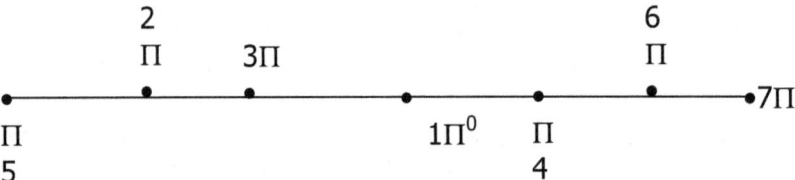

The sphere was a cross with three points stabilizing four points, which was on a line running from one to seven points.

It did not matter the way or from which direction the first spot is numbered because the last number would be seven. The Universe accommodates a flat structure and not one with certain sides removed to serve the explaining of the space. The four moving resurrects the three no moving and the three not moving confirms the four moving where the total rises to the value of seven. If we put the seven in relation to the full compliment of time, then we find we have three more point bringing a total of ten.

```
      2               6
   Π    3Π            Π
 •——•———•————•————•———•—•7Π
   Π           1Π⁰   Π
   5                  4
```

1•Π⁰ Singularity governing

2Π• Singularity in relevance

3Π• Singularity in relevance

•4Π Singularity forming motion to become time

•5Π Singularity forming time distortion becoming space.

•6Π Singularity forming time distortion becoming material in space

•7Π Singularity forming time distortion becoming material ending space

This is where space starts because that is where space started because everything within the Universe is a repeat of what once was. It is not coincidental that only any person can ever view half of the Universe.

Should one ever wish to paint showing that atoms formed as a result of the one mass groping the other mass as the two were clutching one another till dooms day arrive bringing the required coupling by mass coupling then such a venture has to start by providing the incentive to initiate a centre from where the clutching may originate and why such clutching will originate. If mass was first to create such a pulling, then what was before mass to create the pulling that formed the initial mass? What brought about a centre Πr^2 that brought about the mass one went on to groping onto the mass two and why did mass one form as well as mass two formed because the first and foremost forming of mass one and mass two before the groping started had to involve a centre beforehand and afterwards it proved it become groping mass.

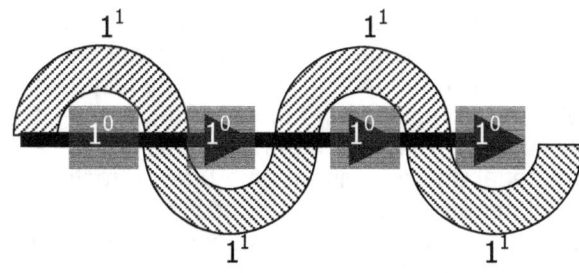

If the Universe did not grow through time as it does, then from where did everything originate? The Universe had to start somewhere to have accumulated what it achieved this far and all that can add to the Universe is what time deposits as space. There is a differentiation in relevancy between what is hot and what is cold and that forms the ever-growing Universe. In that we have infinity being in moving relevancy with eternity by creating a Universe in between 1^0 and 1^1.

But before even being a mass it had to have a centre to secure that mass because having mass means whatever has mass is pulling towards a centre. What brought about such centre into the cosmos? The first evidence of mass being a factor puts the cosmos in a position that already formed a complete Universe where this going back then stops at 10^{-43} sec. When the Universe was in its beginning it began its first material in whatever form the Newtonian wishes to imagine, then there had to be a centre to start whatever mass was groping another mass that had to have started with a centre. By not providing a centre one drops into a basin fluid representing Newtonian backward nonsense since the initial explaining of what and why and where the very centre arrived at that brought about the groping in the first place the whole notion is a loose unstable not very well thought through blubbering of the ridiculousness. It matches all facts in truthfulness as is found in the realms of Little Red Riding Hood. Why did a centre taken from $F = G \dfrac{M_1 M_2}{r^2}$ where eventually r would be one ($r^0 = 1$) and mass multiplied and divided would be in relation to one A centre must form because if there was one centre that is the Newtonian pulling then that moment mass have also had to come from some other centre where that centre had to establish the mass notwithstanding the size or nature of mass. Mass is material contained in movement by restricting free movement so what contained the first mass before there was any mass to contain around a centre. At least I can bring a centre to the table as I prove with my reasons I give for such a centre forming before even any mass came groping and snaring other mass. I can indicate with a fair amount of logical accuracy why such a centre will come into place. I can show how and why and where motion brought on this very first groping and clutching if there is groping and clutching to begin with. I can show what phenomenon is responsible for this centre to become practical and I can show the phenomenon positioning and locating of such a centre. I can show in the shortest manner by using accepted mathematics in the simplest of forms why space is linked to time and all space revolves around a centre using time to do so. I show what natural occurrences are producing motion because producing motion provides the heat in the space it establishes. Space forming to accommodate heat expanding is providing motion that brings about cooling through duplication that is the result of motion. Moving puts the object in smaller portions distributed in more space in the same time duration and by moving the space the object requires where then by duplicating or moving it is exceeding what it was before. By then distributing the space over a wider area in smaller quantities the heat is distributed over a wider area making the general applying heat lower. Cooling establishes duplication, which provides reducing of heat as duplicating reduces the product and in that duplicating the doubling of space is halving the heat in the space. It is a natural process that can be tested anywhere in the cosmos. Heat something and the space grows bigger whereas cooling the something will naturally reduce the space it holds. But the cooling can only be when the air moves rapid over the warm object or the object moves rapid within the air. If you wish to cool then movement is the only way to do it

Hydrogen is as much a liquid as iron is a gas and neon is a solid. It depends on the element relating to the space/heat in the circumstances surrounding the substance at that very precise instant in time. We have to stop telling the cosmos to show us what we wish to find and start accepting what the cosmos is telling us to find. No element is either a gas or is a fluid or is a solid. We arrange the elements in such a manner, but that is only applying to the situation the Earth grants the elements. That is why the Sun freezes outer space to a liquid around the Sun and that I explain in much better detail in The Cosmic Code.
When an element freezes it is solid notwithstanding…
When an element melts it becomes a liquid
When an element boils it is a gas again notwithstanding.
By having movement it cools as by standing still it overheats and that is why solid stars are frozen space and outer space is overheating into the oblivious…

Gravity is the duplicating of material by the dismissing and removing of space that leads to the concentration of heat. Gravity cools space down to liquid stored in concentrated space and further freeze space to turn liquids into solids. By duplicating the protons bring about cooling that that freezes space-time into the single dimension of singularity by removing space-time altogether.

But as it is impotent to realise the above, it is just as important to realise that heat is another form of material and a separate form of material. The two developed on equal basis and as a result of the other. The one produced to save the other and what the one produced saved the other. The one principle brought the incentive for motion while the other took the incentive by providing the motion. The one produced what the other captured and the one retained what the other delivered. Eventually the motion did not bring the required relief and another form of substance had to be devised. By overheating and increasing space it counteracted overheating and by removing the expanded material and retaining it onto the contracting of the other did the two form a synopses where by all received benefits in the form of cooling. Only when further requirements develop did the need arise for more to be made available. The first demand on motion asked no further changes because one change brought on satisfaction to all that suited all. The second was more general and on an ad hoc basis that was established to fit the need of individual places and not groupings in the broader perspective to fit individuals at large. At first the establishing of motion set a trend that brought on required results but afterwards the space required in which to move became a demanding issue as the heat levels required out of control.

The heat had to be stored in space by becoming space to retain heat for later consumption. The number in ratio that produced the heat providing particles that offered to release their form in contribution the have those that retained form do so to save those others in retaining form. But those on offer became those ones that became the danger of destroying Creation instead of saving Creation. There might even be some areas and regions in far off places in our modern day where an imbalance may evolve and some particles become unsuccessful to save those more successful. Those we named Super Nova stars. By going less successful the singularity places a demand on another bringing about the command on space-time so that support can be accomplished to save singularity. Therefore by losing density is a way of gaining security to survive as part of a bigger relative. Density is the distributing of heat in specific relative space and by having less material in more space the density is the offering for the common survival of the lot. But the relevancy brings a contribution in whatever role to secure the survival of the lot in relations. No relevancy therefore can be "nothing" notwithstanding Newton's opinion about the matter as Newton had the opinion a relevancy acquired by rotation brought about an accumulation resulting in nothing.

Seeing that I bring a new aproach to cosmology there inevitebly has to be a new concepts arriving and these concepts will be accompenied by new termenology.

Reading this book re-affirm that that is not my view and I dispute that view in all my heart.

Aanplasing, verplasing, versnelling and inperking
As this book is a translation from Afrikaans originally, some terminology and expressions I had to revise to accommodate my ideas. Where I could I used modified English words to express a new thought to introduce the new idea. One such a term is gravity, which I wanted to change to time because gravity is actually time in the process of establishing space. I had so much criticism about changing the word, which I feel I do not deserve I had to drop the notion. I wanted the change because there is a certain notion clinging to the idea represented by gravity. Gravity links to a force that is all compelling, but I do not agree with such a compelling force, such as the word gravity implies. Gravity I introduce, works on two principles, but gravity to Newtonian standards is a single force. When I refer to gravity, the normal reaction is that I am referring to the force which the existence thereof I deny. By declaring that gravity the force controlling the entire Universe as a standard constant that concept is non-existent. However by changing the concept I bring the wroth of the scientific world upon me, not that I do not do it in any case, but there is a lot more resentment and misunderstanding when I say gravity does not exist. When I make the statement that there is no gravity, every person considers me mentally unstable because they feel gravity in their everyday life. Of course there is a movement of energy keeping all objects attached to the Earth, but gravity implies work, and with that work principle I disagree, because that is not work, that is a cosmic balance that started at a point where eternity split from infinity and will end at a point where eternity meets again with infinity. Only life standing alone and detached from the cosmos being an energy source not part of the normal cosmic time flow is the energy that is able to manipulate space-time. In mimicking cosmic time life can commit work in the sense of being a force. Life can achieve movement not conscripted to cosmic time flow and being a force in that sense it achieves a duplication of cosmic time running concurrent with cosmic time from beginning of life to the end of life and this is then running concurrent with the rest of the Universe where the rest of the Universe is in a balance. The Universe is setting time forming space to which life must adapt and adopt but in that science have an un comprehendible inability to realise the difference between life being a

cosmic addition and what we find is a natural time flow in the cosmos. Life is within the cosmos and the cosmos is without life. In the entire Universe there is no work, it is a balance running concurrent through time forming space. The balance shift in some cases to favour space and in other cases to favour time more but in all of that shifting, a continuous balance strikes every aspect of space-time. This applies new ideas that were never yet brought to light before and the new concepts clashes with the conventional names that science applies to current ideas. I had to divorce the normally used science ideas from those I introduce and the only way was with introducing new terminology. I have to start implementing the newly created terminology, which will apply to the rest of this book. Gravity can reduce the relation of flow by employing the concept of space. However what I write comes from what I wrote in Afrikaans and the ideas was born in Afrikaans. This stems from my lack in ability to find suitable words in the English language that would define the concepts as they are, in order to establish the difference in meaning from the current words, which convey the existing misinterpretations (or if you wish, to my view incorrect applications).

This applies new ideas never brought to light before and the new concepts clashes with the conventional names that science applies to current ideas. I had to divorce the science ideas from those I introduce and the only way was by introducing new terminology to the English language since the concepts could not be correctly described when using normal English. I have to start implementing the newly created terminology, which will apply to the rest of my explaining detail in my more advanced books. This stems from my lack in ability to find suitable words in the English language that would define the concepts as they are, in order to establish the difference in meaning from the current words, which convey the existing misinterpretations (or if you wish, to my view incorrect applications).

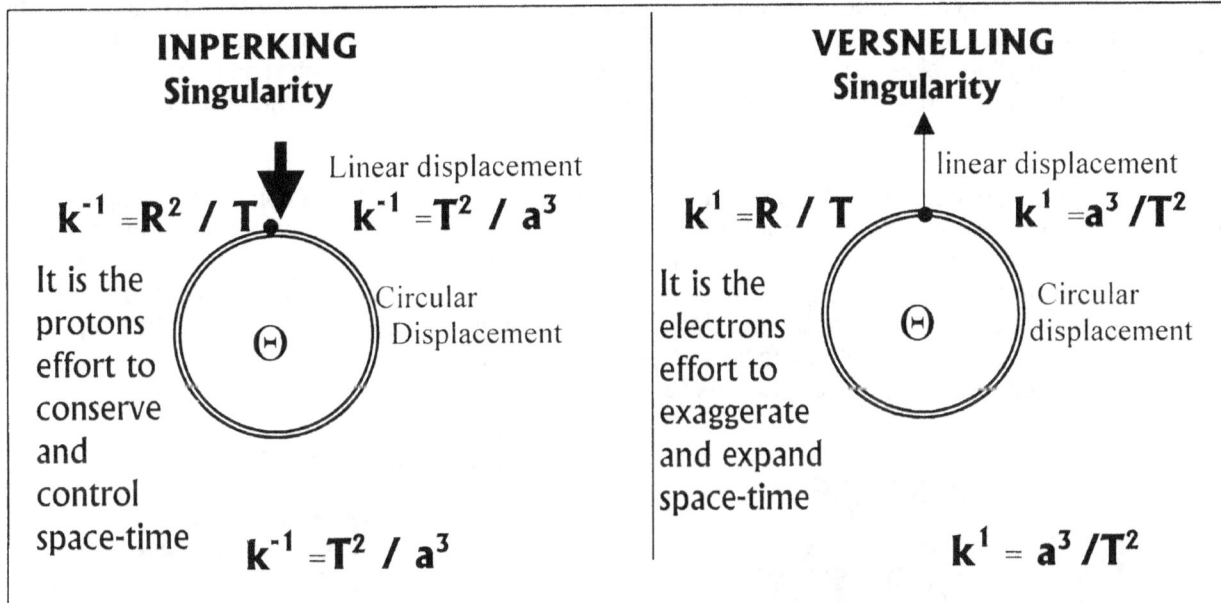
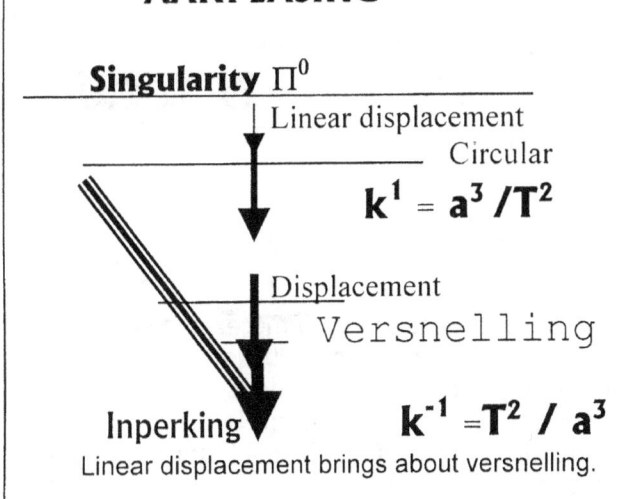
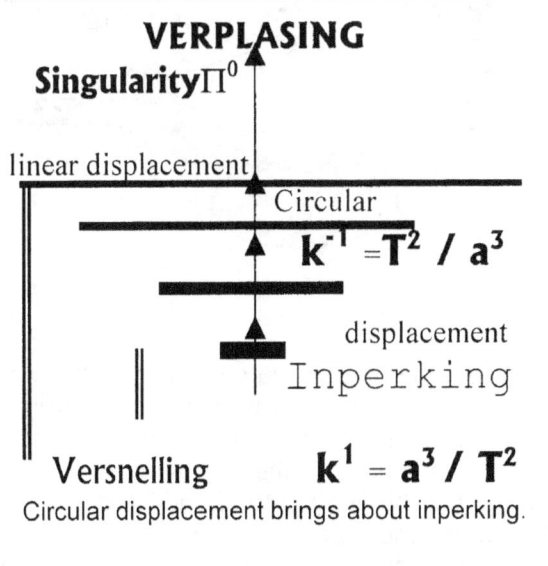

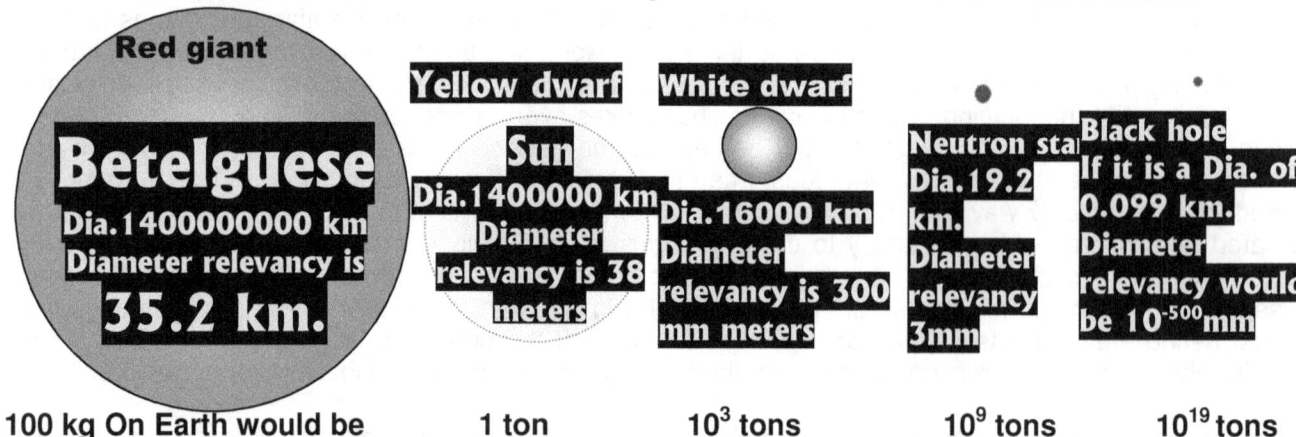

Firstly, we start with the word *densified*, which is not a normal English word, but a word I had to produce in order to make a comprehensible statement. When an atom is in a larger star then to what we see is that the atom shrinks in size but it does not shrink in size because it becomes denser and therefore gravity the atom is densified. The correct word that applies is concentrated, that much I do know about the English language. But concentrated has not the correct meaning or the expression that I would like to bring over. Concentrated can apply to any substance, be it gas, liquid or solids where one of the ingredients become more than the rest of the ingredients. In that way, matter as a solid substance produced from the eternal substance which is heat, cannot be concentrated. Nothing in the entire Universe can compare with the density of pure heat that spins at a rate in which that very heat can produce a value and which has a density far beyond anything else. Therefore, I chose to use the concept of concentration in a position where it makes a lot more sense and when gas is liquefies it is concentrated but when gravity reduces the "size" of the atom it is densified.

A star is concentrated space-time, but there is a huge difference between a star's concentrated space-time and the value of pure matter. When a star does therefore become densified space-time, it can only be at the end of the Big Crunch eternity, witch I prefer to call moment Omega; that is when space becomes infinite once again matching time being eternal. In this light I chose to call matter densified space-time. Densified space-time should therefore be in a definition where matter or substance has reached a point in density that will last one eternity, but has no limit. Concentrated space-time, on the other hand does have a limit, which is at the point where it becomes densified space-time.

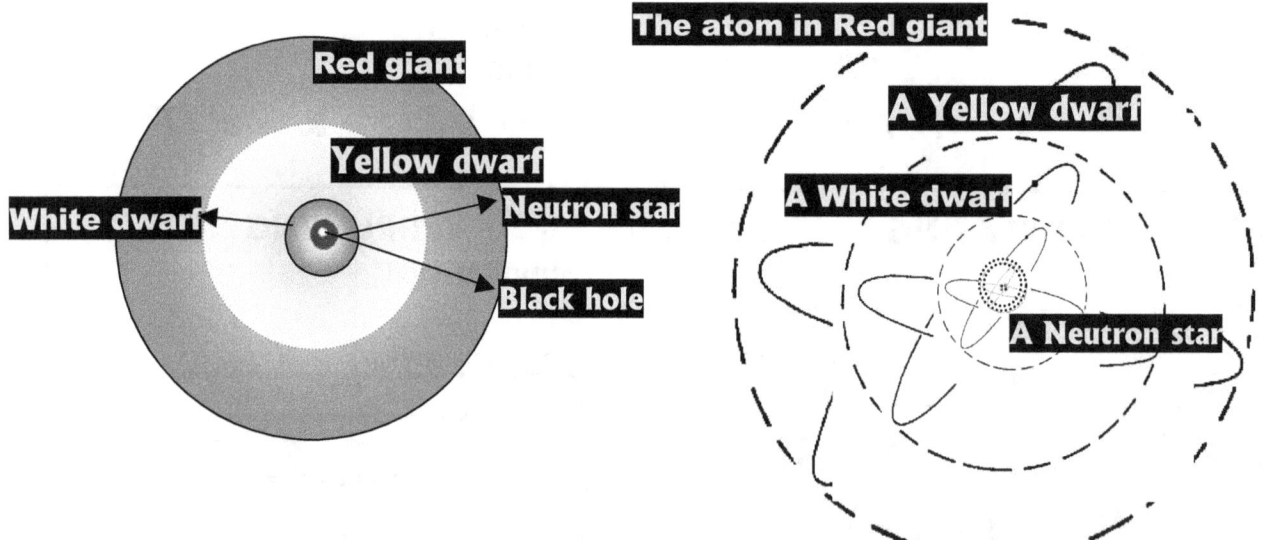

Newtonians dump everything concerning gravity on mass although according to the examples taken from their files they know pretty well that increasing in gravity brings enormous reductions in atomic sizes, increases in mass which means increasing gravity must bring about incredibly denser atoms which are much more compact and the gravity must bring about enormous speed increasers in gravity spin, yet to stick to the bullshit and not to offend Newton they bring no changes to the technical terms accompanying the acknowledgment of changes coming about in stars due to gravity increasing. The Newtonian has to realise that increasing mass by volumetric size does not increase gravity; however reducing volumetric size increases gravity radically. In the more sophisticated books I have no choice but to refer to the terms I now introduce because there is just no such terms available as references.

The second word I created is **Aanplasing**, which is the ongoing redirection of heat as in matter to heat as in time and that connects to a circular deepening of the separation that matter undergo transforming to time as it discards heat for the cold of fusion. Later (I hope) it will be clear enough for every reader to comprehend and to distinguish between the various factors that bring about **aanplasing** as should the reasons be clear why I prefer to have created this new word.

The Atom's gravity process employing aanplasing, verplasing, versnelling and inperking

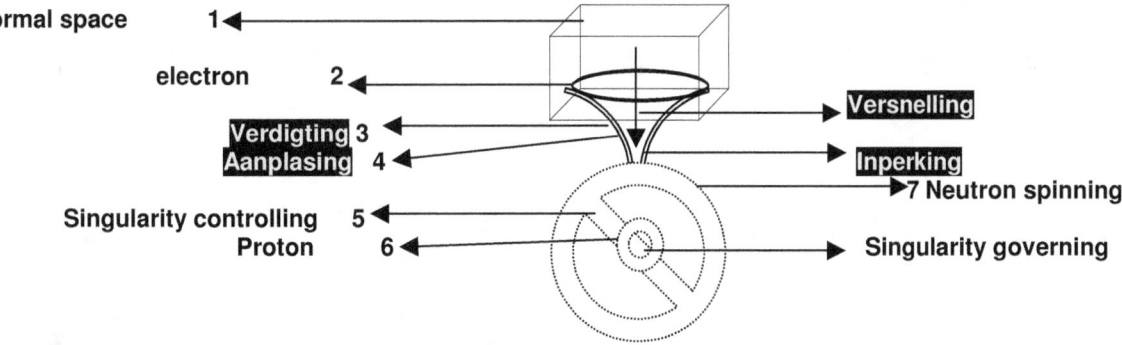

Aanplasing is the way gravity not only takes the object to the centre of the star but also increases the spinning speed while it reduces the actual size and increases the applying density.

In this case however, there was no English word to my liking that could merely be altered and then be re-applied. With a choice of no choice to my disposal I chose to alter an English word as was possible with densified. A more suitable word that relates to a better meaning in the case where I brought in **"densified"** would have been the Afrikaans word **"verdigting"**, where "verdigting" stands in relation to "konsentrasie" (concentrated). The fact of the matter is that I am not wilfully forcing Afrikaans down the throat of the Anglo-American and in the case of density I was able to adapt and modify a known English word that could adopt a new concept. Unfortunately in the case of aanplasing using an English word would mean that there is no liberation from the "misleading" focus that depends on gravity, nor can it liberate the feature of this "misconception".

As for the Afrikaans words: **aanplasing, verplasing, versnelling and inperking**: there are no such words or concepts in existence that the precise meaning can derive from the English written or spoken language. Should any such words exist, the misconceptions that remains connected to the original English words, would not bring justice to the concept which I wish to apply to convey the meaning that lies behind the correct value of the thought. If I stuck to the word "gravity", the concept I wanted to introduce would forever remain confused with Newtonian application. To that end the new realization would then never come across in the way I intend it to be.

The R that I use in the formula has nothing to do with Radius as a term, except when used, to calculate the value of a circle or a sphere. The R is derived from the Afrikaans word **Ruimte**, which means space. In the Afrikaans word: **"Ruimte"** the **"u"** and the **"i"** is used in conjunction, which is pronounced the same way as " ai" is used in English words such as in **p**ain, dr**ai**n, tr**ai**n, v**ai**n, r**ai**n, etc. So spelled incorrectly it should be pronounced as **"Raimte" and the "te" is pronounced "huh"**. The T stands for the word tyd, **which incidentally is time.**

This is what I named negative space-time displacement, which results in **verplasing. The word verplasing is pronounce FHERPLHASHING (FHER – PL –HA – S – H-ING)**, which means to "relocate" without destroying or changing the composition in any way, as the object is moving away from a certain position.

The word **aanplasing is pronounce AHNPLHASHING (AHN –PL – HAS – HING)** and literally means to relocate without damaging or destroying the composure or structure of the objects, as the object is moving toward a certain position. This very same value was previously mistakenly confused as being gravity. It is the effect on matter where space-time is in motion and matter is relatively motionless.

Both **aanplasing** as well as **verplasing** cause time differentiation and matters structural re-valuation. That means the duration of time is re-valued and the space compromised. The excelling of the time factor is; versnelling and the reduction of space are: inperking.

The **word versnelling is pronouncing FHERSNELHING (FHER –SNEL – H-ING)** and means to speed up. If one is placed in a star it would seem as if time inside the star is accelerated while time on the outside of the star would come to a standstill. This concept is explained in far more detail, in a later stage in the book. The **word inperking means reduction or containment of the structure or scaling it down to a different size without penalizing or altering the shape in any way.**

Both inperking and versnelling is how matter relates to change **in space (inperking)** and **time (versnelling)**. The generation of the heat is within the structure and relate to time in space.

Inperking: This stands apart from the idea of curtailment because in curtailing. The **In** part is pronounced as in English where the **per** one pronounced in the same fashion as the sound an English sheep makes BHE placing a <u>H</u> sound before the <u>E</u> with **king** already explained. Sorry but that is as far as the Afrikaans lesson goes for the day.

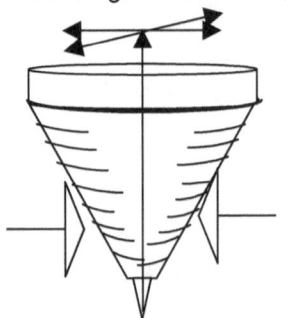 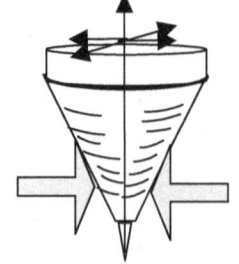 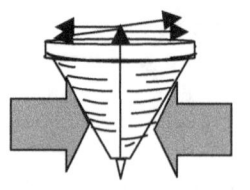 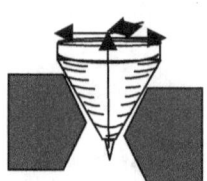

A top in outer space **A top on the surface of the Sun** **A top in the middle Layers of the Sun** **A top in the very core of the Sun**

Inperking: This stands apart from the idea of curtailment because in curtailing something or someone, means that object's or person's movement or moveable motion is deprived. This then brings over the misconception in the accepted notion of an expanding Universe. In due course I shall explain the concept, but inperking involves the same value that was there at first and will be there in the end, only the location in the balance shifts to favour one or the other part of the same coin. Because of the fact that none such a thing applies when space-time "accelerates" (versnel), and where this brings about inperking, it does not apply. Instead, all functions and factors still apply when inperking becomes valid, therefore the meaning of inperking becomes more applicable and this word describes the process much better. Inperking relates to time, where the duration of time extends, but not the value of time as such, as time applies in the cosmic sense.

One should realize that the entire atom, as well as its surroundings including all other surrounding atoms are reduced in space-time volume, so the atom is not actually curtailed, nor is its surrounding which means the word curtailment does not really apply. All aspects of occupied and unoccupied space-time are in reality, re-focused down in the true sense and above all, remains to the precise relative relation value it had for one entire eternity where the relation between such times, only refocuses. However, scaled down would neither apply, because that would not refer to the time involvement, which lies at the hart of this revaluation. Where less time applies, inperking would be more severe and where more time applies, less inperking will apply.

In this reference to time, one second would remain one second to matter inside the star but the duration of that second, compared to geodesic time validation, would appear to stretch enormously. All words in the English language by implying its dictionary meaning, will inevitably lead to more language confusion, seeing that the explanatory meaning does not cover time enhancement and space reduction, heating and slowing of time lapse. By introducing a new word to the reader, I hope to screen out any misconceptions. Hand in hand with inperking, goes versnelling. When the reader encounters the concept of inperking, it should accompany the idea of versnelling.

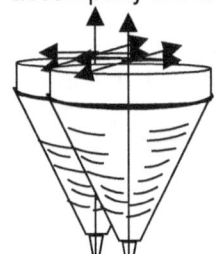  

The faster it moves in massive stars

The more it duplicates

...into smaller fragmented but much denser particles and that is why in truly massive gravity the atoms literally disappear into singularity

Movement by duplication increases because gravity is movement acceleration. That is **<u>Versnelling</u>**

<u>Versnelling</u>: It carries exactly the same meaning as acceleration, but the meaning or concept connected to acceleration applies to matter as the matter increases its own positional change in space and time. That is not the impression I wish to relay when referring to versnelling, because it is exactly the opposite of that meaning. In this, the actual meaning is more applicable to the true connection. These I must explain carefully, not to convey confusion. When a person stands outside an explosion of some sort, the time laps seems instantaneous, quicker than the senses can relate to. However, inside the explosion, time is almost standing still. By creating more space, time reduces in its instant component. That means the more space time is relevant too, the slower the instant would become and the longer it would take to gave the following instant too arrive.

Any person, who is inside such an explosion, would relate to time on the outside as being enduring long. Any person on the outside of the explosion would relive the explosion occurring as instantly. Whether this statement is accepted or not, the truth is that a person in an explosion takes a duration of time that freezes to die, although his body is shattered in a million pieces. Persons dying on the electric chair will find that time increases so much while having the current flow through their body that the time in the body concerning life increase to eternity and the time it takes for the person to die is stretched to a point where the person can't die. It takes to long to die. It would be much better do give many short jolts where in-between the flow of current time could be restored to a normal body experience and then increase to the value electricity places on it. A person experiencing the explosion coming from an atom or nuclear bomb will find the person will never reach the other side because time freezes life to the space within the cell. All this information and much more I intend to explain in the book I aim to write after completing this series. I hope to name it **_The Absolute Relevancy of Singularity in Terms of Life._** A person experiencing a nuclear destruction person is sealed in a period separated from the period he and we lives in. In the nuclear detonation the connection between infinity ands eternity is demolished as the atom that places the distinction in place is destroyed into liquid. This I aim to explain in **_The Absolute Relevancy of Singularity in Terms of Life._** The time duration slows down immensely, but from the outside, it accelerates immensely bercuase the differentiation between eternity and infinity grows immensely. Therefore, time versnel to the outside of where ever one relates to.

I wish to bring over the fact, as just been said, that the concept we have, is quite the opposite. Versnelling implies that the motional increase lies with the transfer of space-time, regardless whether matter occupies it or not. As aanplasing (not gravity) and versnelling bring about inperking (not curtailment as the body remains free to do as it wishes) the space-time that the body occupies and the surrounding sphere are in constant state of versnelling. The increase in motion has an effect on the matter, but the matter stands weightless as its specific density applies the time in that particular space. By moving through much denser space it will take the body longer to go through a shorter distance.

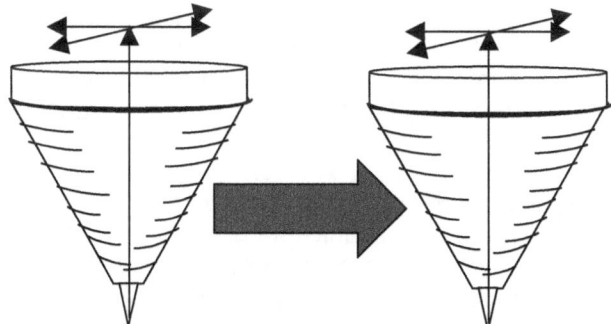

Verplasing is moving by time or gravity from the one point or location to the next point or location. This is in a star normally down and around but could also be up and around.

Verplasing: This word is preferred to that of displacement, because although the matter in motion is displaced, time and space are implicated in the process. Verplasing is in fact the transferring of newly created magnetic space-time by matter, as a body composed of atoms has to replenish the space-time it occupies in order to maintain its position, place and structure in space-time in time in space, according to its geodesic positional allocation within the star's space and in time. Verplasing comes in effect as matter progresses in position, but the time-affect of verplasing that it has on matter, comes into real effect when an object reaches Mach$_3$ depending on its shape and altitude. In short: **_Aanplasing_** is relatively where matter is in a geodesic motionless position as space-time carries the motion component of the two values. This means that aanplasing is relative to positive space-time displacement.

Verplasing on the other hand has to do with the motion being with the newly created space-time in relation with the matter and the geodesic space-time remains relatively motionless. In both cases **_inperking_** and **_versnelling_** is a consequential result of the process. The difference is in the application of the time component itself.

A practical example of the difference between aanplasing and versnelling is as such: a body in **_aanplasing_** is in example where a skydiver is falling towards the Earth and **_verplasing_** is where a body, such as a rocket is on a trajectory path as it fires into space. Both bodies will comply with the linear and circular displacement, but the circular displacement will relate oppositely in each event.

This aspect, Newtonian science disregards, as much as they disregard that there is any connection between the atom and the cosmos, just because Newton never concluded this idea (I suppose).

I deny the fact that a star can have winds, although winds are as close to that concept which the Earth can provide. Winds are gravity but in stars that which moves have so much density one can hardly compare that which move to winds that move. As you will later see in **The Absolute Relevancy of Singularity in Terms The Cosmic Code.**, winds are the transferring of heat, but so is electricity and lightning, and one cannot call lightning wind. Neither can one call lightning electricity. It is altogether different product of the same transformation of heat, but the applied principle separating the products of heat transmitting stand in total different areas where gravity too is the transmitting of heat from a less dense to a denser confinement. As far as ordinary physics go, nothing changes at all because everything is the transfer of heat and that even includes thoughts that are the transfer of heat or electric waves patterns subject to gravity applying.

Every aspect of physics remains the same, except the way science view the cosmos. The formulas I show, has NO CALCULATION ABILITY, although they represent what is said to explain what was never before explained.

The only value in the exercise is proving what no person ever proved before, AND THAT IS THE INFLUENCE THE ATOM HOLDS ON THE Universe, AS THE ATOM INFLUENCE EXTENDS BEYOND ALL COSMIC BOUNDARIES.

To me everything makes perfect sense and while saying this I do admit full heartedly that I am not a Master such as yourself with the knowledge you possess. In that light, should you feel there are aspects I do not explain to a sufficient standard, I am willing to work on it. **This aspect, Newtonian science disregard in, as much as they disregard that there is any connection between the atom and the cosmos**

A star is spinning in liquid forming a solid in motion having liquid flowing around a solid centre. That explanation also covers the explaining of gravity! All gravity is a product of the Coanda principal where motion of liquids creates a governing singularity centre with in the very centre of the star. Electricity, gravity the atomic gravity and the flow of light are amongst many other forms the transmitting or the normal flow of heat. Generating electricity is the same process the Earth and other cosmic structures use to generate gravity and there are no simple one force pulling as gravity. As gravity is, so is electricity and lightning concepts of the same principle where one may be stronger in dynamics when compared to others being weaker in dynamics. All principles are relevancies where one statement only finds value when compared to another forming a relevancy by borders.

What is the Universe? This is such a simple question that every one and every person gets wrong because of the relevancy we humans place on the Universe and the relevancy what the Universe truly is. SUPER – EDUCATED- WIZARDS really get tide in knots with making all about nothing so complicated it absorbs everything holding back nothing. It is so embarrassing simple even I can understand what the Universe is and the Universe is not what science says the Universe is. The Universe is a hollow sphere formed by illusions in light only because Π holds Π^2 at the very end of space and time. By holding a specific centre every centre forms the centre of the Universe and that makes every spot in the Universe a sphere that becomes the strongest form that any object can have. The sphere is without any doubt the favourite choice coming about as the natural form formed by gravity in form of being committed only to gravity. Where gravity has the last say without other influences changing possibilities as collisions leaving debris in space or natural out burst like Super Nova explosions, gravity will enforce the sphere to be the form taken by the particle. But there is no evidence of particles of similar size joining in matrimony through gravity being the shotgun at the wedding. There is no proof that mass does unite by the pulling power of gravity. There is more than abundant evidence that gravity conforms by Π while gravity always confirms Π. In cases where there is a mismatch of size outside any proportions of equality there is then a contracting of the lesser by the greater. In such cases the lesser is not qualifying as material (and that I prove in **The Absolute Relevancy of Singularity in Terms The Cosmic Code**) but the greater consider all the lesser to be heat. It is humans bringing distinction to matter in form.

Two objects where the one is small and the other is large is falling down to the Earth by implication of size holding mass should have their own value of gravity and gravitons and in comparison with the gravitons of the Earth; the mass putting the gravitons at work has apparent insignificant and an unrelated value.

However, these two objects are in their own individual deuce to see who reaches the Earth first. Let's compare an iron ball in matching size and compare such a fall to a wooden ball falling the same distance under the same conditions. It stands to reason that the iron ball's gravitons should give it a superior advantage just because of superior numbers working if mass was capable of producing gravity. This comes about because the two objects are in a position where they compare in relation to one another and share a common second factor, which is the Earth. In relation to the Earth, the gravitons of the two balls do not come into consideration, but this do not play a part since the Earth is a common factor. The balls, however, is put in a situation where they stand in relation to each other. When compared to one another, the gravitons should give the heavier ball a sizable advantage with the heavier ball having more gravitons to bring on more gravity. As much as this is nonsense I have heard experts say on television that the heavier object will fall quicker than the lighter object! I almost swallowed my tongue in disgust to here an expert get away with that nonsense. Galileo said there is no heavy or light big or small since all object sharing similar conditions fall equally. Galileo was the first to indicate space-time but all failed to notice that Galileo's pendulum is going through space during equal time so then pendulum is indicating space (the swing) time (which is what the pendulum is keeping) and Galileo was the first to prove gravity is simply equal motion to all but that fact be it as obvious as it is every one including Newton, which is supposedly the master on motion, failed to see. The sensible example one can show to prove that where some matching structures in size in the cosmos come into conflict by coming to close to each other there is a process coming about where occupied space sharing one of the structures are turned to heat in space by the other and larger structure. The larger object is not pulling the lesser object closer as Newton confirmed by suggesting it happens, but it is literally compressing the air (including the space in which the falling object is) into the lesser space that the compressed air will have.

Even more astonishing is facts about the Binary star system that is seldom to never mentioned. Try and associate what happens in the Roche limit, which is what truly happens in the cosmos to what Newtonians confirmed suggestion about what happens when conflict of space arrive. The Official Policy Protectors never tries to explain the relation between Newton's laws as mentioned above, and the binary star system forming the principle we know as the Roche limit. The binary stars are systems where two stars spin around each other and never collide. These stars are sometimes smaller but mostly many times much larger the size of our Sun. When one applies the same Newtonian formula as given above, these massive giants must crash into each other, destroying themselves in the process. The enormous mystery is not in the apparent misbehaviour of these giants, but the fact that this behaviour is known to science since the previous century. Relate the binary once again to the comet / Sun relation and there is a distinct similarity.

With the comet, the Newtonians regards a force that attaches the Sun and the comet in some way where this force pulls the comet towards the Sun. At the same time another force joins in that pulls the Sun closer to the comet, and is in ratio and resulting because of the mass but such is the mass difference between the Sun and the comet, the force the comet applies never realizes as a force able to move the Sun. In view of this, only the force the Sun applies, comes into effect. The comet proves the reality of this force by speeding up its movement as it comes closer to the Sun. By asking the correct question to the point more proof is uncovered about reality. If the force did not become greater as the radius reduces, why would then miss the Sun and disappear into the dark yonder after circling around the Sun? The Sun started pulling the comet when the comet was hiding in the cold and darkness, but then discovered its location by applying detective gravity and collected a force second to none that collected this hiding comet by force and pulled it along in the direction of the centre of the Sun. Gravity is about circles and about Π and mass plays no part in gravity applying.

Stars will never collide because stars can never collide. To prevent stars colliding we have the Roche limit in place.

The only absence in the cosmos is zero and without zero there cannot be an end to eternity but only an everlasting cycle that breaks to start one more eternity now and then. With the cosmos created instant by instant from where no space is but space forms within the cosmic centre, the cosmos is ruled from a position where everything excluding nothing is outside the area holding space which is what we can see but is holding an infinite position in what we can realise and that is a position also that we know where we know God must be. By accepting singularity as well as the undetectable nature of singularity while being utmost prominent and the rule there of brings into the cosmos things physics are unable to explain, mathematics are unable to calculate and man is unable to dismiss. If you accept physics you have to accept God because you cannot accept one fact that is proving singularity without the other coming through singularity. Singularity proves that what forms the Universe is never part of the Universe and what governs the Universe by control is never within the Universe being controlled.

www.ingramcontent.com/pod-product-compliance
Lightning Source LLC
Chambersburg PA
CBHW080652190526
45169CB00006B/2082